Student Solutions Manual
Calculus
Single Variable
Sixth Edition

Robert A. Adams
University of British Columbia

PEARSON

Addison
Wesley

Toronto

0-321-30715-1

Acquisitions Editor: Kelly Torrance
Developmental Editor: Remo Celio
Production Editor: Mary Ann McCutcheon
Production Coordinator: Andrea Falkenberg

1 2 3 4 5   10 09 08 07 06

Printed and bound in Canada.

PEARSON
Addison
Wesley

FOREWORD

These solutions are provided for the benefit of students using the textbook *Single-Variable Calculus (Sixth Edition)* by R. A. Adams, published by Pearson Education Canada. For the most part, the solutions are detailed, especially in exercises on core material and techniques. Occasionally some details are omitted — for example, in exercises on applications of integration, the evaluation of the integrals encountered is not always given with the same degree of detail as the evaluation of integrals found in those exercises dealing specifically with techniques of integration. With a few exceptions, only the even-numbered exercises in the text are solved in this Manual.

As a student of Calculus, you should use this Manual with caution. It is always more beneficial to attempt exercises and problems on your own, before you look at solutions prepared by others. If you use these solutions as "study material" prior to attempting the exercises, you can lose much of the benefit that can follow from diligent attempts to develop your own analytical powers. When you have tried unsuccessfully to solve a problem, then look at these solutions to try to get a "hint" for a second attempt.

May, 2005.

R. A. Adams
Department of Mathematics
The University of British Columbia
Vancouver, B.C., Canada. V6T 1Z2
adms@math.ubc.ca

CONTENTS

CHAPTER P.   PRELIMINARIES

Section P.1   Real Numbers and the Real Line (page 10)

2. $\dfrac{1}{11} = 0.09090909\cdots = 0.\overline{09}$

4. If $x = 3.277777\cdots$, then $10x - 32 = 0.77777\cdots$ and $100x - 320 = 7 + (10x - 32)$, or $90x = 295$. Thus $x = 295/90 = 59/18$.

6. Two different decimal expansions can represent the same number. For instance, both $0.999999\cdots = 0.\overline{9}$ and $1.000000\cdots = 1.\overline{0}$ represent the number 1.

8. $x < 2$ and $x \ge -3$ define the interval $[-3, 2)$.

10. $x \le -1$ defines the interval $(-\infty, -1]$.

12. $x < 4$ or $x \ge 2$ defines the interval $(-\infty, \infty)$, that is, the whole real line.

14. If $3x + 5 \le 8$, then $3x \le 8 - 5 - 3$ and $x \le 1$. Solution: $(-\infty, 1]$

16. If $\dfrac{6-x}{4} \ge \dfrac{3x-4}{2}$, then $6 - x \ge 6x - 8$. Thus $14 \ge 7x$ and $x \le 2$. Solution: $(-\infty, 2]$

18. If $x^2 < 9$, then $|x| < 3$ and $-3 < x < 3$. Solution: $(-3, 3)$

20. Given: $(x + 1)/x \ge 2$.
CASE I. If $x > 0$, then $x + 1 \ge 2x$, so $x \le 1$.
CASE II. If $x < 0$, then $x + 1 \le 2x$, so $x \ge 1$. (not possible)
Solution: $(0, 1]$.

22. Given $6x^2 - 5x \le -1$, then $(2x - 1)(3x - 1) \le 0$, so either $x \le 1/2$ and $x \ge 1/3$, or $x \le 1/3$ and $x \ge 1/2$. The latter combination is not possible. The solution set is $[1/3, 1/2]$.

24. Given $x^2 - x \le 2$, then $x^2 - x - 2 \le 0$ so $(x - 2)(x + 1) \le 0$. This is possible if $x \le 2$ and $x \ge -1$ or if $x \ge 2$ and $x \le -1$. The latter situation is not possible. The solution set is $[-1, 2]$.

26. Given: $\dfrac{3}{x-1} < \dfrac{2}{x+1}$.
CASE I. If $x > 1$ then $(x - 1)(x + 1) > 0$, so that $3(x + 1) < 2(x - 1)$. Thus $x < -5$. There are no solutions in this case.
CASE II. If $-1 < x < 1$, then $(x - 1)(x + 1) < 0$, so $3(x + 1) > 2(x - 1)$. Thus $x > -5$. In this case all numbers in $(-1, 1)$ are solutions.
CASE III. If $x < -1$, then $(x - 1)(x + 1) > 0$, so that $3(x + 1) < 2(x - 1)$. Thus $x < -5$. All numbers $x < -5$ are solutions.
Solutions: $(-\infty, -5) \cup (-1, 1)$.

28. If $|x - 3| = 7$, then $x - 3 = \pm 7$, so $x = -4$ or $x = 10$.

30. If $|1 - t| = 1$, then $1 - t = \pm 1$, so $t = 0$ or $t = 2$.

32. If $\left|\dfrac{s}{2} - 1\right| = 1$, then $\dfrac{s}{2} - 1 = \pm 1$, so $s = 0$ or $s = 4$.

34. If $|x| \le 2$, then x is in $[-2, 2]$.

36. If $|t + 2| < 1$, then $-2 - 1 < t < -2 + 1$, so t is in $(-3, -1)$.

38. If $|2x + 5| < 1$, then $-5 - 1 < 2x < -5 + 1$, so x is in $(-3, -2)$.

40. If $\left|2 - \dfrac{x}{2}\right| < \dfrac{1}{2}$, then $x/2$ lies between $2 - (1/2)$ and $2 + (1/2)$. Thus x is in $(3, 5)$.

42. $|x - 3| < 2|x| \Leftrightarrow x^2 - 6x + 9 = (x - 3)^2 < 4x^2$ $\Leftrightarrow 3x^2 + 6x - 9 > 0 \Leftrightarrow 3(x + 3)(x - 1) > 0$. This inequality holds if $x < -3$ or $x > 1$.

44. The equation $|x - 1| = 1 - x$ holds if $|x - 1| = -(x - 1)$, that is, if $x - 1 < 0$, or, equivalently, if $x < 1$.

Section P.2   Cartesian Coordinates in the Plane   (page 16)

2. From $A(-1, 2)$ to $B(4, -10)$, $\Delta x = 4 - (-1) = 5$ and $\Delta y = -10 - 2 = -12$. $|AB| = \sqrt{5^2 + (-12)^2} = 13$.

4. From $A(0.5, 3)$ to $B(2, 3)$, $\Delta x = 2 - 0.5 = 1.5$ and $\Delta y = 3 - 3 = 0$. $|AB| = 1.5$.

6. Arrival point: $(-2, -2)$. Increments $\Delta x = -5$, $\Delta y = 1$. Starting point was $(-2 - (-5), -2 - 1)$, that is, $(3, -3)$.

8. $x^2 + y^2 = 2$ represents a circle of radius $\sqrt{2}$ centred at the origin.

10. $x^2 + y^2 = 0$ represents the origin.

12. $y < x^2$ represents all points lying below the parabola $y = x^2$.

14. The vertical line through $(\sqrt{2}, -1.3)$ is $x = \sqrt{2}$; the horizontal line through that point is $y = -1.3$.

16. Line through $(-2, 2)$ with slope $m = 1/2$ is $y = 2 + (1/2)(x + 2)$, or $x - 2y = -6$.

18. Line through $(a, 0)$ with slope $m = -2$ is $y = 0 - 2(x - a)$, or $y = 2a - 2x$.

20. At $x = 3$, the height of the line $x - 4y = 7$ is $y = (3 - 7)/4 = -1$. Thus $(3, -1)$ lies on the line.

22. The line through $(-2, 1)$ and $(2, -2)$ has slope $m = (-2 - 1)/(2 + 2) = -3/4$ and equation $y = 1 - (3/4)(x + 2)$ or $3x + 4y = -2$.

24. The line through $(-2, 0)$ and $(0, 2)$ has slope $m = (2 - 0)/(0 + 2) = 1$ and equation $y = 2 + x$.

26. If $m = -1/2$ and $b = -3$, then the line has equation $y = -(1/2)x - 3$, or $x + 2y = -6$.

28. $x + 2y = -4$ has x-intercept $a = -4$ and y-intercept $b = -4/2 = -2$. Its slope is $-b/a = 2/(-4) = -1/2$.

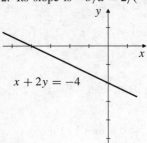

Fig. P.2.28

30. $1.5x - 2y = -3$ has x-intercept $a = -3/1.5 = -2$ and y-intercept $b = -3/(-2) = 3/2$. Its slope is $-b/a = 3/4$.

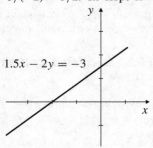

Fig. P.2.30

32. line through $(-2, 2)$ parallel to $2x + y = 4$ is $2x + y = -2$; line perpendicular to $2x + y = 4$ is $x - 2y = -6$.

34. We have

$$2x + y = 8 \quad \Longrightarrow \quad 14x + 7y = 56$$
$$5x - 7y = 1 \quad\quad\quad\quad 5x - 7y = 1.$$

Adding these equations gives $19x = 57$, so $x = 3$ and $y = 8 - 2x = 2$. The intersection point is $(3, 2)$.

36. The line $(x/2) - (y/3) = 1$ has x-intercept $a = 2$, and y-intercept $b = -3$.

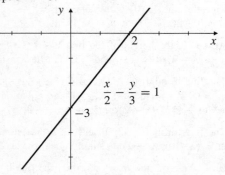

Fig. P.2.36

38. The line through $(-2, 5)$ and $(k, 1)$ has x-intercept 3, so also passes through $(3, 0)$. Its slope m satisfies

$$\frac{1 - 0}{k - 3} = m = \frac{0 - 5}{3 + 2} = -1.$$

Thus $k - 3 = -1$, and so $k = 2$.

40. $-40°$ and $-40°$ is the same temperature on both the Fahrenheit and Celsius scales.

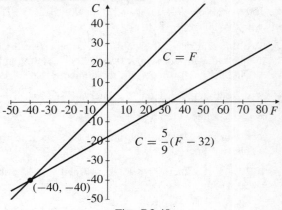

Fig. P.2.40

42. $A = (0, 0), \quad B = (1, \sqrt{3}), \quad C = (2, 0)$

$$|AB| = \sqrt{(1 - 0)^2 + (\sqrt{3} - 0)^2} = \sqrt{4} = 2$$
$$|AC| = \sqrt{(2 - 0)^2 + (0 - 0)^2} = \sqrt{4} = 2$$
$$|BC| = \sqrt{(2 - 1)^2 + (0 - \sqrt{3})^2} = \sqrt{4} = 2.$$

Since $|AB| = |AC| = |BC|$, triangle ABC is equilateral.

44. If $M = (x_m, y_m)$ is the midpoint of $P_1 P_2$, then the displacement of M from P_1 equals the displacement of P_2 from M:

$$x_m - x_1 = x_2 - x_m, \quad y_m - y_1 = y_2 - y_m.$$

Thus $x_m = (x_1 + x_2)/2$ and $y_m = (y_1 + y_2)/2$.

46. Let the coordinates of P be $(x, 0)$ and those of Q be $(X, -2X)$. If the midpoint of PQ is $(2, 1)$, then

$$(x + X)/2 = 2, \quad (0 - 2X)/2 = 1.$$

The second equation implies that $X = -1$, and the second then implies that $x = 5$. Thus P is $(5, 0)$.

48. $\sqrt{(x - 2)^2 + y^2} = \sqrt{x^2 + (y - 2)^2}$ says that (x, y) is equidistant from $(2, 0)$ and $(0, 2)$. Thus (x, y) must lie on the line that is the right bisector of the line from $(2, 0)$ to $(0, 2)$. A simpler equation for this line is $x = y$.

50. For any value of k, the coordinates of the point of intersection of $x + 2y = 3$ and $2x - 3y = -1$ will also satisfy the equation

$$(x + 2y - 3) + k(2x - 3y + 1) = 0$$

because they cause both expressions in parentheses to be 0. The equation above is linear in x and y, and so represents a straight line for any choice of k. This line will pass through $(1, 2)$ provided $1 + 4 - 3 + k(2 - 6 + 1) = 0$, that is, if $k = 2/3$. Therefore, the line through the point of intersection of the two given lines and through the point $(1, 2)$ has equation

$$x + 2y - 3 + \frac{2}{3}(2x - 3y + 1) = 0,$$

or, on simplification, $x = 1$.

Section P.3 Graphs of Quadratic Equations (page 22)

2. $x^2 + (y - 2)^2 = 4$, or $x^2 + y^2 - 4y = 0$

4. $(x - 3)^2 + (y + 4)^2 = 25$, or $x^2 + y^2 - 6x + 8y = 0$.

6. $x^2 + y^2 + 4y = 0$
$x^2 + y^2 + 4y + 4 = 4$
$x^2 + (y + 2)^2 = 4$
centre: $(0, -2)$; radius 2.

8. $x^2 + y^2 - 2x - y + 1 = 0$
$x^2 - 2x + 1 + y^2 - y + \frac{1}{4} = \frac{1}{4}$
$(x - 1)^2 + \left(y - \frac{1}{2}\right)^2 = \frac{1}{4}$
centre: $(1, 1/2)$; radius $1/2$.

10. $x^2 + y^2 < 4$ represents the open disk consisting of all points lying inside the circle of radius 2 centred at the origin.

12. $x^2 + (y - 2)^2 \le 4$ represents the closed disk consisting of all points lying inside or on the circle of radius 2 centred at the point $(0, 2)$.

14. Together, $x^2 + y^2 \le 4$ and $(x + 2)^2 + y^2 \le 4$ represent the region consisting of all points that are inside or on both the circle of radius 2 centred at the origin and the circle of radius 2 centred at $(-2, 0)$.

16. $x^2 + y^2 - 4x + 2y > 4$ can be rewritten $(x - 2)^2 + (y + 1)^2 > 9$. This equation, taken together with $x + y > 1$, represents all points that lie both outside the circle of radius 3 centred at $(2, -1)$ and above the line $x + y = 1$.

18. The exterior of the circle with centre $(2, -3)$ and radius 4 is given by $(x - 2)^2 + (y + 3)^2 > 16$, or $x^2 + y^2 - 4x + 6y > 3$.

20. $x^2 + y^2 > 4$, $(x - 1)^2 + (y - 3)^2 < 10$

22. The parabola with focus $(0, -1/2)$ and directrix $y = 1/2$ has equation $x^2 = -2y$.

24. The parabola with focus $(-1, 0)$ and directrix $x = 1$ has equation $y^2 = -4x$.

26. $y = -x^2$ has focus $(0, -1/4)$ and directrix $y = 1/4$.

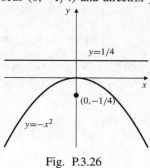

Fig. P.3.26

28. $x = y^2/16$ has focus $(4, 0)$ and directrix $x = -4$.

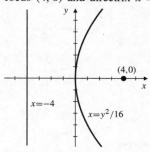

Fig. P.3.28

30.
a) If $y = mx$ is shifted to the right by amount x_1, the equation $y = m(x - x_1)$ results. If (a, b) satisfies this equation, then $b = m(a - x_1)$, and so $x_1 = a - (b/m)$. Thus the shifted equation is
$y = m(x - a + (b/m)) = m(x - a) + b$.

b) If $y = mx$ is shifted vertically by amount y_1, the equation $y = mx + y_1$ results. If (a, b) satisfies this equation, then $b = ma + y_1$, and so $y_1 = b - ma$. Thus the shifted equation is $y = mx + b - ma = m(x - a) + b$, the same equation obtained in part (a).

32. $4y = \sqrt{x + 1}$

34. $(y/2) = \sqrt{4x + 1}$

36. $x^2 + y^2 = 5$ shifted up 2, left 4 gives $(x + 4)^2 + (y - 2)^2 = 5$.

38. $y = \sqrt{x}$ shifted down 2, left 4 gives $y = \sqrt{x + 4} - 2$.

40. $y = x^2 - 6$, $y = 4x - x^2$. Subtracting these equations gives
$2x^2 - 4x - 6 = 0$, or $2(x - 3)(x + 1) = 0$. Thus $x = 3$ or $x = -1$. The corresponding values of y are 3 and -5. The intersection points are $(3, 3)$ and $(-1, -5)$.

42. $2x^2 + 2y^2 = 5$, $xy = 1$. The second equation says that $y = 1/x$. Substituting this into the first equation gives $2x^2 + (2/x^2) = 5$, or $2x^4 - 5x^2 + 2 = 0$. This equation factors to $(2x^2 - 1)(x^2 - 2) = 0$, so its solutions are $x = \pm 1/\sqrt{2}$ and $x = \pm \sqrt{2}$. The corresponding values of y are given by $y = 1/x$. Therefore, the intersection points are $(1/\sqrt{2}, \sqrt{2})$, $(-1/\sqrt{2}, -\sqrt{2})$, $(\sqrt{2}, 1/\sqrt{2})$, and $(-\sqrt{2}, -1/\sqrt{2})$.

44. $9x^2 + 16y^2 = 144$ is an ellipse with major axis between $(-4, 0)$ and $(4, 0)$ and minor axis between $(0, -3)$ and $(0, 3)$.

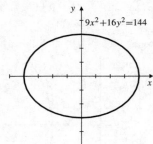

Fig. P.3.44

46. $(x - 1)^2 + \dfrac{(y + 1)^2}{4} = 4$ is an ellipse with centre at $(1, -1)$, major axis between $(1, -5)$ and $(1, 3)$ and minor axis between $(-1, -1)$ and $(3, -1)$.

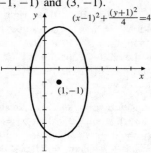

Fig. P.3.46

48. $x^2 - y^2 = -1$ is a rectangular hyperbola with centre at the origin and passing through $(0, \pm 1)$. Its asymptotes are $y = \pm x$.

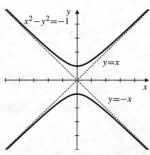

Fig. P.3.48

50. $(x - 1)(y + 2) = 1$ is a rectangular hyperbola with centre at $(1, -2)$ and passing through $(2, -1)$ and $(0, -3)$. Its asymptotes are $x = 1$ and $y = -2$.

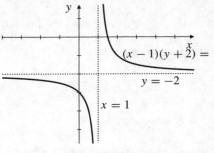

Fig. P.3.50

52. Replacing x with $-x$ and y with $-y$ reflects the graph in both axes. This is equivalent to rotating the graph $180°$ about the origin.

Section P.4   Functions and Their Graphs (page 31)

2. $f(x) = 1 - \sqrt{x}$; domain $[0, \infty)$, range $(-\infty, 1]$

4. $F(x) = 1/(x - 1)$; domain $(-\infty, 1) \cup (1, \infty)$, range $(-\infty, 0) \cup (0, \infty)$

6. $g(x) = \dfrac{1}{1 - \sqrt{x - 2}}$; domain $(2, 3) \cup (3, \infty)$, range $(-\infty, 0) \cup (0, \infty)$. The equation $y = g(x)$ can be solved for $x = 2 - (1 - (1/y))^2$ so has a real solution provided $y \neq 0$.

8.

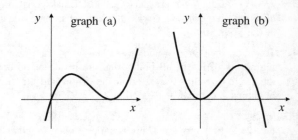

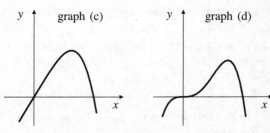

Fig. P.4.8

a) is the graph of $x(1-x)^2$, which is positive for $x > 0$.

b) is the graph of $x^2 - x^3 = x^2(1-x)$, which is positive if $x < 1$.

c) is the graph of $x - x^4$, which is positive if $0 < x < 1$ and behaves like x near 0.

d) is the graph of $x^3 - x^4$, which is positive if $0 < x < 1$ and behaves like x^3 near 0.

10.

x	$f(x) = x^{2/3}$
0	0
± 0.5	0.62996
± 1	1
± 1.5	1.3104
± 2	1.5874

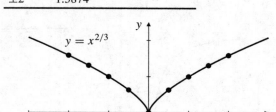

Fig. P.4.10

12. $f(x) = x^3 + x$ is odd: $f(-x) = -f(x)$

14. $f(x) = \dfrac{1}{x^2 - 1}$ is even: $f(-x) = f(x)$

16. $f(x) = \dfrac{1}{x+4}$ is odd about $(-4, 0)$:
$f(-4 - x) = -f(-4 + x)$

18. $f(x) = x^3 - 2$ is odd about $(0, -2)$:
$f(-x) + 2 = -(f(x) + 2)$

20. $f(x) = |x + 1|$ is even about $x = -1$:
$f(-1 - x) = f(-1 + x)$

22. $f(x) = \sqrt{(x - 1)^2}$ is even about $x = 1$:
$f(1 - x) = f(1 + x)$

24.

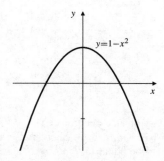

26.

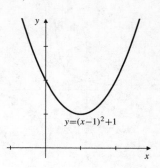

28.

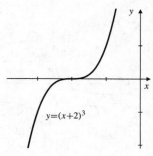

30.

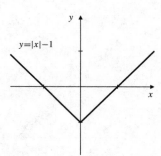

32.

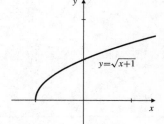

34.

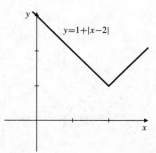

36.

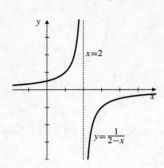

38.

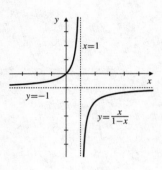

40.

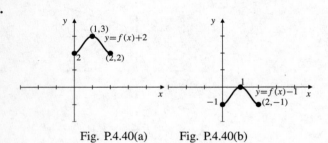

Fig. P.4.40(a)   Fig. P.4.40(b)

42.

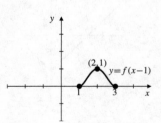

44.

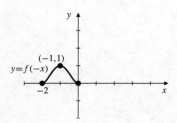

46.

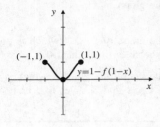

48. Range is approximately $(-\infty, 0.17]$.

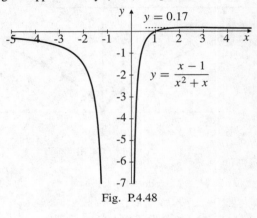

Fig. P.4.48

50.

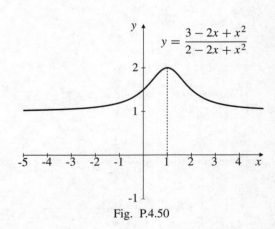

Fig. P.4.50

Apparent symmetry about $x = 1$.
This can be confirmed by calculating $f(2 - x)$, which
turns out to be equal to $f(x)$.

52.

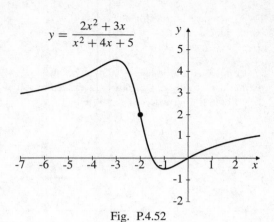

$$y = \frac{2x^2 + 3x}{x^2 + 4x + 5}$$

Fig. P.4.52

Apparent symmetry about $(-2, 2)$.
This can be confirmed by calculating shifting the graph right by 2 (replace x with $x - 2$) and then down 2 (subtract 2). The result is $-5x/(1 + x^2)$, which is odd.

Section P.5   Combining Functions to Make New Functions   (page 37)

2. $f(x) = \sqrt{1 - x}$, $g(x) = \sqrt{1 + x}$.
$\mathcal{D}(f) = (-\infty, 1]$, $\mathcal{D}(g) = [-1, \infty)$.
$\mathcal{D}(f + g) = \mathcal{D}(f - g) = \mathcal{D}(fg) = [-1, 1]$,
$\mathcal{D}(f/g) = (-1, 1]$, $\mathcal{D}(g/f) = [-1, 1)$.
$(f + g)(x) = \sqrt{1 - x} + \sqrt{1 + x}$

$(f - g)(x) = \sqrt{1 - x} - \sqrt{1 + x}$

$(fg)(x) = \sqrt{1 - x^2}$

$(f/g)(x) = \sqrt{(1 - x)/(1 + x)}$

$(g/f)(x) = \sqrt{(1 + x)/(1 - x)}$

4.

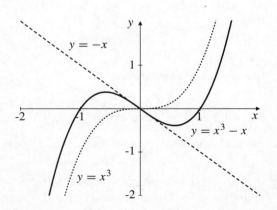

6.

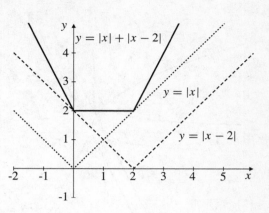

8. $f(x) = 2/x$, $g(x) = x/(1 - x)$.
$f \circ f(x) = 2/(2/x) = x$; $\mathcal{D}(f \circ f) = \{x : x \neq 0\}$
$f \circ g(x) = 2/(x/(1 - x)) = 2(1 - x)/x$;
  $\mathcal{D}(f \circ g) = \{x : x \neq 0, 1\}$
$g \circ f(x) = (2/x)/(1 - (2/x)) = 2/(x - 2)$;
  $\mathcal{D}(g \circ f) = \{x : x \neq 0, 2\}$
$g \circ g(x) = (x/(1 - x))/(1 - (x/(1 - x))) = x/(1 - 2x)$;
  $\mathcal{D}(g \circ g) = \{x : x \neq 1/2, 1\}$

10. $f(x) = (x + 1)/(x - 1) = 1 + 2/(x - 1)$, $g(x) = \text{sgn}\,(x)$.
$f \circ f(x) = 1 + 2/(1 + (2/(x - 1) - 1)) = x$;
$\mathcal{D}(f \circ f) = \{x : x \neq 1\}$
$f \circ g(x) = \dfrac{\text{sgn}\,x + 1}{\text{sgn}\,x - 1} = 0$; $\mathcal{D}(f \circ g) = (-\infty, 0)$

$g \circ f(x) = \text{sgn}\left(\dfrac{x + 1}{x - 1}\right) = \begin{cases} 1 & \text{if } x < -1 \text{ or } x > 1 \\ -1 & \text{if } -1 < x < 1 \end{cases}$;

$\mathcal{D}(g \circ f) = \{x : x \neq -1, \ 1\}$
$g \circ g(x) = \text{sgn}\,(\text{sgn}\,(x)) = \text{sgn}\,(x)$; $\mathcal{D}(g \circ g) = \{x : x \neq 0\}$

	$f(x)$	$g(x)$	$f \circ g(x)$
11.	x^2	$x + 1$	$(x + 1)^2$
12.	$x - 4$	$x + 4$	x
13.	$\sqrt{x}$	x^2	$|x|$
14.	$2x^3 + 3$	$x^{1/3}$	$2x + 3$
15.	$(x + 1)/x$	$1/(x - 1)$	x
16.	$1/(x + 1)^2$	$x - 1$	$1/x^2$

18.

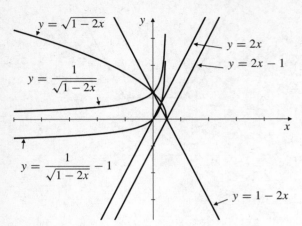

Fig. P.5.18

20.

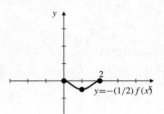

22.

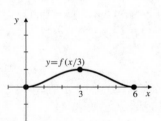

24.

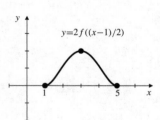

26.

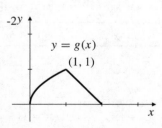

28. $\lfloor x \rfloor = 0$ for $0 \le x < 1$; $\lceil x \rceil = 0$ for $-1 \le x < 0$.

30. $\lceil -x \rceil = -\lfloor x \rfloor$ is true for all real x; if $x = n + y$ where n is an integer and $0 \le y < 1$, then $-x = -n - y$, so that $\lceil -x \rceil = -n$ and $\lfloor x \rfloor = n$.

32. $f(x)$ is called the integer part of x because $|f(x)|$ is the largest integer that does not exceed x; i.e. $|x| = |f(x)| + y$, where $0 \le y < 1$.

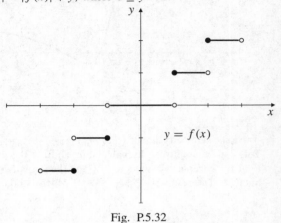

Fig. P.5.32

34. f even $\Leftrightarrow f(-x) = f(x)$
f odd $\Leftrightarrow f(-x) = -f(x)$
f even and odd $\Rightarrow f(x) = -f(x) \Rightarrow 2f(x) = 0$
$\Rightarrow f(x) = 0$

Section P.6   Polynomials and Rational Functions   (page 43)

2. $x^2 - 3x - 10 = (x - 5)(x + 2)$
The roots are 5 and -2.

4. Rather than use the quadratic formula this time, let us complete the square.

$$\begin{aligned} x^2 - 6x + 13 &= x^2 - 6x + 9 + 4 \\ &= (x - 3)^2 + 2^2 \\ &= (x - 3 - 2i)(x - 3 + 2i). \end{aligned}$$

The roots are $3 + 2i$ and $3 - 2i$.

6. $x^4 + 6x^3 + 9x^2 = x^2(x^2 + 6x + 9) = x^2(x + 3)^2$. There are two double roots, 0 and -3.

8. $x^4 - 1 = (x^2 - 1)(x^2 + 1) = (x - 1)(x + 1)(x - i)(x + i)$.
The roots are 1, -1, i, and $-i$.

10. $x^5 - x^4 - 16x + 16 = (x - 1)(x^4 - 16)$
$$\begin{aligned} &= (x - 1)(x^2 - 4)(x^4 + 4) \\ &= (x - 1)(x - 2)(x + 2)(x - 2i)(x + 2i). \end{aligned}$$

The roots are 1, 2, -2, $2i$, and $-2i$.

12. $x^9 - 4x^7 - x^6 + 4x^4 = x^4(x^5 - x^2 - 4x^3 + 4)$
$$= x^4(x^3 - 1)(x^2 - 4)$$
$$= x^4(x - 1)(x - 2)(x + 2)(x^2 + x + 1).$$

Seven of the nine roots are: 0 (with multiplicity 4), 1, 2, and -2. The other two roots are solutions of $x^2 + x + 1 = 0$, namely

$$x = \frac{-1 \pm \sqrt{1 - 4}}{2} = -\frac{1}{2} \pm \frac{\sqrt{3}}{2} i.$$

The required factorization of $x^9 - 4x^7 - x^6 + 4x^4$ is

$$x^4(x-1)(x-2)(x+2)\left(x - \frac{1}{2} + \frac{\sqrt{3}}{2} i\right)\left(x - \frac{1}{2} - \frac{\sqrt{3}}{2} i\right).$$

14.
$$\frac{x^2}{x^2 + 5x + 3} = \frac{x^2 + 5x + 3 - 5x - 3}{x^2 + 5x + 3}$$
$$= 1 + \frac{-5x - 3}{x^2 + 5x + 3}.$$

16.
$$\frac{x^4 + x^2}{x^3 + x^2 + 1} = \frac{x(x^3 + x^2 + 1) - x^3 - x + x^2}{x^3 + x^2 + 1}$$
$$= x + \frac{-(x^3 + x^2 + 1) + x^2 + 1 - x + x^2}{x^3 + x^2 + 1}$$
$$= x - 1 + \frac{2x^2 - x + 1}{x^3 + x^2 + 1}.$$

18. Let $P(x) = a_n x^n + a_{n-1} x^{n-1} + \cdots + a_1 x + a_0$, where $n \geq 1$. By the Factor Theorem, $x + 1$ is a factor of $P(x)$ if and only if $P(-1) = 0$, that is, if and only if $a_0 - a_1 + a_2 - a_3 + \cdots + (-1)^n a_n = 0$. This condition says that the sum of the coefficients of even powers is equal to the sum of coefficients of odd powers.

20. By the previous exercise, $\bar{z} = u - iv$ is also a root of P. Therefore $P(x)$ has two linear factors $x - u - iv$ and $x - u + iv$. The product of these factors is the real quadratic factor $(x - u)^2 - i^2 v^2 = x^2 - 2ux + u^2 + v^2$, which must also be a factor of $P(x)$.

Section P.7 The Trigonometric Functions (page 55)

2. $\tan \dfrac{-3\pi}{4} = -\tan \dfrac{3\pi}{4} = -1$

4. $\sin\left(\dfrac{7\pi}{12}\right) = \sin\left(\dfrac{\pi}{4} + \dfrac{\pi}{3}\right)$
$$= \sin \frac{\pi}{4} \cos \frac{\pi}{3} + \cos \frac{\pi}{4} \sin \frac{\pi}{3}$$
$$= \frac{1}{\sqrt{2}} \frac{1}{2} + \frac{1}{\sqrt{2}} \frac{\sqrt{3}}{2} = \frac{1 + \sqrt{3}}{2\sqrt{2}}$$

6. $\sin \dfrac{11\pi}{12} = \sin \dfrac{\pi}{12}$
$$= \sin\left(\frac{\pi}{3} - \frac{\pi}{4}\right)$$
$$= \sin \frac{\pi}{3} \cos \frac{\pi}{4} - \cos \frac{\pi}{3} \sin \frac{\pi}{4}$$
$$= \left(\frac{\sqrt{3}}{2}\right)\left(\frac{1}{\sqrt{2}}\right) - \left(\frac{1}{2}\right)\left(\frac{1}{\sqrt{2}}\right)$$
$$= \frac{\sqrt{3} - 1}{2\sqrt{2}}$$

8. $\sin(2\pi - x) = -\sin x$

10. $\cos\left(\dfrac{3\pi}{2} + x\right) = \cos \dfrac{3\pi}{2} \cos x - \sin \dfrac{3\pi}{2} \sin x$
$$= (-1)(-\sin x) = \sin x$$

12.
$$\frac{\tan x - \cot x}{\tan x + \cot x} = \frac{\left(\dfrac{\sin x}{\cos x} - \dfrac{\cos x}{\sin x}\right)}{\left(\dfrac{\sin x}{\cos x} + \dfrac{\cos x}{\sin x}\right)}$$
$$= \frac{\left(\dfrac{\sin^2 x - \cos^2 x}{\cos x \sin x}\right)}{\left(\dfrac{\sin^2 x + \cos^2 x}{\cos x \sin x}\right)}$$
$$= \sin^2 x - \cos^2 x$$

14. $(1 - \cos x)(1 + \cos x) = 1 - \cos^2 x = \sin^2 x$ implies $\dfrac{1 - \cos x}{\sin x} = \dfrac{\sin x}{1 + \cos x}$. Now
$$\frac{1 - \cos x}{\sin x} = \frac{1 - \cos 2\left(\dfrac{x}{2}\right)}{\sin 2\left(\dfrac{x}{2}\right)}$$
$$= \frac{1 - \left(1 - 2\sin^2\left(\dfrac{x}{2}\right)\right)}{2 \sin \dfrac{x}{2} \cos \dfrac{x}{2}}$$
$$= \frac{\sin \dfrac{x}{2}}{\cos \dfrac{x}{2}} = \tan \frac{x}{2}$$

16.
$$\frac{\cos x - \sin x}{\cos x + \sin x} = \frac{(\cos x - \sin x)^2}{(\cos x + \sin x)(\cos x - \sin x)}$$
$$= \frac{\cos^2 x - 2\sin x \cos x + \sin^2 x}{\cos^2 x - \sin^2 x}$$
$$= \frac{1 - \sin(2x)}{\cos(2x)}$$
$$= \sec(2x) - \tan(2x)$$

18. $\cos 3x = \cos(2x + x)$
$$= \cos 2x \cos x - \sin 2x \sin x$$
$$= (2\cos^2 x - 1)\cos x - 2\sin^2 x \cos x$$
$$= 2\cos^3 x - \cos x - 2(1 - \cos^2 x)\cos x$$
$$= 4\cos^3 x - 3\cos x$$

20. $\sin\dfrac{x}{2}$ has period 4π.

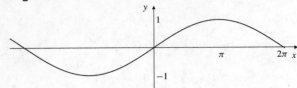

Fig. P.7.20

22. $\cos\dfrac{\pi x}{2}$ has period 4.

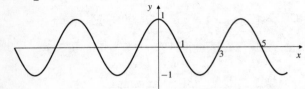

Fig. P.7.22

24.

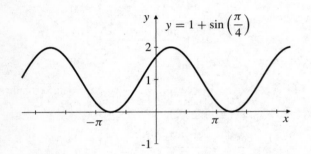

$y = 1 + \sin\left(\dfrac{\pi}{4}\right)$

26. $\tan x = 2$ where x is in $[0, \dfrac{\pi}{2}]$. Then
$\sec^2 x = 1 + \tan^2 x = 1 + 4 = 5$. Hence,
$\sec x = \sqrt{5}$ and $\cos x = \dfrac{1}{\sec x} = \dfrac{1}{\sqrt{5}}$,
$\sin x = \tan x \cos x = \dfrac{2}{\sqrt{5}}$.

28. $\cos x = -\dfrac{5}{13}$ where x is in $\left[\dfrac{\pi}{2}, \pi\right]$. Hence,
$\sin x = \sqrt{1 - \cos^2 x} = \sqrt{1 - \dfrac{25}{169}} = \dfrac{12}{13}$,
$\tan x = -\dfrac{12}{5}$.

30. $\tan x = \dfrac{1}{2}$ where x is in $[\pi, \dfrac{3\pi}{2}]$. Then,
$\sec^2 x = 1 + \dfrac{1}{4} = \dfrac{5}{4}$. Hence,
$\sec x = -\dfrac{\sqrt{5}}{2}$, $\cos x = -\dfrac{2}{\sqrt{5}}$,
$\sin x = \tan x \cos x = -\dfrac{1}{\sqrt{5}}$.

32. $b = 2$, $B = \dfrac{\pi}{3}$

$\dfrac{2}{a} = \tan B = \sqrt{3} \Rightarrow a = \dfrac{2}{\sqrt{3}}$

$\dfrac{2}{c} = \sin B = \dfrac{\sqrt{3}}{2} \Rightarrow c = \dfrac{4}{\sqrt{3}}$

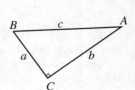

34. $\sin A = \dfrac{a}{c} \Rightarrow a = c \sin A$

36. $\cos B = \dfrac{a}{c} \Rightarrow a = c \cos B$

38. $\sin A = \dfrac{a}{c} \Rightarrow c = \dfrac{a}{\sin A}$

40. $\sin A = \dfrac{a}{c}$

42. $\sin A = \dfrac{a}{c} = \dfrac{a}{\sqrt{a^2 + b^2}}$

44. Given that $a = 2, b = 2, c = 3$.
Since $a^2 = b^2 + c^2 - 2bc \cos A$,
$\cos A = \dfrac{a^2 - b^2 - c^2}{-2bc}$
$= \dfrac{4 - 4 - 9}{-2(2)(3)} = \dfrac{3}{4}$.

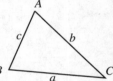

46. Given that $a = 2$, $b = 3$, $C = \dfrac{\pi}{4}$.
$c^2 = a^2 + b^2 - 2ab \cos C = 4 + 9 - 2(2)(3) \cos \dfrac{\pi}{4} = 13 - \dfrac{12}{\sqrt{2}}$.

Hence, $c = \sqrt{13 - \dfrac{12}{\sqrt{2}}} \approx 2.12479$.

48. Given that $a = 2$, $b = 3$, $C = 35°$. Then
$c^2 = 4 + 9 - 2(2)(3) \cos 35°$, hence $c \approx 1.78050$.

50. If $a = 1, b = \sqrt{2}, A = 30°$, then $\dfrac{\sin B}{b} = \dfrac{\sin A}{a} = \dfrac{1}{2}$.

Thus $\sin B = \dfrac{\sqrt{2}}{2} = \dfrac{1}{\sqrt{2}}$, $B = \dfrac{\pi}{4}$ or $\dfrac{3\pi}{4}$, and
$C = \pi - \left(\dfrac{\pi}{4} + \dfrac{\pi}{6}\right) = \dfrac{7\pi}{12}$ or $C = \pi - \left(\dfrac{3\pi}{4} + \dfrac{\pi}{6}\right) = \dfrac{\pi}{12}$.

Thus, $\cos C = \cos \dfrac{7\pi}{12} = \cos \left(\dfrac{\pi}{4} + \dfrac{\pi}{3}\right) = \dfrac{1 - \sqrt{3}}{2\sqrt{2}}$ or
$\cos C = \cos \dfrac{\pi}{12} = \cos \left(\dfrac{\pi}{3} - \dfrac{\pi}{4}\right) = \dfrac{1 + \sqrt{3}}{2\sqrt{2}}$.

Hence,

$$c^2 = a^2 + b^2 - 2ab \cos C$$
$$= 1 + 2 - 2\sqrt{2} \cos C$$
$$= 3 - (1 - \sqrt{3}) \text{ or } 3 - (1 + \sqrt{3})$$
$$= 2 + \sqrt{3} \text{ or } 2 - \sqrt{3}.$$

Hence, $c = \sqrt{2 + \sqrt{3}}$ or $\sqrt{2 - \sqrt{3}}$.

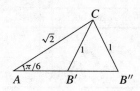

Fig. P.7.50

52. See the following diagram. Since $\tan 40° = h/a$, therefore $a = h/\tan 40°$. Similarly, $b = h/\tan 70°$. Since $a + b = 2$ km, therefore,

$$\frac{h}{\tan 40°} + \frac{h}{\tan 70°} = 2$$

$$h = \frac{2(\tan 40° \tan 70°)}{\tan 70° + \tan 40°} \approx 1.286 \text{ km.}$$

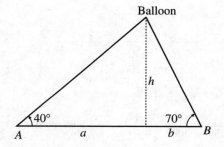

Fig. P.7.52

54. From Exercise 53, area $= \frac{1}{2}ac \sin B$. By Cosine Law, $\cos B = \dfrac{a^2 + c^2 - b^2}{2ac}$. Thus,

$$\sin B = \sqrt{1 - \left(\frac{a^2 + c^2 - b^2}{2ac}\right)^2}$$

$$= \frac{\sqrt{-a^4 - b^4 - c^4 + 2a^2b^2 + 2b^2c^2 + 2a^2c^2}}{2ac}.$$

Hence, Area $= \dfrac{\sqrt{-a^4 - b^4 - c^4 + 2a^2b^2 + 2b^2c^2 + 2a^2c^2}}{4}$ square units. Since,

$$s(s-a)(s-b)(s-c)$$
$$= \frac{b+c+a}{2}\frac{b+c-a}{2}\frac{a-b+c}{2}\frac{a+b-c}{2}$$
$$= \frac{1}{16}\left((b+c)^2 - a^2\right)\left(a^2 - (b-c)^2\right)$$
$$= \frac{1}{16}\left(a^2\left((b+c)^2 + (b-c)^2\right) - a^4 - (b^2 - c^2)^2\right)$$
$$= \frac{1}{16}\left(2a^2b^2 + 2a^2c^2 - a^4 - b^4 - c^4 + 2b^2c^2\right)$$

Thus $\sqrt{s(s-a)(s-b)(s-c)} = $ Area of triangle.

CHAPTER 1. LIMITS AND CONTINUITY

Section 1.1 Examples of Velocity, Growth Rate, and Area (page 61)

2.

h	Avg. vel. over $[2, 2+h]$
1	5.0000
0.1	4.1000
0.01	4.0100
0.001	4.0010
0.0001	4.0001

4. Average volocity on $[2, 2+h]$ is

$$\frac{(2+h)^2 - 4}{(2+h) - 2} = \frac{4 + 4h + h^2 - 4}{h} = \frac{4h + h^2}{h} = 4 + h.$$

As h approaches 0 this average velocity approaches 4 m/s

6. Average velocity over $[t, t+h]$ is

$$\frac{3(t+h)^2 - 12(t+h) + 1 - (3t^2 - 12t + 1)}{(t+h) - t}$$

$$= \frac{6th + 3h^2 - 12h}{h} = 6t + 3h - 12 \text{ m/s}.$$

This average velocity approaches $6t - 12$ m/s as h approaches 0.
At $t = 1$ the velocity is $6 \times 1 - 12 = -6$ m/s.
At $t = 2$ the velocity is $6 \times 2 - 12 = 0$ m/s.
At $t = 3$ the velocity is $6 \times 3 - 12 = 6$ m/s.

8. Average velocity over $[t-k, t+k]$ is

$$\frac{3(t+k)^2 - 12(t+k) + 1 - [3(t-k)^2 - 12(t-k) + 1]}{(t+k) - (t-k)}$$

$$= \frac{1}{2k}\Big(3t^2 + 6tk + 3k^2 - 12t - 12k + 1 - 3t^2 + 6tk - 3k^2$$

$$+ 12t - 12k + 1\Big)$$

$$= \frac{12tk - 24k}{2k} = 6t - 12 \text{ m/s},$$

which is the velocity at time t from Exercise 7.

10. Average velocity over $[1, 1+h]$ is

$$\frac{2 + \dfrac{1}{\pi}\sin\pi(1+h) - \left(2 + \dfrac{1}{\pi}\sin\pi\right)}{h}$$

$$= \frac{\sin(\pi + \pi h)}{\pi h} = \frac{\sin\pi\cos(\pi h) + \cos\pi\sin(\pi h)}{\pi h}$$

$$= -\frac{\sin(\pi h)}{\pi h}.$$

h	Avg. vel. on $[1, 1+h]$
1.0000	0
0.1000	-0.983631643
0.0100	-0.999835515
0.0010	-0.999998355

12. We sketched a tangent line to the graph on page 55 in the text at $t = 20$. The line appeared to pass through the points $(10, 0)$ and $(50, 1)$. On day 20 the biomass is growing at about $(1 - 0)/(50 - 10) = 0.025$ mm^2/d.

14. a)

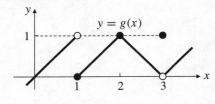

Fig. 1.1.14

b) Average rate of increase in profits between 2002 and 2004 is
$$\frac{174 - 62}{2004 - 2002} = \frac{112}{2} = 56 \text{ (thousand\$/yr)}.$$

c) Drawing a tangent line to the graph in (a) at $t = 2002$ and measuring its slope, we find that the rate of increase of profits in 1992 is about 43 thousand\$/year.

Section 1.2 Limits of Functions (page 68)

2. From inspecting the graph

we see that

$$\lim_{x\to 1} g(x) \text{ does not exist}$$

(left limit is 1, right limit is 0)

$$\lim_{x\to 2} g(x) = 1, \qquad \lim_{x\to 3} g(x) = 0.$$

4. $\displaystyle\lim_{x\to 1+} g(x) = 0$

6. $\lim_{x \to 3-} g(x) = 0$

8. $\lim_{x \to 2} 3(1-x)(2-x) = 3(-1)(2-2) = 0$

10. $\lim_{t \to -4} \dfrac{t^2}{4-t} = \dfrac{(-4)^2}{4+4} = 2$

12. $\lim_{x \to -1} \dfrac{x^2-1}{x+1} = \lim_{x \to -1}(x-1) = -2$

14. $\lim_{x \to -2} \dfrac{x^2+2x}{x^2-4} = \lim_{x \to -2} \dfrac{x}{x-2} = \dfrac{-2}{-4} = \dfrac{1}{2}$

16. $\lim_{h \to 0} \dfrac{3h+4h^2}{h^2-h^3} = \lim_{h \to 0} \dfrac{3+4h}{h-h^2}$ does not exist; denominator approaches 0 but numerator does not approach 0.

18. $\lim_{h \to 0} \dfrac{\sqrt{4+h}-2}{h}$

$= \lim_{h \to 0} \dfrac{4+h-4}{h(\sqrt{4+h}+2)}$

$= \lim_{h \to 0} \dfrac{1}{\sqrt{4+h}+2} = \dfrac{1}{4}$

20. $\lim_{x \to -2} |x-2| = |-4| = 4$

22. $\lim_{x \to 2} \dfrac{|x-2|}{x-2} = \lim_{x \to 2} \begin{cases} 1, & \text{if } x > 2 \\ -1, & \text{if } x < 2. \end{cases}$
Hence, $\lim_{x \to 2} \dfrac{|x-2|}{x-2}$ does not exist.

24. $\lim_{x \to 2} \dfrac{\sqrt{4-4x+x^2}}{x-2}$

$= \lim_{x \to 2} \dfrac{|x-2|}{x-2}$ does not exist.

26. $\lim_{x \to 1} \dfrac{x^2-1}{\sqrt{x+3}-2} = \lim_{x \to 1} \dfrac{(x-1)(x+1)(\sqrt{x+3}+2)}{(x+3)-4}$

$= \lim_{x \to 1}(x+1)(\sqrt{x+3}+2) = (2)(\sqrt{4}+2) = 8$

28. $\lim_{s \to 0} \dfrac{(s+1)^2-(s-1)^2}{s} = \lim_{s \to 0} \dfrac{4s}{s} = 4$

30. $\lim_{x \to -1} \dfrac{x^3+1}{x+1}$

$= \lim_{x \to -1} \dfrac{(x+1)(x^2-x+1)}{x+1} = 3$

32. $\lim_{x \to 8} \dfrac{x^{2/3}-4}{x^{1/3}-2}$

$= \lim_{x \to 8} \dfrac{(x^{1/3}-2)(x^{1/3}+2)}{(x^{1/3}-2)}$

$= \lim_{x \to 8}(x^{1/3}+2) = 4$

34. $\lim_{x \to 2} \left(\dfrac{1}{x-2} - \dfrac{1}{x^2-4} \right)$

$= \lim_{x \to 2} \dfrac{x+2-1}{(x-2)(x+2)}$

$= \lim_{x \to 2} \dfrac{x+1}{(x-2)(x+2)}$ does not exist.

36. $\lim_{x \to 0} \dfrac{|3x-1|-|3x+1|}{x}$

$= \lim_{x \to 0} \dfrac{(3x-1)^2-(3x+1)^2}{x(|3x-1|+|3x+1|)}$

$= \lim_{x \to 0} \dfrac{-12x}{x(|3x-1|+|3x+1|)} = \dfrac{-12}{1+1} = -6$

38. $f(x) = x^3$

$\lim_{h \to 0} \dfrac{f(x+h)-f(x)}{h} = \lim_{h \to 0} \dfrac{(x+h)^3-x^3}{h}$

$= \lim_{h \to 0} \dfrac{3x^2h+3xh^2+h^3}{h}$

$= \lim_{h \to 0} 3x^2+3xh+h^2 = 3x^2$

40. $f(x) = 1/x^2$

$\lim_{h \to 0} \dfrac{f(x+h)-f(x)}{h} = \lim_{h \to 0} \dfrac{\dfrac{1}{(x+h)^2}-\dfrac{1}{x^2}}{h}$

$= \lim_{h \to 0} \dfrac{x^2-(x^2+2xh+h^2)}{h(x+h)^2x^2}$

$= \lim_{h \to 0} -\dfrac{2x+h}{(x+h)^2x^2} = -\dfrac{2x}{x^4} = -\dfrac{2}{x^3}$

42. $f(x) = 1/\sqrt{x}$

$\lim_{h \to 0} \dfrac{f(x+h)-f(x)}{h} = \lim_{h \to 0} \dfrac{\dfrac{1}{\sqrt{x+h}}-\dfrac{1}{\sqrt{x}}}{h}$

$= \lim_{h \to 0} \dfrac{\sqrt{x}-\sqrt{x+h}}{h\sqrt{x}\sqrt{x+h}}$

$= \lim_{h \to 0} \dfrac{x-(x+h)}{h\sqrt{x}\sqrt{x+h}(\sqrt{x}+\sqrt{x+h})}$

$= \lim_{h \to 0} \dfrac{-1}{\sqrt{x}\sqrt{x+h}(\sqrt{x}+\sqrt{x+h})}$

$= \dfrac{-1}{2x^{3/2}}$

44. $\lim_{x \to \pi/4} \cos x = \cos \pi/4 = 1/\sqrt{2}$

46. $\lim_{x \to 2\pi/3} \sin x = \sin 2\pi/3 = \sqrt{3}/2$

48.

x	$(1 - \cos x)/x^2$
± 1.0	0.45969769
± 0.1	0.49958347
± 0.01	0.49999583
± 0.001	0.49999996
0.0001	0.50000000

It appears that $\displaystyle\lim_{x \to 0} \frac{1 - \cos x}{x^2} = \frac{1}{2}$.

50. $\displaystyle\lim_{x \to 2+} \sqrt{2 - x}$ does not exist.

52. $\displaystyle\lim_{x \to -2+} \sqrt{2 - x} = 2$

54. $\displaystyle\lim_{x \to 0-} \sqrt{x^3 - x} = 0$

56. $\displaystyle\lim_{x \to 0+} \sqrt{x^2 - x^4} = 0$

58. $\displaystyle\lim_{x \to a+} \frac{|x - a|}{x^2 - a^2} = \lim_{x \to a+} \frac{x - a}{x^2 - a^2} = \frac{1}{2a}$

60. $\displaystyle\lim_{x \to 2+} \frac{x^2 - 4}{|x + 2|} = \frac{0}{4} = 0$

62. $\displaystyle\lim_{x \to -1+} f(x) = \lim_{x \to -1+} x^2 + 1 = 1 + 1 = 2$

64. $\displaystyle\lim_{x \to 0-} f(x) = \lim_{x \to 0-} x^2 + 1 = 1$

66. If $\lim x \to a\, f(x) = 4$ and $\displaystyle\lim_{x \to a} g(x) = -2$, then

  a) $\displaystyle\lim_{x \to a} \Big(f(x) + g(x) \Big) = 4 + (-2) = 2$

  b) $\displaystyle\lim_{x \to a} f(x) \cdot g(x) = 4 \times (-2) = -8$

  c) $\displaystyle\lim_{x \to a} 4g(x) = 4(-2) = -8$

  d) $\displaystyle\lim_{x \to a} \frac{f(x)}{g(x)} = \frac{4}{-2} = -2$

68. If $\displaystyle\lim_{x \to 0} \frac{f(x)}{x^2} = -2$ then

$\lim_{x \to 0} f(x) = \lim_{x \to 0} x^2 \dfrac{f(x)}{x^2} = 0 \times (-2) = 0,$
and similarly,
$\lim_{x \to 0} \dfrac{f(x)}{x} = \lim_{x \to 0} x \dfrac{f(x)}{x^2} = 0 \times (-2) = 0.$

70.

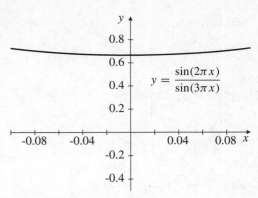

Fig. 1.2.70

$\lim_{x \to 0} \sin(2\pi x)/\sin(3\pi x) = 2/3$

72.

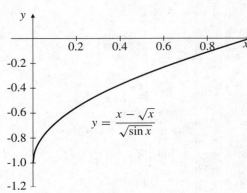

Fig. 1.2.72

$\displaystyle\lim_{x \to 0+} \frac{x - \sqrt{x}}{\sqrt{\sin x}} = -1$

74. Since $\sqrt{5 - 2x^2} \le f(x) \le \sqrt{5 - x^2}$ for $-1 \le x \le 1$, and $\lim_{x \to 0} \sqrt{5 - 2x^2} = \lim_{x \to 0} \sqrt{5 - x^2} = \sqrt{5}$, we have $\lim_{x \to 0} f(x) = \sqrt{5}$ by the squeeze theorem.

76.   a)

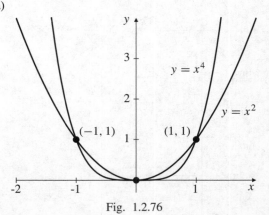

Fig. 1.2.76

b) Since the graph of f lies between those of x^2 and x^4, and since these latter graphs come together at $(\pm 1, 1)$ and at $(0, 0)$, we have $\lim_{x \to \pm 1} f(x) = 1$ and $\lim_{x \to 0} f(x) = 0$ by the squeeze theorem.

78. $f(x) = s \sin \dfrac{1}{x}$ is defined for all $x \neq 0$; its domain is $(-\infty, 0) \cup (0, \infty)$. Since $|\sin t| \leq 1$ for all t, we have $|f(x)| \leq |x|$ and $-|x| \leq f(x) \leq |x|$ for all $x \neq 0$. Since $\lim_{x \to 0} = (-|x|) = 0 = \lim_{x \to 0} |x|$, we have $\lim_{x \to 0} f(x) = 0$ by the squeeze theorem.

Section 1.3 Limits at Infinity and Infinite Limits (page 75)

2. $\displaystyle \lim_{x \to \infty} \frac{x}{x^2 - 4} = \lim_{x \to \infty} \frac{1/x}{1 - (4/x^2)} = \frac{0}{1} = 0$

4. $\displaystyle \lim_{x \to -\infty} \frac{x^2 - 2}{x - x^2}$

$= \displaystyle \lim_{x \to -\infty} \frac{1 - \dfrac{2}{x^2}}{\dfrac{1}{x} - 1} = \frac{1}{-1} = -1$

6. $\displaystyle \lim_{x \to \infty} \frac{x^2 + \sin x}{x^2 + \cos x} = \lim_{x \to \infty} \frac{1 + \dfrac{\sin x}{x^2}}{1 + \dfrac{\cos x}{x^2}} = \frac{1}{1} = 1$

We have used the fact that $\lim_{x \to \infty} \dfrac{\sin x}{x^2} = 0$ (and similarly for cosine) because the numerator is bounded while the denominator grows large.

8. $\displaystyle \lim_{x \to \infty} \frac{2x - 1}{\sqrt{3x^2 + x + 1}}$

$= \displaystyle \lim_{x \to \infty} \frac{x \left(2 - \dfrac{1}{x} \right)}{|x| \sqrt{3 + \dfrac{1}{x} + \dfrac{1}{x^2}}}$ (but $|x| = x$ as $x \to \infty$)

$= \displaystyle \lim_{x \to \infty} \frac{2 - \dfrac{1}{x}}{\sqrt{3 + \dfrac{1}{x} + \dfrac{1}{x^2}}} = \frac{2}{\sqrt{3}}$

10. $\displaystyle \lim_{x \to -\infty} \frac{2x - 5}{|3x + 2|} = \lim_{x \to -\infty} \frac{2x - 5}{-(3x + 2)} = -\frac{2}{3}$

12. $\displaystyle \lim_{x \to 3} \frac{1}{(3 - x)^2} = \infty$

14. $\displaystyle \lim_{x \to 3+} \frac{1}{3 - x} = -\infty$

16. $\displaystyle \lim_{x \to -2/5} \frac{2x + 5}{5x + 2}$ does not exist.

18. $\displaystyle \lim_{x \to -2/5+} \frac{2x + 5}{5x + 2} = \infty$

20. $\displaystyle \lim_{x \to 1-} \frac{x}{\sqrt{1 - x^2}} = \infty$

22. $\displaystyle \lim_{x \to 1-} \frac{1}{|x - 1|} = \infty$

24. $\displaystyle \lim_{x \to 1+} \frac{\sqrt{x^2 - x}}{x - x^2} = \lim_{x \to 1+} \frac{-1}{\sqrt{x^2 - x}} = -\infty$

26. $\displaystyle \lim_{x \to \infty} \frac{x^3 + 3}{x^2 + 2} = \lim_{x \to \infty} \frac{x + \dfrac{3}{x^2}}{1 + \dfrac{2}{x^2}} = \infty$

28. $\displaystyle \lim_{x \to \infty} \left(\frac{x^2}{x + 1} - \frac{x^2}{x - 1} \right) = \lim_{x \to \infty} \frac{-2x^2}{x^2 - 1} = -2$

30. $\displaystyle \lim_{x \to \infty} \left(\sqrt{x^2 + 2x} - \sqrt{x^2 - 2x} \right)$

$= \displaystyle \lim_{x \to \infty} \frac{x^2 + 2x - x^2 + 2x}{\sqrt{x^2 + 2x} + \sqrt{x^2 - 2x}}$

$= \displaystyle \lim_{x \to \infty} \frac{4x}{x \sqrt{1 + \dfrac{2}{x}} + x \sqrt{1 - \dfrac{2}{x}}}$

$= \displaystyle \lim_{x \to \infty} \frac{4}{\sqrt{1 + \dfrac{2}{x}} + \sqrt{1 - \dfrac{2}{x}}} = \frac{4}{2} = 2$

32. $\displaystyle \lim_{x \to -\infty} \frac{1}{\sqrt{x^2 + 2x} - x} = \lim_{x \to -\infty} \frac{1}{|x|(\sqrt{1 + (2/x)} + 1)} = 0$

34. Since $\displaystyle \lim_{x \to \infty} \frac{2x - 5}{|3x + 2|} = \frac{2}{3}$ and $\displaystyle \lim_{x \to -\infty} \frac{2x - 5}{|3x + 2|} = -\frac{2}{3}$, $y = \pm(2/3)$ are horizontal asymptotes of $y = (2x - 5)/|3x + 2|$. The only vertical asymptote is $x = -2/3$, which makes the denominator zero.

36. $\displaystyle \lim_{x \to 1} f(x) = \infty$

38. $\displaystyle \lim_{x \to 2-} f(x) = 2$

40. $\displaystyle \lim_{x \to 3+} f(x) = \infty$

42. $\displaystyle \lim_{x \to 4-} f(x) = 0$

44. $\displaystyle \lim_{x \to 5+} f(x) = 0$

46. horizontal: $y = 1$; vertical: $x = 1$, $x = 3$.

48. $\displaystyle \lim_{x \to 3-} \lfloor x \rfloor = 2$

50. $\displaystyle \lim_{x \to 2.5} \lfloor x \rfloor = 2$

52. $\displaystyle \lim_{x \to -3-} \lfloor x \rfloor = -4$

54. $\lim\limits_{x \to 0+} f(x) = L$
(a) If f is even, then $f(-x) = f(x)$.
Hence, $\lim\limits_{x \to 0-} f(x) = L$.
(b) If f is odd, then $f(-x) = -f(x)$.
Therefore, $\lim\limits_{x \to 0-} f(x) = -L$.

Section 1.4 Continuity (page 85)

2. g has removable discontinuities at $x = -1$ and $x = 2$. Redefine $g(-1) = 1$ and $g(2) = 0$ to make g continuous at those points.

4. Function f is discontinuous at $x = 1$, 2, 3, 4, and 5. f is left continuous at $x = 4$ and right continuous at $x = 2$ and $x = 5$.

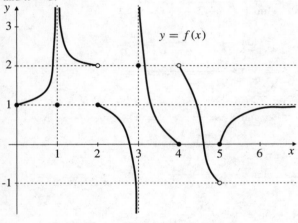

Fig. 1.4.4

6. $\operatorname{sgn} x$ is not defined at $x = 0$, so cannot be either continuous or discontinuous there. (Functions can be continuous or discontinuous only at points in their domains!)

8. $f(x) = \begin{cases} x & \text{if } x < -1 \\ x^2 & \text{if } x \geq -1 \end{cases}$ is continuous everywhere on the real line except at $x = -1$ where it is right continuous, but not left continuous.

$$\lim_{x \to -1-} f(x) = \lim_{x \to -1-} x = -1 \neq 1$$
$$= f(-1) = \lim_{x \to -1+} x^2 = \lim_{x \to -1+} f(x).$$

10. $f(x) = \begin{cases} x^2 & \text{if } x \leq 1 \\ 0.987 & \text{if } x > 1 \end{cases}$ is continuous everywhere except at $x = 1$, where it is left continuous but not right continuous because $0.987 \neq 1$. Close, as they say, but no cigar.

12. $C(t)$ is discontinuous only at the integers. It is continuous on the left at the integers, but not on the right.

14. Since $\dfrac{1+t^3}{1-t^2} = \dfrac{(1+t)(1-t+t^2)}{(1+t)(1-t)} = \dfrac{1-t+t^2}{1-t}$ for $t \neq -1$, we can define the function to be $3/2$ at $t = -1$ to make it continuous there. The continuous extension is $\dfrac{1-t+t^2}{1-t}$.

16. Since
$\dfrac{x^2-2}{x^4-4} = \dfrac{(x-\sqrt{2})(x+\sqrt{2})}{(x-\sqrt{2})(x+\sqrt{2})(x^2+2)} = \dfrac{x+\sqrt{2}}{(x+\sqrt{2})(x^2+2)}$
for $x \neq \sqrt{2}$, we can define the function to be $1/4$ at $x = \sqrt{2}$ to make it continuous there. The continuous extension is $\dfrac{x+\sqrt{2}}{(x+\sqrt{2})(x^2+2)}$. (Note: cancelling the $x + \sqrt{2}$ factors provides a further continuous extension to $x = -\sqrt{2}$.

18. $\lim_{x \to 3-} g(x) = 3 - m$ and $\lim_{x \to 3+} g(x) = 1 - 3m = g(3)$. Thus g will be continuous at $x = 3$ if $3 - m = 1 - 3m$, that is, if $m = -1$.

20. The Max-Min Theorem says that a continuous function defined on a closed, finite interval must have maximum and minimum values. It does not say that other functions cannot have such values. The Heaviside function is not continuous on $[-1, 1]$ (because it is discontinuous at $x = 0$), but it still has maximum and minimum values. Do not confuse a theorem with its converse.

22. Let the numbers be x and y, where $x \geq 0$, $y \geq 0$, and $x + y = 8$. If S is the sum of their squares then
$$S = x^2 + y^2 = x^2 + (8-x)^2$$
$$= 2x^2 - 16x + 64 = 2(x-4)^2 + 32.$$

Since $0 \leq x \leq 8$, the maximum value of S occurs at $x = 0$ or $x = 8$, and is 64. The minimum value occurs at $x = 4$ and is 32.

24. If x desks are shipped, the shipping cost per desk is
$$C = \frac{245x - 30x^2 + x^3}{x} = x^2 - 30x + 245$$
$$= (x-15)^2 + 20.$$

This cost is minimized if $x = 15$. The manufacturer should send 15 desks in each shipment, and the shipping cost will then be \$20 per desk.

26. $f(x) = x^2 + 4x + 3 = (x+1)(x+3)$
$f(x) > 0$ on $(-\infty, -3)$ and $(-1, \infty)$
$f(x) < 0$ on $(-3, -1)$.

28. $f(x) = \dfrac{x^2 + x - 2}{x^3} = \dfrac{(x+2)(x-1)}{x^3}$
$f(x) > 0$ on $(-2, 0)$ and $(1, \infty)$
$f(x) < 0$ on $(-\infty, -2)$ and $(0, 1)$.

30. $f(x) = x^3 - 15x + 1$ is continuous everywhere.
$f(-4) = -3$, $f(-3) = 19$, $f(1) = -13$, $f(4) = 5$.
Because of the sign changes f has a zero between -4 and -3, another zero between -3 and 1, and another between 1 and 4.

32. Let $g(x) = f(x) - x$. Since $0 \le f(x) \le 1$ if $0 \le x \le 1$, therefore, $g(0) \ge 0$ and $g(1) \le 0$. If $g(0) = 0$ let $c = 0$, or if $g(1) = 0$ let $c = 1$. (In either case $f(c) = c$.) Otherwise, $g(0) > 0$ and $g(1) < 0$, and, by IVT, there exists c in $(0, 1)$ such that $g(c) = 0$, i.e., $f(c) = c$.

34. f odd $\Leftrightarrow f(-x) = -f(x)$
f continuous on the right $\Leftrightarrow \lim\limits_{x \to 0+} f(x) = f(0)$
Therefore, letting $t = -x$, we obtain

$$\lim_{x \to 0-} f(x) = \lim_{t \to 0+} f(-t) = \lim_{t \to 0+} -f(t)$$
$$= -f(0) = f(-0) = f(0).$$

Therefore f is continuous at 0 and $f(0) = 0$.

36. max 0.133 at $x = 1.437$; min -0.232 at $x = -1.805$

38. max 1.510 at $x = 0.465$; min 0 at $x = 0$ and $x = 1$

40. root $x = 0.739$

42. roots $x = -0.7244919590$ and $x = 1.220744085$

Section 1.5 The Formal Definition of Limit (page 90)

2. Since 1.2% of 8,000 is 96, we require the edge length x of the cube to satisfy $7904 \le x^3 \le 8096$. It is sufficient that $19.920 \le x \le 20.079$. The edge of the cube must be within 0.079 cm of 20 cm.

4. $4 - 0.1 \le x^2 \le 4 + 0.1$
$1.9749 \le x \le 2.0024$

6. $-2 - 0.01 \le \dfrac{1}{x} \le -2 + 0.01$
$-\dfrac{1}{2.01} \ge x \ge -\dfrac{1}{1.99}$
$-0.5025 \le x \le -0.4975$

8. We need $-0.01 \le \sqrt{2x + 3} - 3 \le 0.01$. Thus

$$2.99 \le \sqrt{2x + 3} \le 3.01$$
$$8.9401 \le 2x + 3 \le 9.0601$$
$$2.97005 \le x \le 3.03005$$
$$3 - 0.02995 \le x - 3 \le 0.03005.$$

Here $\delta = 0.02995$ will do.

10. We need $1 - 0.05 \le 1/(x + 1) \le 1 + 0.05$, or $1.0526 \ge x + 1 \ge 0.9524$. This will occur if $-0.0476 \le x \le 0.0526$. In this case we can take $\delta = 0.0476$.

12. To be proved: $\lim\limits_{x \to 2} (5 - 2x) = 1$.
Proof: Let $\epsilon > 0$ be given. Then $|(5 - 2x) - 1| < \epsilon$ holds if $|2x - 4| < \epsilon$, and so if $|x - 2| < \delta = \epsilon/2$. This confirms the limit.

14. To be proved: $\lim\limits_{x \to 2} \dfrac{x - 2}{1 + x^2} = 0$.
Proof: Let $\epsilon > 0$ be given. Then

$$\left| \frac{x - 2}{1 + x^2} - 0 \right| = \frac{|x - 2|}{1 + x^2} \le |x - 2| < \epsilon$$

provided $|x - 2| < \delta = \epsilon$.

16. To be proved: $\lim\limits_{x \to -2} \dfrac{x^2 + 2x}{x + 2} = -2$.
Proof: Let $\epsilon > 0$ be given. For $x \ne -2$ we have

$$\left| \frac{x^2 + 2x}{x + 2} - (-2) \right| = |x + 2| < \epsilon$$

provided $|x + 2| < \delta = \epsilon$. This completes the proof.

18. To be proved: $\lim\limits_{x \to -1} \dfrac{x + 1}{x^2 - 1} = -\dfrac{1}{2}$.
Proof: Let $\epsilon > 0$ be given. If $x \ne -1$, we have

$$\left| \frac{x + 1}{x^2 - 1} - \frac{1}{2} \right| = \left| \frac{1}{x - 1} - \left(-\frac{1}{2} \right) \right| = \frac{|x + 1|}{2|x - 1|}.$$

If $|x + 1| < 1$, then $-2 < x < 0$, so $-3 < x - 1 < -1$ and $|x - 1| > 1$. Ler $\delta = \min(1, 2\epsilon)$. If $0 < |x - (-1)| < \delta$ then $|x - 1| > 1$ and $|x + 1| < 2\epsilon$. Thus

$$\left| \frac{x + 1}{x^2 - 1} - \frac{1}{2} \right| = \frac{|x + 1|}{2|x - 1|} < \frac{2\epsilon}{2} = \epsilon.$$

This completes the required proof.

20. To be proved: $\lim\limits_{x \to 2} x^3 = 8$.
Proof: Let $\epsilon > 0$ be given. We have $|x^3 - 8| = |x - 2||x^2 + 2x + 4|$. If $|x - 2| < 1$, then $1 < x < 3$ and $x^2 < 9$. Therefore $|x^2 + 2x + 4| \le 9 + 2 \times 3 + 4 = 19$. If $|x - 2| < \delta = \min(1, \epsilon/19)$, then

$$|x^3 - 8| = |x - 2||x^2 + 2x + 4| < \frac{\epsilon}{19} \times 19 = \epsilon.$$

This completes the proof.

22. We say that $\lim_{x \to -\infty} f(x) = L$ if the following condition holds: for every number $\epsilon > 0$ there exists a number $R > 0$, depending on ϵ, such that

$$x < -R \quad \text{implies} \quad |f(x) - L| < \epsilon.$$

24. We say that $\lim_{x \to \infty} f(x) = \infty$ if the following condition holds: for every number $B > 0$ there exists a number $R > 0$, depending on B, such that

$$x > R \quad \text{implies} \quad f(x) > B.$$

26. We say that $\lim_{x \to a-} f(x) = \infty$ if the following condition holds: for every number $B > 0$ there exists a number $\delta > 0$, depending on B, such that

$$a - \delta < x < a \quad \text{implies} \quad f(x) > B.$$

28. To be proved: $\lim_{x \to 1-} \dfrac{1}{x-1} = -\infty$. Proof: Let $B > 0$ be given. We have $\dfrac{1}{x-1} < -B$ if $0 > x - 1 > -1/B$, that is, if $1 - \delta < x < 1$, where $\delta = 1/B$.. This completes the proof.

30. To be proved: $\lim_{x \to \infty} \sqrt{x} = \infty$. Proof: Let $B > 0$ be given. We have $\sqrt{x} > B$ if $x > R$ where $R = B^2$. This completes the proof.

32. To be proved: if $\lim_{x \to a} g(x) = M$, then there exists $\delta > 0$ such that if $0 < |x - a| < \delta$, then $|g(x)| < 1 + |M|$.
Proof: Taking $\epsilon = 1$ in the definition of limit, we obtain a number $\delta > 0$ such that if $0 < |x - a| < \delta$, then $|g(x) - M| < 1$. It follows from this latter inequality that

$$|g(x)| = |(g(x) - M) + M| \le |G(x) - M| + |M| < 1 + |M|.$$

34. To be proved: if $\lim_{x \to a} g(x) = M$ where $M \ne 0$, then there exists $\delta > 0$ such that if $0 < |x - a| < \delta$, then $|g(x)| > |M|/2$.
Proof: By the definition of limit, there exists $\delta > 0$ such that if $0 < |x - a| < \delta$, then $|g(x) - M| < |M|/2$ (since $|M|/2$ is a positive number). This latter inequality implies that

$$|M| = |g(x) + (M - g(x))| \le |g(x)| + |g(x) - M| < |g(x)| + \frac{|M|}{2}.$$

It follows that $|g(x)| > |M| - (|M|/2) = |M|/2$, as required.

36. To be proved: if $\lim_{x \to a} f(x) = L$ and $\lim_{x \to a} f(x) = M \ne 0$, then $\lim_{x \to a} \dfrac{f(x)}{g(x)} = \dfrac{L}{M}$.
Proof: By Exercises 33 and 35 we have

$$\lim_{x \to a} \frac{f(x)}{g(x)} = \lim_{x \to a} f(x) \times \frac{1}{g(x)} = L \times \frac{1}{M} = \frac{L}{M}.$$

38. To be proved: if $f(x) \le g(x) \le h(x)$ in an open interval containing $x = a$ (say, for $a - \delta_1 < x < a + \delta_1$, where $\delta_1 > 0$), and if $\lim_{x \to a} f(x) = \lim_{x \to a} h(x) = L$, then also $\lim_{x \to a} g(x) = L$.
Proof: Let $\epsilon > 0$ be given. Since $\lim_{x \to a} f(x) = L$, there exists $\delta_2 > 0$ such that if $0 < |x - a| < \delta_2$, then $|f(x) - L| < \epsilon/3$. Since $\lim_{x \to a} h(x) = L$, there exists $\delta_3 > 0$ such that if $0 < |x - a| < \delta_3$, then $|h(x) - L| < \epsilon/3$. Let $\delta = \min(\delta_1, \delta_2, \delta_3)$. If $0 < |x - a| < \delta$, then

$$
\begin{aligned}
|g(x) - L| &= |g(x) - f(x) + f(x) - L| \\
&\le |g(x) - f(x)| + |f(x) - L| \\
&\le |h(x) - f(x)| + |f(x) - L| \\
&= |h(x) - L + L - f(x)| + |f(x) - L| \\
&\le |h(x) - L| + |f(x) - L| + |f(x) - L| \\
&< \frac{\epsilon}{3} + \frac{\epsilon}{3} + \frac{\epsilon}{3} = \epsilon.
\end{aligned}
$$

Thus $\lim_{x \to a} g(x) = L$.

Review Exercises 1   (page 91)

2. The average rate of change of $1/x$ over $[-2, -1]$ is

$$\frac{(1/(-1)) - (1/(-2))}{-1 - (-2)} = \frac{-1/2}{1} = -\frac{1}{2}.$$

4. The rate of change of $1/x$ at $x = -3/2$ is

$$
\begin{aligned}
\lim_{h \to 0} \frac{\dfrac{1}{-(3/2) + h} - \left(\dfrac{1}{-3/2}\right)}{h} &= \lim_{h \to 0} \frac{\dfrac{2}{2h - 3} + \dfrac{2}{3}}{h} \\
&= \lim_{h \to 0} \frac{2(3 + 2h - 3)}{3(2h - 3)h} \\
&= \lim_{h \to 0} \frac{4}{3(2h - 3)} = -\frac{4}{9}.
\end{aligned}
$$

6. $\lim_{x \to 2} \dfrac{x^2}{1 - x^2} = \dfrac{2^2}{1 - 2^2} = -\dfrac{4}{3}$

8. $\lim_{x \to 2} \dfrac{x^2 - 4}{x^2 - 5x + 6} = \lim_{x \to 2} \dfrac{(x - 2)(x + 2)}{(x - 2)(x - 3)} = \lim_{x \to 2} \dfrac{x + 2}{x - 3} = -4$

10. $\lim_{x \to 2-} \dfrac{x^2 - 4}{x^2 - 4x + 4} = \lim_{x \to 2-} \dfrac{x + 2}{x - 2} = -\infty$

12. $\lim_{x \to 4} \dfrac{2 - \sqrt{x}}{x - 4} = \lim_{x \to 4} \dfrac{4 - x}{(2 + \sqrt{x})(x - 4)} = -\dfrac{1}{4}$

14. $\lim_{h \to 0} \dfrac{h}{\sqrt{x + 3h} - \sqrt{x}} = \lim_{h \to 0} \dfrac{h(\sqrt{x + 3h} + \sqrt{x})}{(x + 3h) - x}$
$\qquad\qquad = \lim_{h \to 0} \dfrac{\sqrt{x + 3h} + \sqrt{x}}{3} = \dfrac{2\sqrt{x}}{3}$

16. $\lim_{x \to 0} \sqrt{x - x^2}$ does not exist because $\sqrt{x - x^2}$ is not defined for $x < 0$.

18. $\lim\limits_{x \to 1-} \sqrt{x - x^2} = 0$

20. $\lim\limits_{x \to -\infty} \dfrac{2x + 100}{x^2 + 3} = \lim\limits_{x \to -\infty} \dfrac{(2/x) + (100/x^2)}{1 + (3/x^2)} = 0$

22. $\lim\limits_{x \to \infty} \dfrac{x^4}{x^2 - 4} = \lim\limits_{x \to \infty} \dfrac{x^2}{1 - (4/x^2)} = \infty$

24. $\lim\limits_{x \to 1/2} \dfrac{1}{\sqrt{x - x^2}} = \dfrac{1}{\sqrt{1/4}} = 2$

26. $\lim\limits_{x \to \infty} \dfrac{\cos x}{x} = 0$ by the squeeze theorem, since

$$-\frac{1}{x} \le \frac{\cos x}{x} \le \frac{1}{x} \quad \text{for all } x > 0$$

and $\lim_{x \to \infty}(-1/x) = \lim_{x \to \infty}(1/x) = 0$.

28. $\lim\limits_{x \to 0} \sin \dfrac{1}{x^2}$ does not exist; $\sin(1/x^2)$ takes the values -1 and 1 in any interval $(-\delta, \delta)$, where $\delta > 0$, and limits, if they exist, must be unique.

30. $\lim\limits_{x \to \infty} [x + \sqrt{x^2 - 4x + 1}] = \infty + \infty = \infty$

32. $f(x) = \dfrac{x}{x + 1}$ is continuous everywhere on its domain, which consists of all real numbers except $x = -1$. It is discontinuous nowhere.

34. $f(x) = \begin{cases} x^2 & \text{if } x > 1 \\ x & \text{if } x \le 1 \end{cases}$ is defined and continuous everywhere, and so discontinuous nowhere. Observe that $\lim_{x \to 1-} f(x) = 1 = \lim_{x \to 1+} f(x)$.

36. $f(x) = H(9 - x^2) = \begin{cases} 1 & \text{if } -3 \le x \le 3 \\ 0 & \text{if } x < -3 \text{ or } x > 3 \end{cases}$ is defined everywhere and discontinuous at $x = \pm 3$. It is right continuous at -3 and left continuous at 3.

38. $f(x) = \begin{cases} |x|/|x + 1| & \text{if } x \ne -1 \\ 1 & \text{if } x = -1 \end{cases}$ is defined everywhere and discontinuous at $x = -1$ where it is neither left nor right continuous since $\lim_{x \to -1} f(x) = \infty$, while $f(-1) = 1$.

Challenging Problems 1   (page 92)

2. For x near 0 we have $|x - 1| = 1 - x$ and $|x + 1| = x + 1$. Thus

$$\lim_{x \to 0} \frac{x}{|x - 1| - |x + 1|} = \lim_{x \to 0} \frac{x}{(1 - x) - (x + 1)} = -\frac{1}{2}.$$

4. Let $y = x^{1/6}$. Then we have

$$\lim_{x \to 64} \frac{x^{1/3} - 4}{x^{1/2} - 8} = \lim_{y \to 2} \frac{y^2 - 4}{y^3 - 8}$$

$$= \lim_{y \to 2} \frac{(y - 2)(y + 2)}{(y - 2)(y^2 + 2y + 4)}$$

$$= \lim_{y \to 2} \frac{y + 2}{y^2 + 2y + 4} = \frac{4}{12} = \frac{1}{3}.$$

6. $r_+(a) = \dfrac{-1 + \sqrt{1 + a}}{a}, \; r_-(a) = \dfrac{-1 - \sqrt{1 + a}}{a}.$

a) $\lim_{a \to 0} r_-(a)$ does not exist. Observe that the right limit is $-\infty$ and the left limit is ∞.

b) From the following table it appears that $\lim_{a \to 0} r_+(a) = 1/2$, the solution of the linear equation $2x - 1 = 0$ which results from setting $a = 0$ in the quadratic equation $ax^2 + 2x - 1 = 0$.

a	$r_+(a)$
1	0.41421
0.1	0.48810
-0.1	0.51317
0.01	0.49876
-0.01	0.50126
0.001	0.49988
-0.001	0.50013

c) $\lim\limits_{a \to 0} r_+(a) = \lim\limits_{a \to 0} \dfrac{\sqrt{1 + a} - 1}{a}$

$\qquad = \lim\limits_{a \to 0} \dfrac{(1 + a) - 1}{a(\sqrt{1 + a} + 1)}$

$\qquad = \lim\limits_{a \to 0} \dfrac{1}{\sqrt{1 + a} + 1} = \dfrac{1}{2}.$

8. a) To be proved: if f is a continuous function defined on a closed interval $[a, b]$, then the range of f is a closed interval.
Proof: By the Max-Min Theorem there exist numbers u and v in $[a, b]$ such that $f(u) \le f(x) \le f(v)$ for all x in $[a, b]$. By the Intermediate-Value Theorem, $f(x)$ takes on all values between $f(u)$ and $f(v)$ at values of x between u and v, and hence at points of $[a, b]$. Thus the range of f is $[f(u), f(v)]$, a closed interval.

b) If the domain of the continuous function f is an open interval, the range of f can be any interval (open, closed, half open, finite, or infinite).

10. $f(x) = \dfrac{1}{x - x^2} = \dfrac{1}{\frac{1}{4} - \left(\frac{1}{4} - x + x^2\right)} = \dfrac{1}{\frac{1}{4} - \left(x - \frac{1}{2}\right)^2}.$
Observe that $f(x) \ge f(1/2) = 4$ for all x in $(0, 1)$.

CHAPTER 2. DIFFERENTIATION

Section 2.1 Tangent Lines and Their Slopes (page 98)

2. Since $y = x/2$ is a straight line, its tangent at any point $(a, a/2)$ on it is the same line $y = x/2$.

4. The slope of $y = 6 - x - x^2$ at $x = -2$ is

$$m = \lim_{h \to 0} \frac{6 - (-2 + h) - (-2 + h)^2 - 4}{h}$$
$$= \lim_{h \to 0} \frac{3h - h^2}{h} = \lim_{h \to 0} (3 - h) = 3.$$

The tangent line at $(-2, 4)$ is $y = 3x + 10$.

6. The slope of $y = \dfrac{1}{x^2 + 1}$ at $(0, 1)$ is

$$m = \lim_{h \to 0} \frac{1}{h} \left(\frac{1}{h^2 + 1} - 1 \right) = \lim_{h \to 0} \frac{-h}{h^2 + 1} = 0.$$

The tangent line at $(0, 1)$ is $y = 1$.

8. The slope of $y = \dfrac{1}{\sqrt{x}}$ at $x = 9$ is

$$m = \lim_{h \to 0} \frac{1}{h} \left(\frac{1}{\sqrt{9 + h}} - \frac{1}{3} \right)$$
$$= \lim_{h \to 0} \frac{3 - \sqrt{9 + h}}{3h\sqrt{9 + h}} \cdot \frac{3 + \sqrt{9 + h}}{3 + \sqrt{9 + h}}$$
$$= \lim_{h \to 0} \frac{9 - 9 - h}{3h\sqrt{9 + h}(3 + \sqrt{9 + h})}$$
$$= -\frac{1}{3(3)(6)} = -\frac{1}{54}.$$

The tangent line at $(9, \frac{1}{3})$ is $y = \frac{1}{3} - \frac{1}{54}(x - 9)$, or $y = \frac{1}{2} - \frac{1}{54}x$.

10. The slope of $y = \sqrt{5 - x^2}$ at $x = 1$ is

$$m = \lim_{h \to 0} \frac{\sqrt{5 - (1 + h)^2} - 2}{h}$$
$$= \lim_{h \to 0} \frac{5 - (1 + h)^2 - 4}{h \left(\sqrt{5 - (1 + h)^2} + 2 \right)}$$
$$= \lim_{h \to 0} \frac{-2 - h}{\sqrt{5 - (1 + h)^2} + 2} = -\frac{1}{2}.$$

The tangent line at $(1, 2)$ is $y = 2 - \frac{1}{2}(x - 1)$, or $y = \frac{5}{2} - \frac{1}{2}x$.

12. The slope of $y = \dfrac{1}{x}$ at $(a, \frac{1}{a})$ is

$$m = \lim_{h \to 0} \frac{1}{h} \left(\frac{1}{a + h} + \frac{1}{a} \right) = \lim_{h \to 0} \frac{a - a - h}{h(a + h)(a)} = -\frac{1}{a^2}.$$

The tangent line at $(a, \frac{1}{a})$ is $y = \frac{1}{a} - \frac{1}{a^2}(x - a)$, or $y = \frac{2}{a} - \frac{x}{a^2}$.

14. The slope of $f(x) = (x - 1)^{4/3}$ at $x = 1$ is

$$m = \lim_{h \to 0} \frac{(1 + h - 1)^{4/3} - 0}{h} = \lim_{h \to 0} h^{1/3} = 0.$$

The graph of f has a tangent line with slope 0 at $x = 1$. Since $f(1) = 0$, the tangent has equation $y = 0$

16. The slope of $f(x) = |x^2 - 1|$ at $x = 1$ is
$$m = \lim_{h \to 0} \frac{|(1 + h)^2 - 1| - |1 - 1|}{h} = \lim_{h \to 0} \frac{|2h + h^2|}{h},$$
which does not exist, and is not $-\infty$ or ∞. The graph of f has no tangent at $x = 1$.

18. The slope of $y = x^2 - 1$ at $x = x_0$ is

$$m = \lim_{h \to 0} \frac{[(x_0 + h)^2 - 1] - (x_0^2 - 1)}{h}$$
$$= \lim_{h \to 0} \frac{2x_0 h + h^2}{h} = 2x_0.$$

If $m = -3$, then $x_0 = -\frac{3}{2}$. The tangent line with slope $m = -3$ at $(-\frac{3}{2}, \frac{5}{4})$ is $y = \frac{5}{4} - 3(x + \frac{3}{2})$, that is, $y = -3x - \frac{13}{4}$.

20. The slope of $y = x^3 - 3x$ at $x = a$ is

$$m = \lim_{h \to 0} \frac{1}{h} \left[(a + h)^3 - 3(a + h) - (a^3 - 3a) \right]$$
$$= \lim_{h \to 0} \frac{1}{h} \left[a^3 + 3a^2 h + 3ah^2 + h^3 - 3a - 3h - a^3 + 3a \right]$$
$$= \lim_{h \to 0} [3a^2 + 3ah + h^2 - 3] = 3a^2 - 3.$$

At points where the tangent line is parallel to the x-axis, the slope is zero, so such points must satisfy $3a^2 - 3 = 0$. Thus, $a = \pm 1$. Hence, the tangent line is parallel to the x-axis at the points $(1, -2)$ and $(-1, 2)$.

22. The slope of the curve $y = 1/x$ at $x = a$ is

$$m = \lim_{h \to 0} \frac{\dfrac{1}{a + h} - \dfrac{1}{a}}{h} = \lim_{h \to 0} \frac{a - (a + h)}{ah(a + h)} = -\frac{1}{a^2}.$$

The tangent at $x = a$ is perpendicular to the line $y = 4x - 3$ if $-1/a^2 = -1/4$, that is, if $a = \pm 2$. The corresponding points on the curve are $(-2, -1/2)$ and $(2, 1/2)$.

24. The curves $y = kx^2$ and $y = k(x - 2)^2$ intersect at $(1, k)$. The slope of $y = kx^2$ at $x = 1$ is

$$m_1 = \lim_{h \to 0} \frac{k(1 + h)^2 - k}{h} = \lim_{h \to 0} (2 + h)k = 2k.$$

The slope of $y = k(x - 2)^2$ at $x = 1$ is

$$m_2 = \lim_{h \to 0} \frac{k(2 - (1 + h))^2 - k}{h} = \lim_{h \to 0} (-2 + h)k = -2k.$$

The two curves intersect at right angles if $2k = -1/(-2k)$, that is, if $4k^2 = 1$, which is satisfied if $k = \pm 1/2$.

26. Horizontal tangent at $(-1, 8)$ and $(2, -19)$.

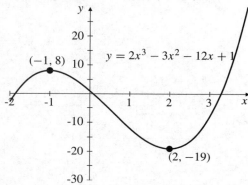

$y = 2x^3 - 3x^2 - 12x + 1$

$(-1, 8)$

$(2, -19)$

Fig. 2.1.26

28. Horizontal tangent at $(a, 2)$ and $(-a, -2)$ for all $a > 1$. No tangents at $(1, 2)$ and $(-1, -2)$.

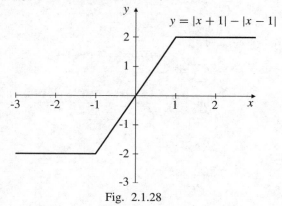

$y = |x + 1| - |x - 1|$

Fig. 2.1.28

30. Horizontal tangent at $(0, 1)$. No tangents at $(-1, 0)$ and $(1, 0)$.

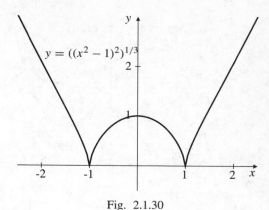

$y = ((x^2 - 1)^2)^{1/3}$

Fig. 2.1.30

32. The slope of $P(x)$ at $x = a$ is

$$m = \lim_{h \to 0} \frac{P(a + h) - P(a)}{h}.$$

Since $P(a + h) = a_0 + a_1 h + a_2 h^2 + \cdots + a_n h^n$ and $P(a) = a_0$, the slope is

$$m = \lim_{h \to 0} \frac{a_0 + a_1 h + a_2 h^2 + \cdots + a_n h^n - a_0}{h}$$
$$= \lim_{h \to 0} a_1 + a_2 h + \cdots + a_n h^{n-1} = a_1.$$

Thus the line $y = \ell(x) = m(x - a) + b$ is tangent to $y = P(x)$ at $x = a$ if and only if $m = a_1$ and $b = a_0$, that is, if and only if

$$P(x) - \ell(x) = a_2(x - a)^2 + a_3(x - a)^3 + \cdots + a_n(x - a)^n$$
$$= (x - a)^2 \left[a_2 + a_3(x - a) + \cdots + a_n(x - a)^{n-2} \right]$$
$$= (x - a)^2 Q(x)$$

where Q is a polynomial.

Section 2.2 The Derivative (page 105)

2.

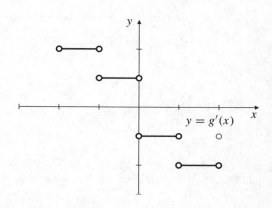

$y = g'(x)$

4.

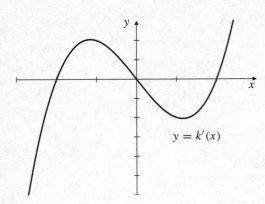

$y = k'(x)$

$y = f(x) = |x^2 - 1| - |x^2 - 4|$

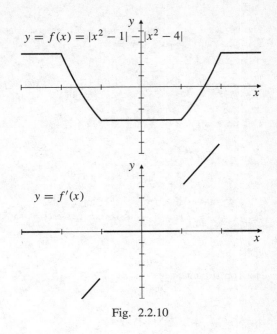

$y = f'(x)$

Fig. 2.2.10

6. Assuming the tick marks are spaced 1 unit apart, the function g is differentiable on the intervals $(-2, -1)$, $(-1, 0)$, $(0, 1)$, and $(1, 2)$.

8. $y = f(x)$ has horizontal tangents at the points near $1/2$ and $3/2$ where $f'(x) = 0$

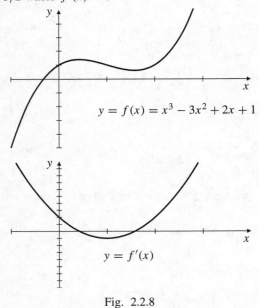

$y = f(x) = x^3 - 3x^2 + 2x + 1$

$y = f'(x)$

Fig. 2.2.8

10. $y = f(x)$ is constant on the intervals $(-\infty, -2)$, $(-1, 1)$, and $(2, \infty)$. It is not differentiable at $x = \pm 2$ and $x = \pm 1$.

12. $f(x) = 1 + 4x - 5x^2$

$$f'(x) = \lim_{h \to 0} \frac{1 + 4(x + h) - 5(x + h)^2 - (1 + 4x - 5x^2)}{h}$$

$$= \lim_{h \to 0} \frac{4h - 10xh - 5h^2}{h} = 4 - 10x$$

14. $s = \dfrac{1}{3 + 4t}$

$$\frac{ds}{dt} = \lim_{h \to 0} \frac{1}{h} \left[\frac{1}{3 + 4(t + h)} - \frac{1}{3 + 4t} \right]$$

$$= \lim_{h \to 0} \frac{3 + 4t - 3 - 4t - 4h}{h(3 + 4t)[3 + (4t + h)]} = -\frac{4}{(3 + 4t)^2}$$

16. $f(x) = \frac{3}{4} \sqrt{2 - x}$

$$f'(x) = \lim_{h \to 0} \frac{\frac{3}{4} \sqrt{2 - (x + h)} - \frac{3}{4} \sqrt{2 - x}}{h}$$

$$= \lim_{h \to 0} \frac{3}{4} \left[\frac{2 - x - h - 2 + x}{h(\sqrt{2 - (x + h)} + \sqrt{2 - x})} \right]$$

$$= -\frac{3}{8\sqrt{2 - x}}$$

18. $z = \dfrac{s}{1 + s}$

$$\frac{dz}{ds} = \lim_{h \to 0} \frac{1}{h} \left[\frac{s + h}{1 + s + h} - \frac{s}{1 + s} \right]$$

$$= \lim_{h \to 0} \frac{(s + h)(1 + s) - s(1 + s + h)}{h(1 + s)(1 + s + h)} = \frac{1}{(1 + s)^2}$$

20. $y = \dfrac{1}{x^2}$

$$y' = \lim_{h \to 0} \frac{1}{h} \left[\frac{1}{(x+h)^2} - \frac{1}{x^2} \right]$$

$$= \lim_{h \to 0} \frac{x^2 - (x+h)^2}{hx^2(x+h)^2} = -\frac{2}{x^3}$$

22. $f(t) = \dfrac{t^2 - 3}{t^2 + 3}$

$$f'(t) = \lim_{h \to 0} \frac{1}{h} \left(\frac{(t+h)^2 - 3}{(t+h)^2 + 3} - \frac{t^2 - 3}{t^2 + 3} \right)$$

$$= \lim_{h \to 0} \frac{[(t+h)^2 - 3](t^2 + 3) - (t^2 - 3)[(t+h)^2 + 3]}{h(t^2 + 3)[(t+h)^2 + 3]}$$

$$= \lim_{h \to 0} \frac{12th + 6h^2}{h(t^2 + 3)[(t+h)^2 + 3]} = \frac{12t}{(t^2 + 3)^2}$$

24. Since $g(x) = x^2 \operatorname{sgn} x = x|x| = \begin{cases} x^2 & \text{if } x > 0 \\ -x^2 & \text{if } x < 0 \end{cases}$, g will become continuous and differentiable at $x = 0$ if we define $g(0) = 0$.

26. $y = x^3 - 2x$

x	$\dfrac{f(x) - f(1)}{x - 1}$	x	$\dfrac{f(x) - f(1)}{x - 1}$
0.9	0.71000	1.1	1.31000
0.99	0.97010	1.01	1.03010
0.999	0.99700	1.001	1.00300
0.9999	0.99970	1.0001	1.00030

$$\frac{d}{dx}(x^3 - 2x)\bigg|_{x=1} = \lim_{h \to 0} \frac{(1+h)^3 - 2(1+h) - (-1)}{h}$$

$$= \lim_{h \to 0} \frac{h + 3h^2 + h^3}{h}$$

$$= \lim_{h \to 0} 1 + 3h + h^2 = 1$$

28. The slope of $y = 5 + 4x - x^2$ at $x = 2$ is

$$\frac{dy}{dx}\bigg|_{x=2} = \lim_{h \to 0} \frac{5 + 4(2+h) - (2+h)^2 - 9}{h}$$

$$= \lim_{h \to 0} \frac{-h^2}{h} = 0.$$

Thus, the tangent line at $x = 2$ has the equation $y = 9$.

30. The slope of $y = \dfrac{t}{t^2 - 2}$ at $t = -2$ and $y = -1$ is

$$\frac{dy}{dt}\bigg|_{t=-2} = \lim_{h \to 0} \frac{1}{h} \left[\frac{-2+h}{(-2+h)^2 - 2} - (-1) \right]$$

$$= \lim_{h \to 0} \frac{-2 + h + [(-2+h)^2 - 2]}{h[(-2+h)^2 - 2]} = -\frac{3}{2}.$$

Thus, the tangent line has the equation
$y = -1 - \frac{3}{2}(t + 2)$, that is, $y = -\frac{3}{2}t - 4$.

32. $f'(x) = -17x^{-18}$ for $x \neq 0$

34. $\dfrac{dy}{dx} = \dfrac{1}{3} x^{-2/3}$ for $x \neq 0$

36. $\dfrac{d}{dt} t^{-2.25} = -2.25 t^{-3.25}$ for $t > 0$

38. $\dfrac{d}{ds} \sqrt{s}\bigg|_{s=9} = \dfrac{1}{2\sqrt{s}}\bigg|_{s=9} = \dfrac{1}{6}.$

40. $f'(8) = -\dfrac{2}{3} x^{-5/3}\bigg|_{x=8} = -\dfrac{1}{48}$

42. The slope of $y = \sqrt{x}$ at $x = x_0$ is

$$\frac{dy}{dx}\bigg|_{x=x_0} = \frac{1}{2\sqrt{x_0}}.$$

Thus, the equation of the tangent line is

$$y = \sqrt{x_0} + \frac{1}{2\sqrt{x_0}}(x - x_0), \text{ that is, } y = \frac{x + x_0}{2\sqrt{x_0}}.$$

44. The intersection points of $y = x^2$ and $x + 4y = 18$ satisfy

$$4x^2 + x - 18 = 0$$
$$(4x + 9)(x - 2) = 0.$$

Therefore $x = -\dfrac{9}{4}$ or $x = 2$.

The slope of $y = x^2$ is $m_1 = \dfrac{dy}{dx} = 2x$.

At $x = -\dfrac{9}{4}$, $m_1 = -\dfrac{9}{2}$. At $x = 2$, $m_1 = 4$.

The slope of $x + 4y = 18$, i.e. $y = -\dfrac{1}{4}x + \dfrac{18}{4}$, is $m_2 = -\dfrac{1}{4}$.

Thus, at $x = 2$, the product of these slopes is $(4)(-\dfrac{1}{4}) = -1$. So, the curve and line intersect at right angles at that point.

46. The slope of $y = \dfrac{1}{x}$ at $x = a$ is

$$\frac{dy}{dx}\bigg|_{x=a} = -\frac{1}{a^2}.$$

If the slope is -2, then $-\dfrac{1}{a^2} = -2$, or $a = \pm\dfrac{1}{\sqrt{2}}$.

Therefore, the equations of the two straight lines are

$$y = \sqrt{2} - 2\left(x - \frac{1}{\sqrt{2}}\right) \text{ and } y = -\sqrt{2} - 2\left(x + \frac{1}{\sqrt{2}}\right),$$

or $y = -2x \pm 2\sqrt{2}$.

48. If a line is tangent to $y = x^2$ at (t, t^2), then its slope is $\dfrac{dy}{dx}\Big|_{x=t} = 2t$. If this line also passes through (a, b), then its slope satisfies

$$\frac{t^2 - b}{t - a} = 2t, \quad \text{that is } t^2 - 2at + b = 0.$$

Hence $t = \dfrac{2a \pm \sqrt{4a^2 - 4b}}{2} = a \pm \sqrt{a^2 - b}$.

If $b < a^2$, i.e. $a^2 - b > 0$, then $t = a \pm \sqrt{a^2 - b}$ has two real solutions. Therefore, there will be two distinct tangent lines passing through (a, b) with equations $y = b + 2\left(a \pm \sqrt{a^2 - b}\right)(x - a)$. If $b = a^2$, then $t = a$. There will be only one tangent line with slope $2a$ and equation $y = b + 2a(x - a)$.

If $b > a^2$, then $a^2 - b < 0$. There will be no real solution for t. Thus, there will be no tangent line.

50. Let $f(x) = x^{-n}$. Then

$$
\begin{aligned}
f'(x) &= \lim_{h \to 0} \frac{(x+h)^{-n} - x^{-n}}{h} \\
&= \lim_{h \to 0} \frac{1}{h}\left(\frac{1}{(x+h)^n} - \frac{1}{x^n}\right) \\
&= \lim_{h \to 0} \frac{x^n - (x+h)^n}{h x^n (x+h)^n} \\
&= \lim_{h \to 0} \frac{x - (x+h)}{h x^n ((x+h)^n} \times \\
&\quad \left(x^{n-1} + x^{n-2}(x+h) + \cdots + (x+h)^{n-1}\right) \\
&= -\frac{1}{x^{2n}} \times n x^{n-1} = -n x^{-(n+1)}.
\end{aligned}
$$

52. Let $f(x) = x^{1/n}$. Then

$$
\begin{aligned}
f'(x) &= \lim_{h \to 0} \frac{(x+h)^{1/n} - x^{1/n}}{h} \quad (\text{let } x + h = a^n,\ x = b^n) \\
&= \lim_{a \to b} \frac{a - b}{a^n - b^n} \\
&= \lim_{a \to b} \frac{1}{a^{n-1} + a^{n-2}b + a^{n-3}b^2 + \cdots + b^{n-1}} \\
&= \frac{1}{n b^{n-1}} = \frac{1}{n} x^{(1/n)-1}.
\end{aligned}
$$

54. Let

$$
\begin{aligned}
f'(a+) &= \lim_{h \to 0+} \frac{f(a+h) - f(a)}{h} \\
f'(a-) &= \lim_{h \to 0-} \frac{f(a+h) - f(a)}{h}
\end{aligned}
$$

If $f'(a+)$ is finite, call the half-line with equation $y = f(a) + f'(a+)(x - a)$, $(x \geq a)$, the *right tangent line* to the graph of f at $x = a$. Similarly, if $f'(a-)$ is finite, call the half-line $y = f(a) + f'(a-)(x - a)$, $(x \leq a)$, the *left tangent line*. If $f'(a+) = \infty$ (or $-\infty$), the right tangent line is the half-line $x = a$, $y \geq f(a)$ (or $x = a$, $y \leq f(a)$). If $f'(a-) = \infty$ (or $-\infty$), the right tangent line is the half-line $x = a$, $y \leq f(a)$ (or $x = a$, $y \geq f(a)$).

The graph has a tangent line at $x = a$ if and only if $f'(a+) = f'(a-)$. (This includes the possibility that both quantities may be $+\infty$ or both may be $-\infty$.) In this case the right and left tangents are two opposite halves of the same straight line. For $f(x) = x^{2/3}$, $f'(x) = \frac{2}{3}x^{-1/3}$. At $(0, 0)$, we have $f'(0+) = +\infty$ and $f'(0-) = -\infty$. In this case both left and right tangents are the *positive* y-axis, and the curve does not have a tangent line at the origin.

For $f(x) = |x|$, we have

$$f'(x) = \text{sgn}(x) = \begin{cases} 1 & \text{if } x > 0 \\ -1 & \text{if } x < 0. \end{cases}$$

At $(0, 0)$, $f'(0+) = 1$, and $f'(0-) = -1$. In this case the right tangent is $y = x$, $(x \geq 0)$, and the left tangent is $y = -x$, $(x \leq 0)$. There is no tangent line.

Section 2.3 Differentiation Rules (page 113)

2. $y = 4x^{1/2} - \dfrac{5}{x}$, $\quad y' = 2x^{-1/2} + 5x^{-2}$

4. $f(x) = \dfrac{6}{x^3} + \dfrac{2}{x^2} - 2$, $\quad f'(x) = -\dfrac{18}{x^4} - \dfrac{4}{x^3}$

6. $y = x^{45} - x^{-45}$ $\quad y' = 45x^{44} + 45x^{-46}$

8. $y = 3\sqrt[3]{t^2} - \dfrac{2}{\sqrt{t^3}} = 3t^{2/3} - 2t^{-3/2}$

$\dfrac{dy}{dt} = 2t^{-1/3} + 3t^{-5/2}$

10. $F(x) = (3x - 2)(1 - 5x)$

$F'(x) = 3(1 - 5x) + (3x - 2)(-5) = 13 - 30x$

12. $g(t) = \dfrac{1}{2t - 3}$, $\quad g'(t) = -\dfrac{2}{(2t - 3)^2}$

14. $y = \dfrac{4}{3 - x}$, $\quad y' = \dfrac{4}{(3 - x)^2}$

16. $g(y) = \dfrac{2}{1 - y^2}$, $\quad g'(y) = \dfrac{4y}{(1 - y^2)^2}$

18. $g(u) = \dfrac{u\sqrt{u} - 3}{u^2} = u^{-1/2} - 3u^{-2}$

$g'(u) = -\dfrac{1}{2}u^{-3/2} + 6u^{-3} = \dfrac{12 - u\sqrt{u}}{2u^3}$

20.
$$z = \frac{x-1}{x^{2/3}} = x^{1/3} - x^{-2/3}$$
$$\frac{dz}{dx} = \frac{1}{3}x^{-2/3} + \frac{2}{3}x^{-5/3} = \frac{x+2}{3x^{5/3}}$$

22.
$$z = \frac{t^2 + 2t}{t^2 - 1}$$
$$z' = \frac{(t^2-1)(2t+2) - (t^2+2t)(2t)}{(t^2-1)^2}$$
$$= -\frac{2(t^2 + t + 1)}{(t^2-1)^2}$$

24.
$$f(x) = \frac{x^3 - 4}{x + 1}$$
$$f'(x) = \frac{(x+1)(3x^2) - (x^3-4)(1)}{(x+1)^2}$$
$$= \frac{2x^3 + 3x^2 + 4}{(x+1)^2}$$

26.
$$F(t) = \frac{t^2 + 7t - 8}{t^2 - t + 1}$$
$$F'(t) = \frac{(t^2-t+1)(2t+7) - (t^2+7t-8)(2t-1)}{(t^2-t+1)^2}$$
$$= \frac{-8t^2 + 18t - 1}{(t^2-t+1)^2}$$

28.
$$f(r) = (r^{-2} + r^{-3} - 4)(r^2 + r^3 + 1)$$
$$f'(r) = (-2r^{-3} - 3r^{-4})(r^2 + r^3 + 1)$$
$$\qquad\qquad + (r^{-2} + r^{-3} - 4)(2r + 3r^2)$$

or

$$f(r) = -2 + r^{-1} + r^{-2} + r^{-3} + r - 4r^2 - 4r^3$$
$$f'(r) = -r^{-2} - 2r^{-3} - 3r^{-4} + 1 - 8r - 12r^2$$

30.
$$y = \frac{(x^2+1)(x^3+2)}{(x^2+2)(x^3+1)}$$
$$= \frac{x^5 + x^3 + 2x^2 + 2}{x^5 + 2x^3 + x^2 + 2}$$
$$y' = \frac{(x^5 + 2x^3 + x^2 + 2)(5x^4 + 3x^2 + 4x)}{(x^5 + 2x^3 + x^2 + 2)^2}$$
$$\quad - \frac{(x^5 + x^3 + 2x^2 + 2)(5x^4 + 6x^2 + 2x)}{(x^5 + 2x^3 + x^2 + 2)^2}$$
$$= \frac{2x^7 - 3x^6 - 3x^4 - 6x^2 + 4x}{(x^5 + 2x^3 + x^2 + 2)^2}$$
$$= \frac{2x^7 - 3x^6 - 3x^4 - 6x^2 + 4x}{(x^2+2)^2(x^3+1)^2}$$

32.
$$f(x) = \frac{(\sqrt{x}-1)(2-x)(1-x^2)}{\sqrt{x}(3+2x)}$$
$$= \left(1 - \frac{1}{\sqrt{x}}\right) \cdot \frac{2 - x - 2x^2 + x^3}{3 + 2x}$$
$$f'(x) = \left(\frac{1}{2}x^{-3/2}\right)\frac{2 - x - 2x^2 + x^3}{3 + 2x} + \left(1 - \frac{1}{\sqrt{x}}\right)$$
$$\times \frac{(3+2x)(-1 - 4x + 3x^2) - (2 - x - 2x^2 + x^3)(2)}{(3+2x)^2}$$
$$= \frac{(2-x)(1-x^2)}{2x^{3/2}(3+2x)}$$
$$\quad + \left(1 - \frac{1}{\sqrt{x}}\right)\frac{4x^3 + 5x^2 - 12x - 7}{(3+2x)^2}$$

34.
$$\frac{d}{dx}\left(\frac{f(x)}{x^2}\right)\Bigg|_{x=2} = \frac{x^2 f'(x) - 2xf(x)}{x^4}\Bigg|_{x=2}$$
$$= \frac{4f'(2) - 4f(2)}{16} = \frac{4}{16} = \frac{1}{4}$$

36.
$$\frac{d}{dx}\left(\frac{f(x)}{x^2 + f(x)}\right)\Bigg|_{x=2}$$
$$= \frac{(x^2 + f(x))f'(x) - f(x)(2x + f'(x))}{(x^2 + f(x))^2}\Bigg|_{x=2}$$
$$= \frac{(4 + f(2))f'(2) - f(2)(4 + f'(2))}{(4 + f(2))^2} = \frac{18 - 14}{6^2} = \frac{1}{9}$$

38.
$$\frac{d}{dt}\left[\frac{t(1+\sqrt{t})}{5-t}\right]\Bigg|_{t=4}$$
$$= \frac{d}{dt}\left[\frac{t + t^{3/2}}{5-t}\right]\Bigg|_{t=4}$$
$$= \frac{(5-t)(1 + \frac{3}{2}t^{1/2}) - (t + t^{3/2})(-1)}{(5-t)^2}\Bigg|_{t=4}$$
$$= \frac{(1)(4) - (12)(-1)}{(1)^2} = 16$$

40.
$$\frac{d}{dt}[(1+t)(1+2t)(1+3t)(1+4t)]\Bigg|_{t=0}$$
$$= (1)(1+2t)(1+3t)(1+4t) + (1+t)(2)(1+3t)(1+4t) +$$
$$(1+t)(1+2t)(3)(1+4t) + (1+t)(1+2t)(1+3t)(4)\Bigg|_{t=0}$$
$$= 1 + 2 + 3 + 4 = 10$$

42. For $y = \dfrac{x+1}{x-1}$ we calculate

$$y' = \frac{(x-1)(1) - (x+1)(1)}{(x-1)^2} = -\frac{2}{(x-1)^2}.$$

At $x = 2$ we have $y = 3$ and $y' = -2$. Thus, the equation of the tangent line is $y = 3 - 2(x-2)$, or $y = -2x + 7$. The normal line is $y = 3 + \frac{1}{2}(x-2)$, or $y = \frac{1}{2}x + 2$.

44. If $y = x^2(4 - x^2)$, then

$$y' = 2x(4 - x^2) + x^2(-2x) = 8x - 4x^3 = 4x(2 - x^2).$$

The slope of a horizontal line must be zero, so $4x(2 - x^2) = 0$, which implies that $x = 0$ or $x = \pm\sqrt{2}$. At $x = 0$, $y = 0$ and at $x = \pm\sqrt{2}$, $y = 4$. Hence, there are two horizontal lines that are tangent to the curve. Their equations are $y = 0$ and $y = 4$.

46. If $y = \dfrac{x+1}{x+2}$, then

$$y' = \frac{(x+2)(1) - (x+1)(1)}{(x+2)^2} = \frac{1}{(x+2)^2}.$$

In order to be parallel to $y = 4x$, the tangent line must have slope equal to 4, i.e.,

$$\frac{1}{(x+2)^2} = 4, \qquad \text{or } (x+2)^2 = \tfrac{1}{4}.$$

Hence $x + 2 = \pm\tfrac{1}{2}$, and $x = -\tfrac{3}{2}$ or $-\tfrac{5}{2}$. At $x = -\tfrac{3}{2}$, $y = -1$, and at $x = -\tfrac{5}{2}$, $y = 3$.
Hence, the tangent is parallel to $y = 4x$ at the points $\left(-\tfrac{3}{2}, -1\right)$ and $\left(-\tfrac{5}{2}, 3\right)$.

48. Since $\dfrac{1}{\sqrt{x}} = y = x^2 \Rightarrow x^{5/2} = 1$, therefore $x = 1$ at the intersection point. The slope of $y = x^2$ at $x = 1$ is $2x\Big|_{x=1} = 2$. The slope of $y = \dfrac{1}{\sqrt{x}}$ at $x = 1$ is

$$\frac{dy}{dx}\Big|_{x=1} = -\frac{1}{2}x^{-3/2}\Big|_{x=1} = -\frac{1}{2}.$$

The product of the slopes is $(2)\left(-\tfrac{1}{2}\right) = -1$. Hence, the two curves intersect at right angles.

50. The tangent to $y = x^2/(x - 1)$ at $(a, a^2/(a-1))$ has slope

$$m = \frac{(x-1)2x - x^2(1)}{(x-1)^2}\Big|_{x=a} = \frac{a^2 - 2a}{(a-1)^2}.$$

The equation of the tangent is

$$y - \frac{a^2}{a-1} = \frac{a^2 - 2a}{(a-1)^2}(x - a).$$

This line passes through $(2, 0)$ provided

$$0 - \frac{a^2}{a-1} = \frac{a^2 - 2a}{(a-1)^2}(2 - a),$$

or, upon simplification, $3a^2 - 4a = 0$. Thus we can have either $a = 0$ or $a = 4/3$. There are two tangents through $(2, 0)$. Their equations are $y = 0$ and $y = -8x + 16$.

52. $f(x) = |x^3| = \begin{cases} x^3 & \text{if } x \geq 0 \\ -x^3 & \text{if } x < 0 \end{cases}$. Therefore f is differentiable everywhere except *possibly* at $x = 0$, However,

$$\lim_{h\to 0+} \frac{f(0+h) - f(0)}{h} = \lim_{h\to 0+} h^2 = 0$$

$$\lim_{h\to 0-} \frac{f(0+h) - f(0)}{h} = \lim_{h\to 0-} (-h^2) = 0.$$

Thus $f'(0)$ exists and equals 0. We have

$$f'(x) = \begin{cases} 3x^2 & \text{if } x \geq 0 \\ -3x^2 & \text{if } x < 0. \end{cases}$$

54. To be proved:

$$(f_1 f_2 \cdots f_n)'$$
$$= f_1' f_2 \cdots f_n + f_1 f_2' \cdots f_n + \cdots + f_1 f_2 \cdots f_n'$$

Proof: The case $n = 2$ is just the Product Rule. Assume the formula holds for $n = k$ for some integer $k > 2$. Using the Product Rule and this hypothesis we calculate

$$(f_1 f_2 \cdots f_k f_{k+1})'$$
$$= [(f_1 f_2 \cdots f_k) f_{k+1}]'$$
$$= (f_1 f_2 \cdots f_k)' f_{k+1} + (f_1 f_2 \cdots f_k) f_{k+1}'$$
$$= (f_1' f_2 \cdots f_k + f_1 f_2' \cdots f_k + \cdots + f_1 f_2 \cdots f_k') f_{k+1}$$
$$\quad + (f_1 f_2 \cdots f_k) f_{k+1}'$$
$$= f_1' f_2 \cdots f_k f_{k+1} + f_1 f_2' \cdots f_k f_{k+1} + \cdots$$
$$\quad + f_1 f_2 \cdots f_k' f_{k+1} + f_1 f_2 \cdots f_k f_{k+1}'$$

so the formula is also true for $n = k + 1$. The formula is therefore for all integers $n \geq 2$ by induction.

Section 2.4   The Chain Rule   (page 118)

2. $y = \left(1 - \dfrac{x}{3}\right)^{99}$

$$y' = 99\left(1 - \frac{x}{3}\right)^{98}\left(-\frac{1}{3}\right) = -33\left(1 - \frac{x}{3}\right)^{98}$$

4. $\dfrac{dy}{dx} = \dfrac{d}{dx}\sqrt{1 - 3x^2} = \dfrac{-6x}{2\sqrt{1 - 3x^2}} = -\dfrac{3x}{\sqrt{1 - 3x^2}}$

6. $z = (1 + x^{2/3})^{3/2}$

$$z' = \tfrac{3}{2}(1 + x^{2/3})^{1/2}(\tfrac{2}{3}x^{-1/3}) = x^{-1/3}(1 + x^{2/3})^{1/2}$$

8. $y = (1 - 2t^2)^{-3/2}$

$$y' = -\tfrac{3}{2}(1 - 2t^2)^{-5/2}(-4t) = 6t(1 - 2t^2)^{-5/2}$$

10. $f(t) = |2 + t^3|$

$$f'(t) = [\text{sgn}\,(2 + t^3)](3t^2) = \frac{3t^2(2 + t^3)}{|2 + t^3|}$$

12. $\quad y = (2 + |x|^3)^{1/3}$

$\quad y' = \frac{1}{3}(2 + |x|^3)^{-2/3}(3|x|^2)\text{sgn}\,(x)$

$\quad\quad = |x|^2(2 + |x|^3)^{-2/3}\left(\dfrac{x}{|x|}\right) = x|x|(2 + |x|^3)^{-2/3}$

14. $\quad f(x) = \left(1 + \sqrt{\dfrac{x-2}{3}}\right)^4$

$\quad f'(x) = 4\left(1 + \sqrt{\dfrac{x-2}{3}}\right)^3 \left(\dfrac{1}{2}\sqrt{\dfrac{3}{x-2}}\right)\left(\dfrac{1}{3}\right)$

$\quad\quad = \dfrac{2}{3}\sqrt{\dfrac{3}{x-2}}\left(1 + \sqrt{\dfrac{x-2}{3}}\right)^3$

16. $\quad y = \dfrac{x^5\sqrt{3 + x^6}}{(4 + x^2)^3}$

$\quad y' = \dfrac{1}{(4+x^2)^6}\left((4+x^2)^3\left[5x^4\sqrt{3+x^6} + x^5\left(\dfrac{3x^5}{\sqrt{3+x^6}}\right)\right]\right.$

$\quad\quad \left. - x^5\sqrt{3+x^6}\left[3(4+x^2)^2(2x)\right]\right)$

$\quad = \dfrac{(4+x^2)\left[5x^4(3+x^6) + 3x^{10}\right] - x^5(3+x^6)(6x)}{(4+x^2)^4\sqrt{3+x^6}}$

$\quad = \dfrac{60x^4 - 3x^6 + 32x^{10} + 2x^{12}}{(4+x^2)^4\sqrt{3+x^6}}$

18.

20. $\quad \dfrac{d}{dx}x^{3/4} = \dfrac{d}{dx}\sqrt{x\sqrt{x}} = \dfrac{1}{2\sqrt{x\sqrt{x}}}\left(\sqrt{x} + \dfrac{x}{2\sqrt{x}}\right) = \dfrac{3}{4}x^{-1/4}$

22. $\quad \dfrac{d}{dt}f(2t + 3) = 2f'(2t + 3)$

24. $\quad \dfrac{d}{dx}\left[f\left(\dfrac{2}{x}\right)\right]^3 = 3\left[f\left(\dfrac{2}{x}\right)\right]^2 f'\left(\dfrac{2}{x}\right)\left(\dfrac{-2}{x^2}\right)$

$\quad\quad = -\dfrac{2}{x^2}f'\left(\dfrac{2}{x}\right)\left[f\left(\dfrac{2}{x}\right)\right]^2$

26. $\quad \dfrac{d}{dt}f(\sqrt{3 + 2t}) = f'(\sqrt{3 + 2t})\dfrac{2}{2\sqrt{3+2t}}$

$\quad\quad = \dfrac{1}{\sqrt{3 + 2t}}f'(\sqrt{3 + 2t})$

28. $\quad \dfrac{d}{dt}f\Big(2f\big(3f(x)\big)\Big)$

$\quad = f'\Big(2f\big(3f(x)\big)\Big)\cdot 2f'\big(3f(x)\big)\cdot 3f'(x)$

$\quad = 6f'(x)f'\big(3f(x)\big)f'\Big(2f\big(3f(x)\big)\Big)$

30. $\quad \dfrac{d}{dx}\left(\dfrac{\sqrt{x^2-1}}{x^2+1}\right)\bigg|_{x=-2}$

$\quad = \dfrac{(x^2+1)\dfrac{x}{\sqrt{x^2-1}} - \sqrt{x^2-1}(2x)}{(x^2+1)^2}\bigg|_{x=-2}$

$\quad = \dfrac{(5)\left(-\dfrac{2}{\sqrt{3}}\right) - \sqrt{3}(-4)}{25} = \dfrac{2}{25\sqrt{3}}$

32. $\quad f(x) = \dfrac{1}{\sqrt{2x + 1}}$

$\quad f'(4) = -\dfrac{1}{(2x+1)^{3/2}}\bigg|_{x=4} = -\dfrac{1}{27}$

34. $\quad F(x) = (1 + x)(2 + x)^2(3 + x)^3(4 + x)^4$

$\quad F'(x) = (2 + x)^2(3 + x)^3(4 + x)^4 +$

$\quad\quad 2(1 + x)(2 + x)(3 + x)^3(4 + x)^4 +$

$\quad\quad 3(1 + x)(2 + x)^2(3 + x)^2(4 + x)^4 +$

$\quad\quad 4(1 + x)(2 + x)^2(3 + x)^3(4 + x)^3$

$\quad F'(0) = (2^2)(3^3)(4^4) + 2(1)(2)(3^3)(4^4) +$

$\quad\quad 3(1)(2^2)(3^2)(4^4) + 4(1)(2^2)(3^3)(4^3)$

$\quad\quad = 4(2^2 \cdot 3^3 \cdot 4^4) = 110{,}592$

36. The slope of $y = \sqrt{1 + 2x^2}$ at $x = 2$ is

$$\dfrac{dy}{dx}\bigg|_{x=2} = \dfrac{4x}{2\sqrt{1 + 2x^2}}\bigg|_{x=2} = \dfrac{4}{3}.$$

Thus, the equation of the tangent line at $(2, 3)$ is $y = 3 + \frac{4}{3}(x - 2)$, or $y = \frac{4}{3}x + \frac{1}{3}$.

38. The slope of $y = (ax + b)^8$ at $x = \dfrac{b}{a}$ is

$\quad \dfrac{dy}{dx}\bigg|_{x=b/a} = 8a(ax + b)^7\bigg|_{x=b/a} = 1024ab^7.$

The equation of the tangent line at $x = \dfrac{b}{a}$ and $y = (2b)^8 = 256b^8$ is

$y = 256b^8 + 1024ab^7\left(x - \dfrac{b}{a}\right)$, or $y = 2^{10}ab^7x - 3\times2^8 b^8$.

40. Given that $f(x) = (x - a)^m(x - b)^n$ then

$\quad f'(x) = m(x - a)^{m-1}(x - b)^n + n(x - a)^m(x - b)^{n-1}$

$\quad\quad = (x - a)^{m-1}(x - b)^{n-1}(mx - mb + nx - na).$

If $x \ne a$ and $x \ne b$, then $f'(x) = 0$ if and only if

$$mx - mb + nx - na = 0,$$

which is equivalent to

$$x = \frac{n}{m+n}a + \frac{m}{m+n}b.$$

This point lies lies between a and b.

42. $4(7x^4 - 49x^2 + 54)/x^7$

44. $5/8$

46. It may happen that $k = g(x + h) - g(x) = 0$ for values of h arbitrarily close to 0 so that the division by k in the "proof" is not justified.

Section 2.5   Derivatives of Trigonometric Functions   (page 123)

2. $\dfrac{d}{dx}\cot x = \dfrac{d}{dx}\dfrac{\cos x}{\sin x} = \dfrac{-\cos^2 x - \sin^2 x}{\sin^2 x} = -\csc^2 x$

4. $y = \sin\dfrac{x}{5}, \quad y' = \dfrac{1}{5}\cos\dfrac{x}{5}.$

6. $y = \sec ax, \quad y' = a\sec ax \tan ax.$

8. $\dfrac{d}{dx}\sin\dfrac{\pi - x}{3} = -\dfrac{1}{3}\cos\dfrac{\pi - x}{3}$

10. $y = \sin(Ax + B), \quad y' = A\cos(Ax + B)$

12. $\dfrac{d}{dx}\cos(\sqrt{x}) = -\dfrac{1}{2\sqrt{x}}\sin(\sqrt{x})$

14. $\dfrac{d}{dx}\sin(2\cos x) = \cos(2\cos x)(-2\sin x)$
$$= -2\sin x\cos(2\cos x)$$

16. $g(\theta) = \tan(\theta\sin\theta)$
$g'(\theta) = (\sin\theta + \theta\cos\theta)\sec^2(\theta\sin\theta)$

18. $y = \sec(1/x), \quad y' = -(1/x^2)\sec(1/x)\tan(1/x)$

20. $G(\theta) = \dfrac{\sin a\theta}{\cos b\theta}$
$G'(\theta) = \dfrac{a\cos b\theta\cos a\theta + b\sin a\theta\sin b\theta}{\cos^2 b\theta}.$

22. $\dfrac{d}{dx}(\cos^2 x - \sin^2 x) = \dfrac{d}{dx}\cos(2x)$
$$= -2\sin(2x) = -4\sin x\cos x$$

24. $\dfrac{d}{dx}(\sec x - \csc x) = \sec x\tan x + \csc x\cot x$

26. $\dfrac{d}{dx}\tan(3x)\cot(3x) = \dfrac{d}{dx}(1) = 0$

28. $\dfrac{d}{dt}(t\sin t + \cos t) = \sin t + t\cos t - \sin t = t\cos t$

30. $\dfrac{d}{dx}\dfrac{\cos x}{1 + \sin x} = \dfrac{(1 + \sin x)(-\sin x) - \cos(x)(\cos x)}{(1 + \sin x)^2}$
$$= \dfrac{-\sin x - 1}{(1 + \sin x)^2} = \dfrac{-1}{1 + \sin x}$$

32. $g(t) = \sqrt{(\sin t)/t}$
$g'(t) = \dfrac{1}{2\sqrt{(\sin t)/t}} \times \dfrac{t\cos t - \sin t}{t^2}$
$$= \dfrac{t\cos t - \sin t}{2t^{3/2}\sqrt{\sin t}}$$

34. $z = \dfrac{\sin\sqrt{x}}{1 + \cos\sqrt{x}}$
$z' = \dfrac{(1 + \cos\sqrt{x})(\cos\sqrt{x}/2\sqrt{x}) - (\sin\sqrt{x})(-\sin\sqrt{x}/2\sqrt{x})}{(1 + \cos\sqrt{x})^2}$
$$= \dfrac{1 + \cos\sqrt{x}}{2\sqrt{x}(1 + \cos\sqrt{x})^2} = \dfrac{1}{2\sqrt{x}(1 + \cos\sqrt{x})}$$

36. $f(s) = \cos(s + \cos(s + \cos s))$
$f'(s) = -[\sin(s + \cos(s + \cos s))]$
$$\times [1 - (\sin(s + \cos s))(1 - \sin s)]$$

38. Differentiate both sides of $\cos(2x) = \cos^2 x - \sin^2 x$ and divide by -2 to get $\sin(2x) = 2\sin x\cos x$.

40. The slope of $y = \tan(2x)$ at $(0, 0)$ is $2\sec^2(0) = 2$. Therefore the tangent and normal lines to $y = \tan(2x)$ at $(0, 0)$ have equations $y = 2x$ and $y = -x/2$, respectively.

42. The slope of $y = \cos^2 x$ at $(\pi/3, 1/4)$ is $-\sin(2\pi/3) = -\sqrt{3}/2$. Therefore the tangent and normal lines to $y = \tan(2x)$ at $(0, 0)$ have equations $y = (1/4) - (\sqrt{3}/2)(x - (\pi/3))$ and $y = (1/4) + (2/\sqrt{3})(x - (\pi/3))$, respectively.

44. For $y = \sec(x°) = \sec\left(\dfrac{x\pi}{180}\right)$ we have
$$\dfrac{dy}{dx} = \dfrac{\pi}{180}\sec\left(\dfrac{x\pi}{180}\right)\tan\left(\dfrac{x\pi}{180}\right).$$
At $x = 60$ the slope is $\dfrac{\pi}{180}(2\sqrt{3}) = \dfrac{\pi\sqrt{3}}{90}$.
Thus, the normal line has slope $-\dfrac{90}{\pi\sqrt{3}}$ and has equation
$$y = 2 - \dfrac{90}{\pi\sqrt{3}}(x - 60).$$

46. The slope of $y = \tan(2x)$ at $x = a$ is $2\sec^2(2a)$. The tangent there is normal to $y = -x/8$ if $2\sec^2(2a) = 8$, or $\cos(2a) = \pm 1/2$. The only solutions in $(-\pi/4, \pi/4)$ are $a = \pm\pi/6$. The corresponding points on the graph are $(\pi/6, \sqrt{3})$ and $(-\pi/6, -\sqrt{3})$.

48. $\dfrac{d}{dx}\tan x = \sec^2 x = 0$ nowhere.
$\dfrac{d}{dx}\cot x = -\csc^2 x = 0$ nowhere.
Thus neither of these functions has a horizontal tangent.

50. $y = 2x + \sin x$ has no horizontal tangents because $dy/dx = 2 + \cos x \geq 1$ everywhere.

52. $y = x + 2\cos x$ has horizontal tangents at $x = \pi/6$ and $x = 5\pi/6$ because $dy/dx = 1 - 2\sin x = 0$ at those points.

54. $\displaystyle\lim_{x\to\pi} \sec(1 + \cos x) = \sec(1 - 1) = \sec 0 = 1$

56. $\displaystyle\lim_{x\to 0} \cos\left(\frac{\pi - \pi\cos^2 x}{x^2}\right) = \lim_{x\to 0}\cos\pi\left(\frac{\sin x}{x}\right)^2 = \cos\pi = -1$

58. f will be differentiable at $x = 0$ if

$$2\sin 0 + 3\cos 0 = b, \qquad \text{and}$$
$$\frac{d}{dx}(2\sin x + 3\cos x)\Big|_{x=0} = a.$$

Thus we need $b = 3$ and $a = 2$.

60. 1

62. a) As suggested by the figure in the problem, the square of the length of chord AP is $(1 - \cos\theta)^2 + (0 - \sin\theta)^2$, and the square of the length of arc AP is θ^2. Hence

$$(1 + \cos\theta)^2 + \sin^2\theta < \theta^2,$$

and, since squares cannot be negative, each term in the sum on the left is less than θ^2. Therefore

$$0 \le |1 - \cos\theta| < |\theta|, \quad 0 \le |\sin\theta| < |\theta|.$$

Since $\lim_{\theta\to 0}|\theta| = 0$, the squeeze theorem implies that

$$\lim_{\theta\to 0} 1 - \cos\theta = 0, \quad \lim_{\theta\to 0}\sin\theta = 0.$$

From the first of these, $\lim_{\theta\to 0}\cos\theta = 1$.

b) Using the result of (a) and the addition formulas for cosine and sine we obtain

$$\lim_{h\to 0}\cos(\theta_0 + h) = \lim_{h\to 0}(\cos\theta_0\cos h - \sin\theta_0\sin h) = \cos\theta_0$$
$$\lim_{h\to 0}\sin(\theta_0 + h) = \lim_{h\to 0}(\sin\theta_0\cos h + \cos\theta_0\sin h) = \sin\theta_0.$$

This says that cosine and sine are continuous at any point θ_0.

Section 2.6   The Mean-Value Theorem (page 131)

2. If $f(x) = \dfrac{1}{x}$, and $f'(x) = -\dfrac{1}{x^2}$ then

$$\frac{f(2) - f(1)}{2 - 1} = \frac{1}{2} - 1 = -\frac{1}{2} = -\frac{1}{c^2} = f'(c)$$

where $c = \sqrt{2}$ lies between 1 and 2.

4. If $f(x) = \cos x + (x^2/2)$, then $f'(x) = x - \sin x > 0$ for $x > 0$. By the MVT, if $x > 0$, then $f(x) - f(0) = f'(c)(x - 0)$ for some $c > 0$, so $f(x) > f(0) = 1$. Thus $\cos x + (x^2/2) > 1$ and $\cos x > 1 - (x^2/2)$ for $x > 0$. Since both sides of the inequality are even functions, it must hold for $x < 0$ as well.

6. Let $f(x) = (1 + x)^r - 1 - rx$ where $r > 1$.
Then $f'(x) = r(1 + x)^{r-1} - r$.
If $-1 \le x < 0$ then $f'(x) < 0$; if $x > 0$, then $f'(x) > 0$.
Thus $f(x) > f(0) = 0$ if $-1 \le x < 0$ or $x > 0$.
Thus $(1 + x)^r > 1 + rx$ if $-1 \le x < 0$ or $x > 0$.

8. If $f(x) = x^2 + 2x + 2$ then $f'(x) = 2x + 2 = 2(x + 1)$.
Evidently, $f'(x) > 0$ if $x > -1$ and $f'(x) < 0$ if $x < -1$.
Therefore, f is increasing on $(-1, \infty)$ and decreasing on $(-\infty, -1)$.

10. If $f(x) = x^3 + 4x + 1$, then $f'(x) = 3x^2 + 4$. Since $f'(x) > 0$ for all real x, hence $f(x)$ is increasing on the whole real line, i.e., on $(-\infty, \infty)$.

12. If $f(x) = \dfrac{1}{x^2 + 1}$ then $f'(x) = \dfrac{-2x}{(x^2 + 1)^2}$. Evidently, $f'(x) > 0$ if $x < 0$ and $f'(x) < 0$ if $x > 0$. Therefore, f is increasing on $(-\infty, 0)$ and decreasing on $(0, \infty)$.

14. If $f(x) = x - 2\sin x$, then $f'(x) = 1 - 2\cos x = 0$ at $x = \pm\pi/3 + 2n\pi$ for $n = 0, \pm 1, \pm 2, \dots$.
f is decreasing on $(-\pi/3 + 2n\pi, \pi + 2n\pi)$.
f is increasing on $(\pi/3 + 2n\pi, -\pi/3 + 2(n + 1)\pi)$ for integers n.

16. If $x_1 < x_2 < \dots < x_n$ belong to I, and $f(x_i) = 0$, $(1 \le i \le n)$, then there exists y_i in (x_i, x_{i+1}) such that $f'(y_i) = 0$, $(1 \le i \le n - 1)$ by MVT.

18. For $x \ne 0$, we have $f'(x) = 2x\sin(1/x) - \cos(1/x)$ which has no limit as $x \to 0$. However, $f'(0) = \lim_{h\to 0} f(h)/h = \lim_{h\to 0} h\sin(1/h) = 0$ does exist even though f' cannot be continuous at 0.

20. $f(x) = \begin{cases} x + 2x^2\sin(1/x) & \text{if } x \ne 0 \\ 0 & \text{if } x = 0. \end{cases}$

a) $f'(0) = \displaystyle\lim_{h\to 0}\frac{f(0 + h) - f(0)}{h}$
$= \displaystyle\lim_{h\to 0}\frac{h + 2h^2\sin(1/h)}{h}$
$= \displaystyle\lim_{h\to 0}(1 + 2h\sin(1/h)) = 1,$
because $|2h\sin(1/h)| \le 2|h| \to 0$ as $h \to 0$.

b) For $x \ne 0$, we have

$$f'(x) = 1 + 4x\sin(1/x) - 2\cos(1/x).$$

There are numbers x arbitrarily close to 0 where $f'(x) = -1$; namely, the numbers $x = \pm 1/(2n\pi)$, where $n = 1, 2, 3, \ldots$. Since $f'(x)$ is continuous at every $x \neq 0$, it is negative in a small interval about every such number. Thus f cannot be increasing on any interval containing $x = 0$.

Section 2.7 Using Derivatives (page 136)

2. If $y = 1/x$, then $\Delta y \approx (-1/x^2)\,\Delta x$. If $\Delta x = (2/100)x$, then $\Delta y \approx (-2/100)/x = (-2/100)y$, so y decreases by about 2%.

4. If $y = x^3$, then $\Delta y \approx 3x^2\,\Delta x$. If $\Delta x = (2/100)x$, then $\Delta y \approx (6/100)x^3 = (6/100)y$, so y increases by about 6%.

6. If $y = x^{-2/3}$, then $\Delta y \approx (-2/3)x^{-5/3}\,\Delta x$. If $\Delta x = (2/100)x$, then $\Delta y \approx (-4/300)x^{2/3} = (-4/300)y$, so y decreases by about 1.33%.

8. If V is the volume and x is the edge length of the cube then $V = x^3$. Thus $\Delta V \approx 3x^2\,\Delta x$. $\Delta V = -(6/100)V$, then $-6x^3/100 = 3x^2\,\Delta x$, so $\Delta x \approx -(2/100)x$. The edge of the cube decreases by about 2%.

10. If $A = s^2$, then $s = \sqrt{A}$ and $ds/dA = 1/(2\sqrt{A})$. If $A = 16$ m^2, then the side is changing at rate $ds/dA = 1/8$ m/m^2.

12. Since $A = \pi D^2/4$, the rate of change of area with respect to diameter is $dA/dD = \pi D/2$ square units per unit.

14. Let A be the area of a square, s be its side length and L be its diagonal. Then, $L^2 = s^2 + s^2 = 2s^2$ and $A = s^2 = \frac{1}{2}L^2$, so $\dfrac{dA}{dL} = L$. Thus, the rate of change of the area of a square with respect to its diagonal L is L.

16. Let s be the side length and V be the volume of a cube. Then $V = s^3 \Rightarrow s = V^{1/3}$ and $\dfrac{ds}{dV} = \frac{1}{3}V^{-2/3}$. Hence, the rate of change of the side length of a cube with respect to its volume V is $\frac{1}{3}V^{-2/3}$.

18. If $f(x) = x^3 - 12x + 1$, then $f'(x) = 3(x^2 - 4)$. The critical points of f are $x = \pm 2$. f is increasing on $(-\infty, -2)$ and $(2, \infty)$ where $f'(x) > 0$, and is decreasing on $(-2, 2)$ where $f'(x) < 0$.

20. If $y = 1 - x - x^5$, then $y' = -1 - 5x^4 < 0$ for all x. Thus y has no critical points and is decreasing on the whole real line.

22. If $f(x) = x + 2\sin x$, then $f'(x) = 1 + 2\cos x > 0$ if $\cos x > -1/2$. Thus f is increasing on the intervals $(-(4\pi/3) + 2n\pi, (4\pi/3) + 2n\pi)$ where n is any integer.

24. CPs $x = -1.366025$ and $x = 0.366025$

26. CP $x = 0.521350$

28. Flow rate $F = kr^4$, so $\Delta F \approx 4kr^3\,\Delta r$. If $\Delta F = F/10$, then

$$\Delta r \approx \frac{F}{40kr^3} = \frac{kr^4}{40kr^3} = 0.025r.$$

The flow rate will increase by 10% if the radius is increased by about 2.5%.

30. If price $=$ \$$p$, then revenue is \$$R = 4{,}000p - 10p^2$.

a) Sensitivity of R to p is $dR/dp = 4{,}000 - 20p$. If $p = 100, 200$, and 300, this sensitivity is $2{,}000$ \$/\$, 0 \$/\$, and $-2{,}000$ \$/\$ respectively.

b) The distributor should charge \$200. This maximizes the revenue.

32. Daily profit if production is x sheets per day is \$$P(x)$ where

$$P(x) = 8x - 0.005x^2 - 1{,}000.$$

a) Marginal profit $P'(x) = 8 - 0.01x$. This is positive if $x < 800$ and negative if $x > 800$.

b) To maximize daily profit, production should be 800 sheets/day.

34. Daily profit $P = 13x - Cx = 13x - 10x - 20 - \dfrac{x^2}{1000}$

$$= 3x - 20 - \frac{x^2}{1000}$$

Graph of P is a parabola opening downward. P will be maximum where the slope is zero:

$$0 = \frac{dP}{dx} = 3 - \frac{2x}{1000} \quad \text{so } x = 1500$$

Should extract 1500 tonnes of ore per day to maximize profit.

36. If $y = Cp^{-r}$, then the elasticity of y is

$$-\frac{p}{y}\frac{dy}{dp} = -\frac{p}{Cp^{-r}}(-r)Cp^{-r-1} = r.$$

Section 2.8 Higher-Order Derivatives (page 140)

2. $y = x^2 - \dfrac{1}{x}$ 　　　 $y'' = 2 - \dfrac{2}{x^3}$

$y' = 2x + \dfrac{1}{x^2}$ 　　　 $y''' = \dfrac{6}{x^4}$

4. $y = \sqrt{ax + b}$ 　　　 $y'' = -\dfrac{a^2}{4(ax + b)^{3/2}}$

$y' = \dfrac{a}{2\sqrt{ax + b}}$ 　　　 $y''' = \dfrac{3a^3}{8(ax + b)^{5/2}}$

6. $y = x^{10} + 2x^8$ 　　　 $y'' = 90x^8 + 112x^6$

$y' = 10x^9 + 16x^7$ 　　　 $y''' = 720x^7 + 672x^5$

8. $\quad y = \dfrac{x-1}{x+1} \qquad y'' = -\dfrac{4}{(x+1)^3}$

$\quad y' = \dfrac{2}{(x+1)^2} \qquad y''' = \dfrac{12}{(x+1)^4}$

10. $\quad y = \sec x \qquad\qquad y'' = \sec x \tan^2 x + \sec^3 x$

$\quad y' = \sec x \tan x \qquad y''' = \sec x \tan^3 x + 5 \sec^3 x \tan x$

12. $\quad y = \dfrac{\sin x}{x}$

$\quad y' = \dfrac{\cos x}{x} - \dfrac{\sin x}{x^2}$

$\quad y'' = \dfrac{(2-x^2)\sin x}{x^3} - \dfrac{2\cos x}{x^2}$

$\quad y''' = \dfrac{(6-x^2)\cos x}{x^3} + \dfrac{3(x^2-2)\sin x}{x^4}$

14. $\quad f(x) = \dfrac{1}{x^2} = x^{-2}$

$\quad f'(x) = -2x^{-3}$

$\quad f''(x) = -2(-3)x^{-4} = 3!x^{-4}$

$\quad f^{(3)}(x) = -2(-3)(-4)x^{-5} = -4!x^{-5}$

Conjecture:

$$f^{(n)}(x) = (-1)^n (n+1)! x^{-(n+2)} \qquad \text{for } n = 1,\,2,\,3,\,\dots$$

Proof: Evidently, the above formula holds for $n = 1$, 2 and 3. Assume it holds for $n = k$, i.e., $f^{(k)}(x) = (-1)^k (k+1)! x^{-(k+2)}$. Then

$$\begin{aligned}
f^{(k+1)}(x) &= \frac{d}{dx} f^{(k)}(x) \\
&= (-1)^k (k+1)![(-1)(k+2)]x^{-(k+2)-1} \\
&= (-1)^{k+1}(k+2)! x^{-[(k+1)+2]}.
\end{aligned}$$

Thus, the formula is also true for $n = k+1$. Hence it is true for $n = 1$, 2, 3, $\dots$ by induction.

16. $\quad f(x) = \sqrt{x} = x^{1/2}$

$\quad f'(x) = \frac{1}{2}x^{-1/2}$

$\quad f''(x) = \frac{1}{2}\left(-\frac{1}{2}\right)x^{-3/2}$

$\quad f'''(x) = \frac{1}{2}\left(-\frac{1}{2}\right)\left(-\frac{3}{2}\right)x^{-5/2}$

$\quad f^{(4)}(x) = \frac{1}{2}\left(-\frac{1}{2}\right)\left(-\frac{3}{2}\right)\left(-\frac{5}{2}\right)x^{-7/2}$

Conjecture:

$$f^{(n)}(x) = (-1)^{n-1}\frac{1\cdot 3\cdot 5\cdots(2n-3)}{2^n}x^{-(2n-1)/2} \quad (n \geq 2).$$

Proof: Evidently, the above formula holds for $n = 2, 3$ and 4. Assume that it holds for $n = k$, i.e.

$$f^{(k)}(x) = (-1)^{k-1}\frac{1\cdot 3\cdot 5\cdots(2k-3)}{2^k}x^{-(2k-1)/2}.$$

Then

$$f^{(k+1)}(x) = \frac{d}{dx} f^{(k)}(x)$$

$$= (-1)^{k-1}\frac{1\cdot 3\cdot 5\cdots(2k-3)}{2^k}\cdot\left[\frac{-(2k-1)}{2}\right]x^{-[(2k-1)/2]-1}$$

$$= (-1)^{(k+1)-1}\frac{1\cdot 3\cdot 5\cdots(2k-3)[2(k+1)-3]}{2^{k+1}}x^{-[2(k+1)-1]/2}.$$

Thus, the formula is also true for $n = k+1$. Hence, it is true for $n \geq 2$ by induction.

18. $\quad f(x) = x^{2/3}$

$\quad f'(x) = \frac{2}{3}x^{-1/3}$

$\quad f''(x) = \frac{2}{3}\left(-\frac{1}{3}\right)x^{-4/3}$

$\quad f'''(x) = \frac{2}{3}\left(-\frac{1}{3}\right)\left(-\frac{4}{3}\right)x^{-7/3}$

Conjecture:

$$f^{(n)}(x) = 2(-1)^{n-1}\frac{1\cdot 4\cdot 7\cdots(3n-5)}{3^n}x^{-(3n-2)/3} \text{ for}$$
$n \geq 2$.

Proof: Evidently, the above formula holds for $n = 2$ and 3. Assume that it holds for $n = k$, i.e.

$$f^{(k)}(x) = 2(-1)^{k-1}\frac{1\cdot 4\cdot 7\cdots(3k-5)}{3^k}x^{-(3k-2)/3}.$$

Then,

$$f^{(k+1)}(x) = \frac{d}{dx} f^{(k)}(x)$$

$$= 2(-1)^{k-1}\frac{1\cdot 4\cdot 7\cdots(3k-5)}{3^k}\cdot\left[\frac{-(3k-2)}{3}\right]x^{-[(3k-2)/3]-1}$$

$$= 2(-1)^{(k+1)-1}\frac{1\cdot 4\cdot 7\cdots(3k-5)[3(k+1)-5]}{3^{(k+1)}}x^{-[3(k+1)-2]/3}.$$

Thus, the formula is also true for $n = k+1$. Hence, it is true for $n \geq 2$ by induction.

20. $\quad f(x) = x\cos x$

$\quad f'(x) = \cos x - x\sin x$

$\quad f''(x) = -2\sin x - x\cos x$

$\quad f'''(x) = -3\cos x + x\sin x$

$\quad f^{(4)}(x) = 4\sin x + x\cos x$

This suggests the formula (for $k = 0,\ 1,\ 2,\ \dots$)

$$f^{(n)}(x) = \begin{cases} n\sin x + x\cos x & \text{if } n = 4k \\ n\cos x - x\sin x & \text{if } n = 4k+1 \\ -n\sin x - x\cos x & \text{if } n = 4k+2 \\ -n\cos x + x\sin x & \text{if } n = 4k+3 \end{cases}$$

Differentiating any of these four formulas produces the one for the next higher value of n, so induction confirms the overall formula.

22. $\quad f(x) = \dfrac{1}{|x|} = |x|^{-1}$. Recall that $\dfrac{d}{dx}|x| = \operatorname{sgn} x$, so

$$f'(x) = -|x|^{-2}\operatorname{sgn} x.$$

If $x \neq 0$ we have

$$\frac{d}{dx}\operatorname{sgn} x = 0 \quad \text{and} \quad (\operatorname{sgn} x)^2 = 1.$$

Thus we can calculate successive derivatives of f using the product rule where necessary, but will get only one nonzero term in each case:

$$f''(x) = 2|x|^{-3}(\text{sgn}\,x)^2 = 2|x|^{-3}$$
$$f^{(3)}(x) = -3!|x|^{-4}\text{sgn}\,x$$
$$f^{(4)}(x) = 4!|x|^{-5}.$$

The pattern suggests that

$$f^{(n)}(x) = \begin{cases} -n!|x|^{-(n+1)}\text{sgn}\,x & \text{if } n \text{ is odd} \\ n!|x|^{-(n+1)} & \text{if } n \text{ is even} \end{cases}$$

Differentiating this formula leads to the same formula with n replaced by $n+1$ so the formula is valid for all $n \geq 1$ by induction.

24. If $y = \tan(kx)$, then $y' = k \sec^2(kx)$ and

$$y'' = 2k^2 sec^2(kx)tan(kx)$$
$$= 2k^2(1 + \tan^2(kx))\tan(kx) = 2k^2 y(1 + y^2).$$

26. To be proved: if $f(x) = \sin(ax + b)$, then

$$f^{(n)}(x) = \begin{cases} (-1)^k a^n \sin(ax + b) & \text{if } n = 2k \\ (-1)^k a^n \cos(ax + b) & \text{if } n = 2k + 1 \end{cases}$$

for $k = 0, 1, 2, \ldots$ Proof: The formula works for $k = 0$ ($n = 2 \times 0 = 0$ and $n = 2 \times 0 + 1 = 1$):

$$\begin{cases} f^{(0)}(x) = f(x) = (-1)^0 a^0 \sin(ax + b) = \sin(ax + b) \\ f^{(1)}(x) = f'(x) = (-1)^0 a^1 \cos(ax + b) = a\cos(ax + b) \end{cases}$$

Now assume the formula holds for some $k \geq 0$.
If $n = 2(k + 1)$, then

$$f^{(n)}(x) = \frac{d}{dx} f^{(n-1)}(x) = \frac{d}{dx} f^{(2k+1)}(x)$$
$$= \frac{d}{dx}\left((-1)^k a^{2k+1} \cos(ax + b)\right)$$
$$= (-1)^{k+1} a^{2k+2} \sin(ax + b)$$

and if $n = 2(k + 1) + 1 = 2k + 3$, then

$$f^{(n)}(x) = \frac{d}{dx}\left((-1)^{k+1} a^{2k+2} \sin(ax + b)\right)$$
$$= (-1)^{k+1} a^{2k+3} \cos(ax + b).$$

Thus the formula also holds for $k + 1$. Therefore it holds for all positive integers k by induction.

28. $(fg)'' = (f'g + fg')' = f''g + f'g' + f'g' + fg''$
$$= f''g + 2f'g' + fg''$$

30. Let a, b, and c be three points in I where f vanishes; that is, $f(a) = f(b) = f(c) = 0$. Suppose $a < b < c$. By the Mean-Value Theorem, there exist points r in (a, b) and s in (b, c) such that $f'(r) = f'(s) = 0$. By the Mean-Value Theorem applied to f' on $[r, s]$, there is some point t in (r, s) (and therefore in I) such that $f''(t) = 0$.

32. Given that $f(0) = f(1) = 0$ and $f(2) = 1$:

a) By MVT,

$$f'(a) = \frac{f(2) - f(0)}{2 - 0} = \frac{1 - 0}{2 - 0} = \frac{1}{2}$$

for some a in $(0, 2)$.

b) By MVT, for some r in $(0, 1)$,

$$f'(r) = \frac{f(1) - f(0)}{1 - 0} = \frac{0 - 0}{1 - 0} = 0.$$

Also, for some s in $(1, 2)$,

$$f'(s) = \frac{f(2) - f(1)}{2 - 1} = \frac{1 - 0}{2 - 1} = 1.$$

Then, by MVT applied to f' on the interval $[r, s]$, for some b in (r, s),

$$f''(b) = \frac{f'(s) - f'(r)}{s - r} = \frac{1 - 0}{s - r}$$
$$= \frac{1}{s - r} > \frac{1}{2}$$

since $s - r < 2$.

c) Since $f''(x)$ exists on $[0, 2]$, therefore $f'(x)$ is continuous there. Since $f'(r) = 0$ and $f'(s) = 1$, and since $0 < \frac{1}{7} < 1$, the Intermediate-Value Theorem assures us that $f'(c) = \frac{1}{7}$ for some c between r and s.

Section 2.9 Implicit Differentiation (page 145)

2. $x^3 + y^3 = 1$

$3x^2 + 3y^2 y' = 0$, so $y' = -\dfrac{x^2}{y^2}$.

4. $x^3 y + xy^5 = 2$
$3x^2 y + x^3 y' + y^5 + 5xy^4 y' = 0$
$y' = \dfrac{-3x^2 y - y^5}{x^3 + 5xy^4}$

6. $x^2 + 4(y - 1)^2 = 4$
$2x + 8(y - 1)y' = 0$, so $y' = \dfrac{x}{4(1 - y)}$

8. $x\sqrt{x+y} = 8 - xy$

$\sqrt{x+y} + x\dfrac{1}{2\sqrt{x+y}}(1+y') = -y - xy'$

$2(x+y) + x(1+y') = -2\sqrt{x+y}(y+xy')$

$y' = -\dfrac{3x + 2y + 2y\sqrt{x+y}}{x + 2x\sqrt{x+y}}$

10. $x^2y^3 - x^3y^2 = 12$

$2xy^3 + 3x^2y^2y' - 3x^2y^2 - 2x^3yy' = 0$

At $(-1, 2)$: $-16 + 12y' - 12 + 4y' = 0$, so the slope is

$y' = \dfrac{12 + 16}{12 + 4} = \dfrac{28}{16} = \dfrac{7}{4}$.

Thus, the equation of the tangent line is

$y = 2 + \frac{7}{4}(x+1)$, or $7x - 4y + 15 = 0$.

12. $x + 2y + 1 = \dfrac{y^2}{x-1}$

$1 + 2y' = \dfrac{(x-1)2yy' - y^2(1)}{(x-1)^2}$

At $(2, -1)$ we have $1 + 2y' = -2y' - 1$ so $y' = -\frac{1}{2}$.

Thus, the equation of the tangent is

$y = -1 - \frac{1}{2}(x-2)$, or $x + 2y = 0$.

14. $\tan(xy^2) = (2/\pi)xy$

$(\sec^2(xy^2))(y^2 + 2xyy') = (2/\pi)(y + xy')$.

At $(-\pi, 1/2)$: $2((1/4) - \pi y') = (1/\pi) - 2y'$, so

$y' = (\pi - 2)/(4\pi(\pi - 1))$. The tangent has equation

$$y = \frac{1}{2} + \frac{\pi - 2}{4\pi(\pi - 1)}(x + \pi).$$

16. $\cos\left(\dfrac{\pi y}{x}\right) = \dfrac{x^2}{y} - \dfrac{17}{2}$

$\left[-\sin\left(\dfrac{\pi y}{x}\right)\right]\dfrac{\pi(xy' - y)}{x^2} = \dfrac{2xy - x^2y'}{y^2}$.

At $(3, 1)$: $-\dfrac{\sqrt{3}}{2}\dfrac{\pi(3y' - 1)}{9} = 6 - 9y'$,

so $y' = (108 - \sqrt{3}\pi)/(162 - 3\sqrt{3}\pi)$. The tangent has equation

$$y = 1 + \frac{108 - \sqrt{3}\pi}{162 - 3\sqrt{3}\pi}(x - 3).$$

18. $x^2 + 4y^2 = 4$, $2x + 8yy' = 0$, $2 + 8(y')^2 + 8yy'' = 0$.

Thus, $y' = \dfrac{-x}{4y}$ and

$y'' = \dfrac{-2 - 8(y')^2}{8y} = -\dfrac{1}{4y} - \dfrac{x^2}{16y^3} = \dfrac{-4y^2 - x^2}{16y^3} = -\dfrac{1}{4y^3}$.

20. $x^3 - 3xy + y^3 = 1$

$3x^2 - 3y - 3xy' + 3y^2y' = 0$

$6x - 3y' - 3y' - 3xy'' + 6y(y')^2 + 3y^2y'' = 0$

Thus

$y' = \dfrac{y - x^2}{y^2 - x}$

$y'' = \dfrac{-2x + 2y' - 2y(y')^2}{y^2 - x}$

$= \dfrac{2}{y^2 - x}\left[-x + \left(\dfrac{y - x^2}{y^2 - x}\right) - y\left(\dfrac{y - x^2}{y^2 - x}\right)^2\right]$

$= \dfrac{2}{y^2 - x}\left[\dfrac{-2xy}{(y^2 - x)^2}\right] = \dfrac{4xy}{(x - y^2)^3}$.

22. $Ax^2 + By^2 = C$

$2Ax + 2Byy' = 0 \Rightarrow y' = -\dfrac{Ax}{By}$

$2A + 2B(y')^2 + 2Byy'' = 0$.

Thus,

$y'' = \dfrac{-A - B(y')^2}{By} = \dfrac{-A - B\left(\dfrac{Ax}{By}\right)^2}{By}$

$= \dfrac{-A(By^2 + Ax^2)}{B^2y^3} = -\dfrac{AC}{B^2y^3}$.

24. Maple gives the slope as $\dfrac{206}{55}$.

26. Maple gives the value $-\dfrac{855,000}{371,293}$.

28. The slope of the ellipse $\dfrac{x^2}{a^2} + \dfrac{y^2}{b^2} = 1$ is found from

$$\frac{2x}{a^2} + \frac{2y}{b^2}y' = 0, \quad \text{i.e.} \quad y' = -\frac{b^2x}{a^2y}.$$

Similarly, the slope of the hyperbola $\dfrac{x^2}{A^2} - \dfrac{y^2}{B^2} = 1$ at (x, y) satisfies

$$\frac{2x}{A^2} - \frac{2y}{B^2}y' = 0, \quad \text{or } y' = \frac{B^2x}{A^2y}.$$

If the point (x, y) is an intersection of the two curves, then

$$\frac{x^2}{a^2} + \frac{y^2}{b^2} = \frac{x^2}{A^2} - \frac{y^2}{B^2}$$

$$x^2\left(\frac{1}{A^2} - \frac{1}{a^2}\right) = y^2\left(\frac{1}{B^2} + \frac{1}{b^2}\right).$$

Thus, $\dfrac{x^2}{y^2} = \dfrac{b^2 + B^2}{B^2b^2} \cdot \dfrac{A^2a^2}{a^2 - A^2}$.

Since $a^2 - b^2 = A^2 + B^2$, therefore $B^2 + b^2 = a^2 - A^2$, and $\dfrac{x^2}{y^2} = \dfrac{A^2a^2}{B^2b^2}$. Thus, the product of the slope of the two curves at (x, y) is

$$-\frac{b^2x}{a^2y} \cdot \frac{B^2x}{A^2y} = -\frac{b^2B^2}{a^2A^2} \cdot \frac{A^2a^2}{B^2b^2} = -1.$$

Therefore, the curves intersect at right angles.

30. $\dfrac{x-y}{x+y} = \dfrac{x}{y} + 1 \Leftrightarrow xy - y^2 = x^2 + xy + xy + y^2$

$$\Leftrightarrow x^2 + 2y^2 + xy = 0$$

Differentiate with respect to x:

$$2x + 4yy' + y + xy' = 0 \quad \Rightarrow \quad y' = -\frac{2x+y}{4y+x}.$$

However, since $x^2 + 2y^2 + xy = 0$ can be written

$$x + xy + \frac{1}{4}y^2 + \frac{7}{4}y^2 = 0, \text{ or } (x + \frac{y}{2})^2 + \frac{7}{4}y^2 = 0,$$

the only solution is $x = 0$, $y = 0$, and these values do not satisfy the original equation. There are no points on the given curve.

Section 2.10 Antiderivatives and Initial-Value Problems (page 151)

2. $\displaystyle\int x^2\,dx = \frac{1}{3}x^3 + C$

4. $\displaystyle\int x^{12}\,dx = \frac{1}{13}x^{13} + C$

6. $\displaystyle\int (x + \cos x)\,dx = \frac{x^2}{2} + \sin x + C$

8. $\displaystyle\int \frac{1 + \cos^3 x}{\cos^2 x}\,dx = \int (\sec^2 x + \cos x)\,dx = \tan x + \sin x + C$

10. $\displaystyle\int (A + Bx + Cx^2)\,dx = Ax + \frac{B}{2}x^2 + \frac{C}{3}x^3 + K$

12. $\displaystyle\int \frac{6(x-1)}{x^{4/3}}\,dx = \int (6x^{-1/3} - 6x^{-4/3})\,dx$
$$= 9x^{2/3} + 18x^{-1/3} + C$$

14. $105\displaystyle\int (1 + t^2 + t^4 + t^6)\,dt$
$$= 105(t + \frac{1}{3}t^3 + \frac{1}{5}t^5 + \frac{1}{7}t^7) + C$$
$$= 105t + 35t^3 + 21t^5 + 15t^7 + C$$

16. $\displaystyle\int \sin\left(\frac{x}{2}\right) dx = -2\cos\left(\frac{x}{2}\right) + C$

18. $\displaystyle\int \sec(1-x)\tan(1-x)\,dx = -\sec(1-x) + C$

20. Since $\dfrac{d}{dx}\sqrt{x+1} = \dfrac{1}{2\sqrt{x+1}}$, therefore

$$\int \frac{4}{\sqrt{x+1}}\,dx = 8\sqrt{x+1} + C.$$

22. Since $\dfrac{d}{dx}\sqrt{x^2+1} = \dfrac{x}{\sqrt{x^2+1}}$, therefore

$$\int \frac{2x}{\sqrt{x^2+1}}\,dx = 2\sqrt{x^2+1} + C.$$

24. $\displaystyle\int \sin x \cos x\,dx = \int \frac{1}{2}\sin(2x)\,dx = -\frac{1}{4}\cos(2x) + C$

26. $\displaystyle\int \sin^2 x\,dx = \int \frac{1-\cos(2x)}{2}\,dx = \frac{x}{2} - \frac{\sin(2x)}{4} + C$

28. Given that
$$\begin{cases} y' = x^{-2} - x^{-3} \\ y(-1) = 0, \end{cases}$$
then $y = \displaystyle\int (x^{-2} - x^{-3})\,dx = -x^{-1} + \frac{1}{2}x^{-2} + C$
and $0 = y(-1) = -(-1)^{-1} + \frac{1}{2}(-1)^{-2} + C$ so $C = -\frac{3}{2}$.
Hence, $y(x) = -\dfrac{1}{x} + \dfrac{1}{2x^2} - \dfrac{3}{2}$ which is valid on the interval $(-\infty, 0)$.

30. Given that
$$\begin{cases} y' = x^{1/3} \\ y(0) = 5, \end{cases}$$
then $y = \displaystyle\int x^{1/3}\,dx = \frac{3}{4}x^{4/3} + C$ and $5 = y(0) = C$.
Hence, $y(x) = \frac{3}{4}x^{4/3} + 5$ which is valid on the whole real line.

32. Given that
$$\begin{cases} y' = x^{-9/7} \\ y(1) = -4, \end{cases}$$
then $y = \displaystyle\int x^{-9/7}\,dx = -\frac{7}{2}x^{-2/7} + C$.
Also, $-4 = y(1) = -\frac{7}{2} + C$, so $C = -\frac{1}{2}$. Hence,
$y = -\frac{7}{2}x^{-2/7} - \frac{1}{2}$, which is valid in the interval $(0, \infty)$.

34. For $\begin{cases} y' = \sin(2x) \\ y(\pi/2) = 1 \end{cases}$, we have

$$y = \int \sin(2x)\,dx = -\frac{1}{2}\cos(2x) + C$$
$$1 = -\frac{1}{2}\cos\pi + C = \frac{1}{2} + C \implies C = \frac{1}{2}$$
$$y = \frac{1}{2}\left(1 - \cos(2x)\right) \quad \text{(for all } x\text{).}$$

36. For $\begin{cases} y' = \sec^2 x \\ y(\pi) = 1 \end{cases}$, we have

$$y = \int \sec^2 x\,dx = \tan x + C$$
$$1 = \tan\pi + C = C \implies C = 1$$
$$y = \tan x + 1 \quad \text{(for } \pi/2 < x < 3\pi/2\text{).}$$

38. Given that

$$\begin{cases} y'' = x^{-4} \\ y'(1) = 2 \\ y(1) = 1, \end{cases}$$

then $y' = \int x^{-4}\,dx = -\frac{1}{3}x^{-3} + C$.

Since $2 = y'(1) = -\frac{1}{3} + C$, therefore $C = \frac{7}{3}$,
and $y' = -\frac{1}{3}x^{-3} + \frac{7}{3}$. Thus

$$y = \int\left(-\frac{1}{3}x^{-3} + \frac{7}{3}\right)dx = \frac{1}{6}x^{-2} + \frac{7}{3}x + D,$$

and $1 = y(1) = \frac{1}{6} + \frac{7}{3} + D$, so that $D = -\frac{3}{2}$. Hence,
$y(x) = \frac{1}{6}x^{-2} + \frac{7}{3}x - \frac{3}{2}$, which is valid in the interval
$(0, \infty)$.

40. Given that

$$\begin{cases} y'' = 5x^2 - 3x^{-1/2} \\ y'(1) = 2 \\ y(1) = 0, \end{cases}$$

we have $y' = \int 5x^2 - 3x^{-1/2}\,dx = \frac{5}{3}x^3 - 6x^{1/2} + C$.

Also, $2 = y'(1) = \frac{5}{3} - 6 + C$ so that $C = \dfrac{19}{3}$. Thus,
$y' = \frac{5}{3}x^3 - 6x^{1/2} + \frac{19}{3}$, and

$$y = \int\left(\frac{5}{3}x^3 - 6x^{1/2} + \frac{19}{3}\right)dx = \frac{5}{12}x^4 - 4x^{3/2} + \frac{19}{3}x + D.$$

Finally, $0 = y(1) = \frac{5}{12} - 4 + \frac{19}{3} + D$ so that $D = -\frac{11}{4}$.
Hence, $y(x) = \frac{5}{12}x^4 - 4x^{3/2} + \frac{19}{3}x - \frac{11}{4}$.

42. For $\begin{cases} y'' = x + \sin x \\ y(0) = 2 \\ y'(0) = 0 \end{cases}$ we have

$$y' = \int(x + \sin x)\,dx = \frac{x^2}{2} - \cos x + C_1$$

$$0 = 0 - \cos 0 + C_1 \implies C_1 = 1$$

$$y = \int\left(\frac{x^2}{2} - \cos x + 1\right)dx = \frac{x^3}{6} - \sin x + x + C_2$$

$$2 = 0 - \sin 0 + 0 + C_2 \implies C_2 = 2$$

$$y = \frac{x^3}{6} - \sin x + x + 2.$$

44. Let r_1 and r_2 be distinct rational roots of the equation
$ar(r - 1) + br + c = 0$
Let $y = Ax^{r_1} + Bx^{r_2}$ $(x > 0)$
Then $y' = Ar_1 x^{r_1-1} + Br_2 x^{r_2-1}$,
and $y'' = Ar_1(r_1 - 1)x^{r_1-2} + Br_2(r_2 - 1)x^{r_2-2}$. Thus
$ax^2 y'' + bxy' + cy$

$$= ax^2(Ar_1(r_1 - 1)x^{r_1-2} + Br_2(r_2 - 1)x^{r_2-2}$$
$$\quad + bx(Ar_1 x^{r_1-1} + Br_2 x^{r_2-1}) + c(Ax^{r_1} + Bx^{r_2})$$

$$= A\Big(ar_1(r_1 - 1) + br_1 + c\Big)x^{r_1}$$
$$\quad + B\Big(ar_2(r_2 - 1) + br_2 + c\Big)x^{r_2}$$

$$= 0x^{r_1} + 0x^{r_2} \equiv 0 \quad (x > 0)$$

46. Consider

$$\begin{cases} x^2 y'' - 6y = 0 \\ y(1) = 1 \\ y'(1) = 1. \end{cases}$$

Let $y = x^r$, $y' = rx^{r-1}$, $y'' = r(r - 1)x^{r-2}$. Substituting
these expressions into the differential equation we obtain

$$x^2[r(r - 1)x^{r-2}] - 6x^r = 0$$
$$[r(r - 1) - 6]x^r = 0.$$

Since this equation must hold for all $x > 0$, we must
have

$$r(r - 1) - 6 = 0$$
$$r^2 - r - 6 = 0$$
$$(r - 3)(r + 2) = 0.$$

There are two roots: $r_1 = -2$, and $r_2 = 3$. Thus the
differential equation has solutions of the form
$y = Ax^{-2} + Bx^3$. Then $y' = -2Ax^{-3} + 3Bx^2$. Since
$1 = y(1) = A + B$ and $1 = y'(1) = -2A + 3B$, therefore
$A = \frac{2}{5}$ and $B = \frac{3}{5}$. Hence, $y = \frac{2}{5}x^{-2} + \frac{3}{5}x^3$.

Section 2.11   Velocity and Acceleration (page 157)

2. $x = 4 + 5t - t^2$, $v = 5 - 2t$, $a = -2$.

a) The point is moving to the right if $v > 0$, i.e., when
$t < \frac{5}{2}$.

b) The point is moving to the left if $v < 0$, i.e., when
$t > \frac{5}{2}$.

c) The point is accelerating to the right if $a > 0$, but
$a = -2$ at all t; hence, the point never accelerates to
the right.

d) The point is accelerating to the left if $a < 0$, i.e., for
all t.

e) The particle is speeding up if v and a have the same
sign, i.e., for $t > \frac{5}{2}$.

f) The particle is slowing down if v and a have opposite sign, i.e., for $t < \frac{5}{2}$.

g) Since $a = -2$ at all t, $a = -2$ at $t = \frac{5}{2}$ when $v = 0$.

h) The average velocity over $[0, 4]$ is
$$\frac{x(4) - x(0)}{4} = \frac{8 - 4}{4} = 1.$$

4. $x = \dfrac{t}{t^2 + 1}$,　$v = \dfrac{(t^2 + 1)(1) - (t)(2t)}{(t^2 + 1)^2} = \dfrac{1 - t^2}{(t^2 + 1)^2}$,

$a = \dfrac{(t^2 + 1)^2(-2t) - (1 - t^2)(2)(t^2 + 1)(2t)}{(t^2 + 1)^4} = \dfrac{2t(t^2 - 3)}{(t^2 + 1)^3}.$

a) The point is moving to the right if $v > 0$, i.e., when $1 - t^2 > 0$, or $-1 < t < 1$.

b) The point is moving to the left if $v < 0$, i.e., when $t < -1$ or $t > 1$.

c) The point is accelerating to the right if $a > 0$, i.e., when $2t(t^2 - 3) > 0$, that is, when $t > \sqrt{3}$ or $-\sqrt{3} < t < 0$.

d) The point is accelerating to the left if $a < 0$, i.e., for $t < -\sqrt{3}$ or $0 < t < \sqrt{3}$.

e) The particle is speeding up if v and a have the same sign, i.e., for $t < -\sqrt{3}$, or $-1 < t < 0$ or $1 < t < \sqrt{3}$.

f) The particle is slowing down if v and a have opposite sign, i.e., for $-\sqrt{3} < t < -1$, or $0 < t < 1$ or $t > \sqrt{3}$.

g) $v = 0$ at $t = \pm 1$. At $t = -1$, $a = \dfrac{-2(-2)}{(2)^3} = \dfrac{1}{2}$.

At $t = 1$, $a = \dfrac{2(-2)}{(2)^3} = -\dfrac{1}{2}$.

h) The average velocity over $[0, 4]$ is
$$\frac{x(4) - x(0)}{4} = \frac{\frac{4}{17} - 0}{4} = \frac{1}{17}.$$

6. Given that $y = 100 - 2t - 4.9t^2$, the time t at which the ball reaches the ground is the positive root of the equation $y = 0$, i.e., $100 - 2t - 4.9t^2 = 0$, namely,
$$t = \frac{-2 + \sqrt{4 + 4(4.9)(100)}}{9.8} \approx 4.318 \text{ s}.$$

The average velocity of the ball is $\dfrac{-100}{4.318} = -23.16$ m/s. Since $-23.159 = v = -2 - 9.8t$, then $t \simeq 2.159$ s.

8. Let $y(t)$ be the height of the projectile t seconds after it is fired upward from ground level with initial speed v_0. Then
$$y''(t) = -9.8, \quad y'(0) = v_0, \quad y(0) = 0.$$

Two antidifferentiations give
$$y = -4.9t^2 + v_0 t = t(v_0 - 4.9t).$$

Since the projectile returns to the ground at $t = 10$ s, we have $y(10) = 0$, so $v_0 = 49$ m/s. On Mars, the acceleration of gravity is 3.72 m/s^2 rather than 9.8 m/s^2, so the height of the projectile would be
$$y = -1.86t^2 + v_0 t = t(49 - 1.86t).$$

The time taken to fall back to ground level on Mars would be $t = 49/1.86 \approx 26.3$ s.

10. To get to $3h$ metres above Mars, the ball would have to be thrown upward with speed
$$v_M = \sqrt{6g_M h} = \sqrt{6g_M v_0^2/(2g)} = v_0\sqrt{3g_M/g}.$$

Since $g_M = 3.72$ and $g = 9.80$, we have $v_M \approx 1.067 v_0$ m/s.

12. If the cliff is h ft high, then the height of the rock t seconds after it is thrown down is $y = h - 32t - 16t^2$ ft. The rock hits the ground ($y = 0$) at time
$$t = \frac{-32 + \sqrt{32^2 + 64h}}{32} = -1 + \frac{1}{4}\sqrt{16 + h} \text{ s}.$$

Its speed at that time is
$$v = -32 - 32t = -8\sqrt{16 + h} = -160 \text{ ft/s}.$$

Solving this equation for h gives the height of the cliff as 384 ft.

14. $x = At^2 + Bt + C$, $v = 2At + B$.
The average velocity over $[t_1, t_2]$ is
$$\frac{x(t_2) - x(t_1)}{t_2 - t_1}$$
$$= \frac{At_2^2 + Bt_2 + C - At_1^2 - Bt_1 - C}{t_2 - t_1}$$
$$= \frac{A(t_2^2 - t_1^2) + B(t_2 - t_1)}{(t_2 - t_1)}$$
$$= \frac{A(t_2 + t_1)(t_2 - t_1) + B(t_2 - t_1)}{(t_2 - t_1)}$$
$$= A(t_2 + t_1) + B.$$
The instantaneous velocity at the midpoint of $[t_1, t_2]$ is
$$v\left(\frac{t_2 + t_1}{2}\right) = 2A\left(\frac{t_2 + t_1}{2}\right) + B = A(t_2 + t_1) + B.$$
Hence, the average velocity over the interval is equal to the instantaneous velocity at the midpoint.

16. This exercise and the next three refer to the following figure depicting the velocity of a rocket fired from a tower as a function of time since firing.

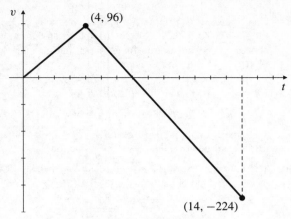

Fig. 2.11.16

The rocket's acceleration while its fuel lasted is the slope of the first part of the graph, namely $96/4 = 24$ ft/s.

18. As suggested in Example 1 on page 154 of the text, the distance travelled by the rocket while it was falling from its maximum height to the ground is the area between the velocity graph and the part of the t-axis where $v < 0$. The area of this triangle is $(1/2)(14 - 7)(224) = 784$ ft. This is the maximum height the rocket achieved.

20. Let $s(t)$ be the distance the car travels in the t seconds after the brakes are applied. Then $s''(t) = -t$ and the velocity at time t is given by

$$s'(t) = \int (-t)\, dt = -\frac{t^2}{2} + C_1,$$

where $C_1 = 20$ m/s (that is, 72km/h) as determined in Example 6. Thus

$$s(t) = \int \left(20 - \frac{t^2}{2}\right) dt = 20t - \frac{t^3}{6} + C_2,$$

where $C_2 = 0$ because $s(0) = 0$. The time taken to come to a stop is given by $s'(t) = 0$, so it is $t = \sqrt{40}$ s. The distance travelled is

$$s = 20\sqrt{40} - \frac{1}{6}40^{3/2} \approx 84.3 \text{ m.}$$

Review Exercises 2   (page 158)

2. $\dfrac{d}{dx}\sqrt{1 - x^2} = \lim_{h \to 0} \dfrac{\sqrt{1 - (x + h)^2} - \sqrt{1 - x^2}}{h}$

$= \lim_{h \to 0} \dfrac{1 - (x + h)^2 - (1 - x^2)}{h(\sqrt{1 - (x + h)^2} + \sqrt{1 - x^2})}$

$= \lim_{h \to 0} \dfrac{-2x - h}{\sqrt{1 - (x + h)^2} + \sqrt{1 - x^2}} = -\dfrac{x}{\sqrt{1 - x^2}}$

4. $g(t) = \dfrac{t - 5}{1 + \sqrt{t}}$

$g'(9) = \lim_{h \to 0} \dfrac{\dfrac{4 + h}{1 + \sqrt{9 + h}} - 1}{h}$

$= \lim_{h \to 0} \dfrac{(3 + h - \sqrt{9 + h})(3 + h + \sqrt{9 + h})}{h(1 + \sqrt{9 + h})(3 + h + \sqrt{9 + h})}$

$= \lim_{h \to 0} \dfrac{9 + 6h + h^2 - (9 + h)}{h(1 + \sqrt{9 + h})(3 + h + \sqrt{9 + h})}$

$= \lim_{h \to 0} \dfrac{5 + h}{(1 + \sqrt{9 + h})(3 + h + \sqrt{9 + h})}$

$= \dfrac{5}{24}$

6. At $x = \pi$ the curve $y = \tan(x/4)$ has slope $(\sec^2(\pi/4))/4 = 1/2$. The normal to the curve there has equation $y = 1 - 2(x - \pi)$.

8. $\dfrac{d}{dx}\dfrac{1 + x + x^2 + x^3}{x^4} = \dfrac{d}{dx}(x^{-4} + x^{-3} + x^{-2} + x^{-1})$

$= -4x^{-5} - 3x^{-4} - 2x^{-3} - x^{-2}$

$= -\dfrac{4 + 3x + 2x^2 + x^3}{x^5}$

10. $\dfrac{d}{dx}\sqrt{2 + \cos^2 x} = \dfrac{-2\cos x \sin x}{2\sqrt{2 + \cos^2 x}} = \dfrac{-\sin x \cos x}{\sqrt{2 + \cos^2 x}}$

12. $\dfrac{d}{dt}\dfrac{\sqrt{1 + t^2} - 1}{\sqrt{1 + t^2} + 1}$

$= \dfrac{(\sqrt{1 + t^2} + 1)\dfrac{t}{\sqrt{1 + t^2}} - (\sqrt{1 + t^2} - 1)\dfrac{t}{\sqrt{1 + t^2}}}{(\sqrt{1 + t^2} + 1)^2}$

$= \dfrac{2t}{\sqrt{1 + t^2}(\sqrt{1 + t^2} + 1)^2}$

14. $\lim_{x \to 2} \dfrac{\sqrt{4x + 1} - 3}{x - 2} = \lim_{h \to 0} 4\dfrac{\sqrt{9 + 4h} - 3}{4h}$

$= \dfrac{d}{dx}4\sqrt{x}\Big|_{x=9} = \dfrac{4}{2\sqrt{9}} = \dfrac{2}{3}$

16. $\lim_{x \to -a} \dfrac{(1/x^2) - (1/a^2)}{x + a} = \lim_{h \to 0} \dfrac{\dfrac{1}{(-a + h)^2} - \dfrac{1}{(-a)^2}}{h}$

$= \dfrac{d}{dx}\dfrac{1}{x^2}\Big|_{x=-a} = \dfrac{2}{a^3}$

18. $\dfrac{d}{dx}[f(\sqrt{x})]^2 = 2f(\sqrt{x})f'(\sqrt{x})\dfrac{1}{2\sqrt{x}} = \dfrac{f(\sqrt{x})f'(\sqrt{x})}{\sqrt{x}}$

20. $\dfrac{d}{dx}\dfrac{f(x)-g(x)}{f(x)+g(x)}$

$= \dfrac{1}{(f(x)+g(x))^2}\Big[f(x)+g(x))(f'(x)-g'(x))$

$\qquad - (f(x)-g(x))(f'(x)+g'(x))\Big]$

$= \dfrac{2(f'(x)g(x)-f(x)g'(x))}{(f(x)+g(x))^2}$

22. $\dfrac{d}{dx}f\left(\dfrac{g(x^2)}{x}\right) = \dfrac{2x^2g'(x^2)-g(x^2)}{x^2}f'\left(\dfrac{g(x^2)}{x}\right)$

24. $\dfrac{d}{dx}\sqrt{\dfrac{\cos f(x)}{\sin g(x)}}$

$= \dfrac{1}{2}\sqrt{\dfrac{\sin g(x)}{\cos f(x)}}$

$\quad\times \dfrac{-f'(x)\sin f(x)\sin g(x)-g'(x)\cos f(x)\cos g(x)}{(\sin g(x))^2}$

26. $3\sqrt{2}x\sin(\pi y)+8y\cos(\pi x)=2$
$3\sqrt{2}\sin(\pi y)+3\pi\sqrt{2}x\cos(\pi y)y'+8y'\cos(\pi x)$
$-8\pi y\sin(\pi x)=0$
At $(1/3, 1/4)$: $3+\pi y'+4y'-\pi\sqrt{3}=0$, so the slope
there is $y' = \dfrac{\pi\sqrt{3}-3}{\pi+4}$.

28. $\displaystyle\int\dfrac{1+x}{\sqrt{x}}\,dx = \int(x^{-1/2}+x^{1/2})\,dx = 2\sqrt{x}+\dfrac{2}{3}x^{3/2}+C$

30. $\displaystyle\int(2x+1)^4\,dx = \int(16x^4+32x^3+24x^2+8x+1)\,dx$

$= \dfrac{16x^5}{5}+8x^4+8x^3+4x^2+x+C$

or, equivalently,

$\displaystyle\int(2x+1)^4\,dx = \dfrac{(2x+1)^5}{10}+C$

32. If $g'(x)=\sin(x/3)+\cos(x/6)$, then

$$g(x)=-3\cos(x/3)+6\sin(x/6)+C.$$

If $(\pi, 2)$ lies on $y=g(x)$, then $-(3/2)+3+C=2$, so
$C=1/2$ and $g(x)=-3\cos(x/3)+6\sin(x/6)+(1/2)$.

34. If $f'(x)=f(x)$ and $g(x)=x\,f(x)$, then

$$g'(x)=f(x)+xf'(x)=(1+x)f(x)$$
$$g''(x)=f(x)+(1+x)f'(x)=(2+x)f(x)$$
$$g'''(x)=f(x)+(2+x)f'(x)=(3+x)f(x)$$

Conjecture: $g^{(n)}(x)=(n+x)f(x)$ for $n=1, 2, 3, \ldots$
Proof: The formula is true for $n=1, 2$, and 3 as shown
above. Suppose it is true for $n=k$; that is, suppose
$g^{(k)}(x)=(k+x)f(x)$. Then

$$g^{(k+1)}(x)=\dfrac{d}{dx}\Big((k+x)f(x)\Big)$$
$$=f(x)+(k+x)f'(x)=((k+1)+x)f(x).$$

Thus the formula is also true for $n=k+1$. It is therefore
true for all positive integers n by induction.

36. The tangent to $y=\sqrt{2+x^2}$ at $x=a$ has slope
$a/\sqrt{2+a^2}$ and equation

$$y=\sqrt{2+a^2}+\dfrac{a}{\sqrt{2+a^2}}(x-a).$$

This line passes through $(0, 1)$ provided

$$1=\sqrt{2+a^2}-\dfrac{a^2}{\sqrt{2+a^2}}$$
$$\sqrt{2+a^2}=2+a^2-a^2=2$$
$$2+a^2=4$$

The possibilities are $a=\pm\sqrt{2}$, and the equations of the
corrresponding tangent lines are $y=1\pm(x/\sqrt{2})$.

38. $\dfrac{d}{dx}\Big(\sin^n x\cos(nx)\Big)$

$= n\sin^{n-1} x\cos x\cos(nx)-n\sin^n x\sin(nx)$
$= n\sin^{n-1} x[\cos x\cos(nx)-\sin x\sin(nx)]$
$= n\sin^{n-1} x\cos((n+1)x)$

$\dfrac{d}{dx}\Big(\cos^n x\sin(nx)\Big)$

$= -n\cos^{n-1} x\sin x\sin(nx)+n\cos^n x\cos(nx)$
$= n\cos^{n-1} x[\cos x\cos(nx)-\sin x\sin(nx)]$
$= n\cos^{n-1} x\cos((n+1)x)$

$\dfrac{d}{dx}\Big(\cos^n x\cos(nx)\Big)$

$= -n\cos^{n-1} x\sin x\cos(nx)-n\cos^n x\sin(nx)$
$= -n\cos^{n-1} x[\sin x\cos(nx)+\cos x\sin(nx)]$
$= -n\cos^{n-1} x\sin((n+1)x)$

40. The average profit per tonne if x tonnes are exported is
$P(x)/x$, that is the slope of the line joining $(x, P(x))$ to
the origin. This slope is maximum if the line is tangent
to the graph of $P(x)$. In this case the slope of the line is
$P'(x)$, the marginal profit.

42. $PV=kT$. Differentiate with respect to P holding T
constant to get

$$V+P\dfrac{dV}{dP}=0$$

Thus the isothermal compressibility of the gas is

$$\dfrac{1}{V}\dfrac{dV}{dP}=\dfrac{1}{V}\left(-\dfrac{V}{P}\right)=-\dfrac{1}{P}.$$

44. The first ball has initial height 60 m and initial velocity 0, so its height at time t is

$$y_1 = 60 - 4.9t^2 \text{ m.}$$

The second ball has initial height 0 and initial velocity v_0, so its height at time t is

$$y_2 = v_0 t - 4.9t^2 \text{ m.}$$

The two balls collide at a height of 30 m (at time T, say). Thus

$$30 = 60 - 4.9T^2$$
$$30 = v_0 T - 4.9T^2.$$

Thus $v_0 T = 60$ and $T^2 = 30/4.9$. The initial upward speed of the second ball is

$$v_0 = \frac{60}{T} = 60\sqrt{\frac{4.9}{30}} \approx 24.25 \text{ m/s.}$$

At time T, the velocity of the first ball is

$$\left.\frac{dy_1}{dt}\right|_{t=T} = -9.8T \approx -24.25 \text{ m/s.}$$

At time T, the velocity of the second ball is

$$\left.\frac{dy_2}{dt}\right|_{t=T} = v_0 - 9.8T = 0 \text{ m/s.}$$

46. $P = 2\pi\sqrt{L/g} = 2\pi L^{1/2} g^{-1/2}$.

a) If L remains constant, then

$$\Delta P \approx \frac{dP}{dg}\Delta g = -\pi L^{1/2} g^{-3/2}\Delta g$$

$$\frac{\Delta P}{P} \approx \frac{-\pi L^{1/2} g^{-3/2}}{2\pi L^{1/2} g^{-1/2}}\Delta g = -\frac{1}{2}\frac{\Delta g}{g}.$$

If g increases by 1%, then $\Delta g/g = 1/100$, and $\Delta P/P = -1/200$. Thus P decreases by 0.5%.

b) If g remains constant, then

$$\Delta P \approx \frac{dP}{dL}\Delta L = \pi L^{-1/2} g^{-1/2}\Delta L$$

$$\frac{\Delta P}{P} \approx \frac{\pi L^{-1/2} g^{-1/2}}{2\pi L^{1/2} g^{-1/2}}\Delta L = \frac{1}{2}\frac{\Delta L}{L}.$$

If L increases by 2%, then $\Delta L/L = 2/100$, and $\Delta P/P = 1/100$. Thus P increases by 1%.

Challenging Problems 2   (page 159)

2. $f'(x) = 1/x$, $f(2) = 9$.

a)
$$\lim_{x\to 2}\frac{f(x^2+5)-f(9)}{x-2} = \lim_{h\to 0}\frac{f(9+4h+h^2)-f(9)}{h}$$
$$= \lim_{h\to 0}\frac{f(9+4h+h^2)-f(9)}{4h+h^2}\times\frac{4h+h^2}{h}$$
$$= \lim_{k\to 0}\frac{f(9+k)-f(9)}{k}\times\lim_{h\to 0}(4+h)$$
$$= f'(9)\times 4 = \frac{4}{9}$$

b)
$$\lim_{x\to 2}\frac{\sqrt{f(x)}-3}{x-2} = \lim_{h\to 0}\frac{\sqrt{f(2+h)}-3}{h}$$
$$= \lim_{h\to 0}\frac{f(2+h)-9}{h}\times\frac{1}{\sqrt{f(2+h)}+3}$$
$$= f'(2)\times\frac{1}{6} = \frac{1}{12}.$$

4. $f(x) = \begin{cases} x & \text{if } x = 1,\ 1/2,\ 1/3,\ \dots \\ x^2 & \text{otherwise} \end{cases}$.

a) f is continuous except at $1/2$, $1/3$, $1/4$, ... It is continuous at $x = 1$ and $x = 0$ (and everywhere else). Note that

$$\lim_{x\to 1} x^2 = 1 = f(1),$$
$$\lim_{x\to 0} x^2 = \lim_{x\to 0} x = 0 = f(0)$$

b) If $a = 1/2$ and $b = 1/3$, then

$$\frac{f(a)+f(b)}{2} = \frac{1}{2}\left(\frac{1}{2}+\frac{1}{3}\right) = \frac{5}{12}.$$

If $1/3 < x < 1/2$, then $f(x) = x^2 < 1/4 < 5/12$. Thus the statement is FALSE.

c) By (a) f cannot be differentiable at $x = 1/2$, $1/2$, ... It is not differentiable at $x = 0$ either, since

$$\lim_{h\to 0} h - 0h = 1 \neq 0 = \lim_{h\to 0}\frac{h^2-0}{h}.$$

f is differentiable elsewhere, including at $x = 1$ where its derivative is 2.

6. Given that $f'(0) = k$, $f(0) \neq 0$, and $f(x+y) = f(x)f(y)$, we have

$$f(0) = f(0+0) = f(0)f(0) \implies f(0) = 0 \text{ or } f(0) = 1.$$

Thus $f(0) = 1$.

$$f'(x) = \lim_{h\to 0}\frac{f(x+h)-f(x)}{h}$$
$$= \lim_{h\to 0}\frac{f(x)f(h)-f(x)}{h} = f(x)f'(0) = kf(x).$$

8.   a) $f'(x) = \lim\limits_{k \to 0} \dfrac{f(x + k) - f(x)}{k}$ (let $k = -h$)

$= \lim\limits_{h \to 0} \dfrac{f(x - h) - f(x)}{-h} = \lim\limits_{h \to 0} \dfrac{f(x) - f(x - h)}{h}.$

$f'(x) = \dfrac{1}{2}\Big(f'(x) + f'(x)\Big)$

$= \dfrac{1}{2}\left(\lim\limits_{h \to 0} \dfrac{f(x + h) - f(x)}{h} \right.$

$\left. + \lim\limits_{h \to 0} \dfrac{f(x) - f(x - h)}{h} \right)$

$= \lim\limits_{h \to 0} \dfrac{f(x + h) - f(x - h)}{2h}.$

b) The change of variables used in the first part of (a) shows that

$$\lim\limits_{h \to 0} \dfrac{f(x + h) - f(x)}{h} \quad and \quad \lim\limits_{h \to 0} \dfrac{f(x) - f(x - h)}{h}$$

are always equal if either exists.

c) If $f(x) = |x|$, then $f'(0)$ does not exist, but

$$\lim\limits_{h \to 0} \dfrac{f(0 + h) - f(0 - h)}{2h} = \lim\limits_{h \to 0} \dfrac{|h| - |h|}{h} = \lim\limits_{h \to 0} \dfrac{0}{h} = 0.$$

10.   By symmetry, any line tangent to both curves must pass through the origin.

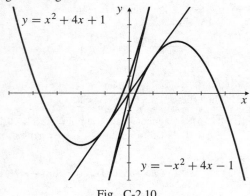

Fig. C-2.10

The tangent to $y = x^2 + 4x + 1$ at $x = a$ has equation

$$y = a^2 + 4a + 1 + (2a + 4)(x - a)$$
$$= (2a + 4)x - (a^2 - 1),$$

which passes through the origin if $a = \pm 1$. The two common tangents are $y = 6x$ and $y = 2x$.

12.   The point $Q = (a, a^2)$ on $y = x^2$ that is closest to $P = (3, 0)$ is such that PQ is normal to $y = x^2$ at Q. Since PQ has slope $a^2/(a - 3)$ and $y = x^2$ has slope $2a$ at Q, we require

$$\frac{a^2}{a - 3} = -\frac{1}{2a},$$

which simplifies to $2a^3 + a - 3 = 0$. Observe that $a = 1$ is a solution of this cubic equation. Since the slope of $y = 2x^3 + x - 3$ is $6x^2 + 1$, which is always positive, the cubic equation can have only one real solution. Thus $Q = (1, 1)$ is the point on $y = x^2$ that is closest to P. The distance from P to the curve is $|PQ| = \sqrt{5}$ units.

14.   Parabola $y = x^2$ has tangent $y = 2ax - a^2$ at (a, a^2). Parabola $y = Ax^2 + Bx + C$ has tangent

$$y = (2Ab + B)x - Ab^2 + C$$

at $(b, Ab^2 + Bb + C)$. These two tangents coincide if

$$2Ab + B = 2a \qquad\qquad (*)$$
$$Ab^2 - C = a^2.$$

The two curves have one (or more) common tangents if $(*)$ has real solutions for a and b. Eliminating a between the two equations leads to

$$(2Ab + B)^2 = 4Ab^2 - 4C,$$

or, on simplification,

$$4A(A - 1)b^2 + 4ABb + (B^2 + 4C) = 0.$$

This quadratic equation in b has discriminant

$$D = 16A^2 B^2 - 16A(A-1)(B^2 + 4C) = 16A(B^2 - 4(A-1)C).$$

There are five cases to consider:

CASE I. If $A = 1$, $B \neq 0$, then $(*)$ gives

$$b = -\frac{B^2 + 4C}{4B}, \quad a = \frac{B^2 - 4C}{4B}.$$

There is a single common tangent in this case.

CASE II. If $A = 1$, $B = 0$, then $(*)$ forces $C = 0$, which is not allowed. There is no common tangent in this case.

CASE III. If $A \neq 1$ but $B^2 = 4(A - 1)C$, then

$$b = \frac{-B}{2(A - 1)} = a.$$

There is a single common tangent, and since the points of tangency on the two curves coincide, the two curves are tangent to each other.

CASE IV. If $A \neq 1$ and $B^2 - 4(A-1)C < 0$, there are no real solutions for b, so there can be no common tangents.

CASE V. If $A \neq 1$ and $B^2 - 4(A - 1)C > 0$, there are two distinct real solutions for b, and hence two common tangent lines.

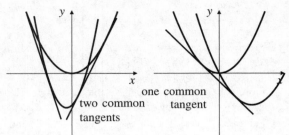

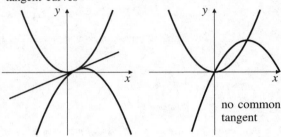

two common tangents

one common tangent

tangent curves

no common tangent

Fig. C-2.14

16.

a) $y = x^4 - 2x^2$ has horizontal tangents at points x satisfying $4x^3 - 4x = 0$, that is, at $x = 0$ and $x = \pm 1$. The horizontal tangents are $y = 0$ and $y = -1$. Note that $y = -1$ is a double tangent; it is tangent at the two points $(\pm 1, -1)$.

b) The tangent to $y = x^4 - 2x^2$ at $x = a$ has equation

$$y = a^4 - 2a^2 + (4a^3 - 4a)(x - a)$$
$$= 4a(a^2 - 1)x - 3a^4 + 2a^2.$$

Similarly, the tangent at $x = b$ has equation

$$y = 4b(b^2 - 1)x - 3b^4 + 2b^2.$$

These tangents are the same line (and hence a double tangent) if

$$4a(a^2 - 1) = 4b(b^2 - 1)$$
$$-3a^4 + 2a^2 = -3b^4 + 2b^2.$$

The second equation says that either $a^2 = b^2$ or $3(a^2 + b^2) = 2$; the first equation says that $a^3 - b^3 = a - b$, or, equivalently, $a^2 + ab + b^2 = 1$. If $a^2 = b^2$, then $a = -b$ ($a = b$ is not allowed). Thus $a^2 = b^2 = 1$ and the two points are $(\pm 1, -1)$ as discovered in part (a).
If $a^2 + b^2 = 2/3$, then $ab = 1/3$. This is not possible since it implies that

$$0 = a^2 + b^2 - 2ab = (a - b)^2 > 0.$$

Thus $y = -1$ is the only double tangent to $y = x^4 - 2x^2$.

c) If $y = Ax + B$ is a double tangent to $y = x^4 - 2x^2 + x$, then $y = (A - 1)x + B$ is a double tangent to $y = x^4 - 2x^2$. By (b) we must have $A - 1 = 0$ and $B = -1$. Thus the only double tangent to $y = x^4 - 2x^2 + x$ is $y = x - 1$.

18.

a) Claim: $\dfrac{d^n}{dx^n} \cos(ax) = a^n \cos\left(ax + \dfrac{n\pi}{2}\right)$.

Proof: For $n = 1$ we have

$$\frac{d}{dx} \cos(ax) = -a\sin(ax) = a\cos\left(ax + \frac{\pi}{2}\right),$$

so the formula above is true for $n = 1$. Assume it is true for $n = k$, where k is a positive integer. Then

$$\frac{d^{k+1}}{dx^{k+1}} \cos(ax) = \frac{d}{dx}\left[a^k \cos\left(ax + \frac{k\pi}{2}\right)\right]$$
$$= a^k\left[-a\sin\left(ax + \frac{k\pi}{2}\right)\right]$$
$$= a^{k+1}\cos\left(ax + \frac{(k+1)\pi}{2}\right).$$

Thus the formula holds for $n = 1, 2, 3, \dots$ by induction.

b) Claim: $\dfrac{d^n}{dx^n} \sin(ax) = a^n \sin\left(ax + \dfrac{n\pi}{2}\right)$.

Proof: For $n = 1$ we have

$$\frac{d}{dx} \sin(ax) = a\cos(ax) = a\sin\left(ax + \frac{\pi}{2}\right),$$

so the formula above is true for $n = 1$. Assume it is true for $n = k$, where k is a positive integer. Then

$$\frac{d^{k+1}}{dx^{k+1}} \sin(ax) = \frac{d}{dx}\left[a^k \sin\left(ax + \frac{k\pi}{2}\right)\right]$$
$$= a^k\left[a\cos\left(ax + \frac{k\pi}{2}\right)\right]$$
$$= a^{k+1}\sin\left(ax + \frac{(k+1)\pi}{2}\right).$$

Thus the formula holds for $n = 1, 2, 3, \dots$ by induction.

c) Note that

$$\frac{d}{dx}(\cos^4 x + \sin^4 x) = -4\cos^3 x \sin x + 4\sin^3 x \cos x$$
$$= -4\sin x \cos x(\cos^2 - \sin^2 x)$$
$$= -2\sin(2x)\cos(2x)$$
$$= -\sin(4x) = \cos\left(4x + \frac{\pi}{2}\right).$$

It now follows from part (a) that

$$\frac{d^n}{dx^n}(\cos^4 x + \sin^4 x) = 4^{n-1}\cos\left(4x + \frac{n\pi}{2}\right).$$

CHAPTER 3.   TRANSCENDENTAL FUNCTIONS

Section 3.1   Inverse Functions   (page 167)

2. $f(x) = 2x - 1$. If $f(x_1) = f(x_2)$, then $2x_1 - 1 = 2x_2 - 1$. Thus $2(x_1 - x_2) = 0$ and $x_1 = x_2$. Hence, f is one-to-one.
Let $y = f^{-1}(x)$. Thus $x = f(y) = 2y - 1$, so $y = \frac{1}{2}(x + 1)$. Thus $f^{-1}(x) = \frac{1}{2}(x + 1)$.
$\mathcal{D}(f) = \mathcal{R}(f^{-1}) = (-\infty, \infty)$.
$\mathcal{R}(f) = \mathcal{D}(f^{-1}) = (-\infty, \infty)$.

4. $f(x) = -\sqrt{x - 1}$ for $x \geq 1$.
If $f(x_1) = f(x_2)$, then $-\sqrt{x_1 - 1} = -\sqrt{x_2 - 1}$ and $x_1 - 1 = x_2 - 1$. Thus $x_1 = x_2$ and f is one-to-one.
Let $y = f^{-1}(x)$. Then $x = f(y) = -\sqrt{y - 1}$ so $x^2 = y - 1$ and $y = x^2 + 1$. Thus, $f^{-1}(x) = x^2 + 1$.
$\mathcal{D}(f) = \mathcal{R}(f^{-1}) = [1, \infty)$. $\mathcal{R}(f) = \mathcal{D}(f^{-1}) = (-\infty, 0]$.

6. $f(x) = 1 + \sqrt[3]{x}$. If $f(x_1) = f(x_2)$, then $1 + \sqrt[3]{x_1} = 1 + \sqrt[3]{x_2}$ so $x_1 = x_2$. Thus, f is one-to-one.
Let $y = f^{-1}(x)$ so that $x = f(y) = 1 + \sqrt[3]{y}$. Thus $y = (x - 1)^3$ and $f^{-1}(x) = (x - 1)^3$.
$\mathcal{D}(f) = \mathcal{R}(f^{-1}) = (-\infty, \infty)$.
$\mathcal{R}(f) = \mathcal{D}(f^{-1}) = (-\infty, \infty)$.

8. $f(x) = (1 - 2x)^3$. If $f(x_1) = f(x_2)$, then $(1 - 2x_1)^3 = (1 - 2x_2)^3$ and $x_1 = x_2$. Thus, f is one-to-one.
Let $y = f^{-1}(x)$. Then $x = f(y) = (1 - 2y)^3$ so $y = \frac{1}{2}(1 - \sqrt[3]{x})$. Thus, $f^{-1}(x) = \frac{1}{2}(1 - \sqrt[3]{x})$.
$\mathcal{D}(f) = \mathcal{R}(f^{-1}) = (-\infty, \infty)$.
$\mathcal{R}(f) = \mathcal{D}(f^{-1}) = (-\infty, \infty)$.

10. $f(x) = \dfrac{x}{1 + x}$. If $f(x_1) = f(x_2)$, then $\dfrac{x_1}{1 + x_1} = \dfrac{x_2}{1 + x_2}$.
Hence $x_1(1 + x_2) = x_2(1 + x_1)$ and, on simplification, $x_1 = x_2$. Thus, f is one-to-one.
Let $y = f^{-1}(x)$. Then $x = f(y) = \dfrac{y}{1 + y}$ and $x(1 + y) = y$. Thus $y = \dfrac{x}{1 - x} = f^{-1}(x)$.
$\mathcal{D}(f) = \mathcal{R}(f^{-1}) = (-\infty, -1) \cup (-1, \infty)$.
$\mathcal{R}(f) = \mathcal{D}(f^{-1}) = (-\infty, 1) \cup (1, \infty)$.

12. $f(x) = \dfrac{x}{\sqrt{x^2 + 1}}$. If $f(x_1) = f(x_2)$, then
$$\dfrac{x_1}{\sqrt{x_1^2 + 1}} = \dfrac{x_2}{\sqrt{x_2^2 + 1}}. \qquad (*)$$
Thus $x_1^2(x_2^2 + 1) = x_2^2(x_1^2 + 1)$ and $x_1^2 = x_2^2$.
From $(*)$, x_1 and x_2 must have the same sign. Hence, $x_1 = x_2$ and f is one-to-one.
Let $y = f^{-1}(x)$. Then $x = f(y) = \dfrac{y}{\sqrt{y^2 + 1}}$, and
$x^2(y^2 + 1) = y^2$. Hence $y^2 = \dfrac{x^2}{1 - x^2}$. Since $f(y)$ and y have the same sign, we must have $y = \dfrac{x}{\sqrt{1 - x^2}}$, so
$f^{-1}(x) = \dfrac{x}{\sqrt{1 - x^2}}$.
$\mathcal{D}(f) = \mathcal{R}(f^{-1}) = (-\infty, \infty)$.
$\mathcal{R}(f) = \mathcal{D}(f^{-1}) = (-1, 1)$.

14. $h(x) = f(2x)$. Let $y = h^{-1}(x)$. Then $x = h(y) = f(2y)$ and $2y = f^{-1}(x)$. Thus $h^{-1}(x) = y = \frac{1}{2}f^{-1}(x)$.

16. $m(x) = f(x - 2)$. Let $y = m^{-1}(x)$. Then $x = m(y) = f(y - 2)$, and $y - 2 = f^{-1}(x)$. Hence $m^{-1}(x) = y = f^{-1}(x) + 2$.

18. $q(x) = \dfrac{f(x) - 3}{2}$ Let $y = q^{-1}(x)$. Then $x = q(y) = \dfrac{f(y) - 3}{2}$ and $f(y) = 2x + 3$. Hence $q^{-1}(x) = y = f^{-1}(2x + 3)$.

20. $s(x) = \dfrac{1 + f(x)}{1 - f(x)}$. Let $y = s^{-1}(x)$.
Then $x = s(y) = \dfrac{1 + f(y)}{1 - f(y)}$. Solving for $f(y)$ we obtain $f(y) = \dfrac{x - 1}{x + 1}$. Hence $s^{-1}(x) = y = f^{-1}\left(\dfrac{x - 1}{x + 1}\right)$.

22. $g(x) = x^3$ if $x \geq 0$, and $g(x) = x^{1/3}$ if $x < 0$.
Suppose $f(x_1) = f(x_2)$. If $x_1 \geq 0$ and $x_2 \geq 0$ then $x_1^3 = x_2^3$ so $x_1 = x_2$.
Similarly, $x_1 = x_2$ if both are negative. If x_1 and x_2 have opposite sign, then so do $g(x_1)$ and $g(x_2)$.
Therefore g is one-to-one. Let $y = g^{-1}(x)$. Then
$x = g(y) = \begin{cases} y^3 & \text{if } y \geq 0 \\ y^{1/3} & \text{if } y < 0. \end{cases}$
Thus $g^{-1}(x) = y = \begin{cases} x^{1/3} & \text{if } x \geq 0 \\ x^3 & \text{if } x < 0. \end{cases}$

24. $y = f^{-1}(x) \Leftrightarrow x = f(y) = y^3 + y$. To find $y = f^{-1}(2)$ we solve $y^3 + y = 2$ for y. Evidently $y = 1$ is the only solution, so $f^{-1}(2) = 1$.

26. $h(x) = -3$ if $x|x| = -4$, that is, if $x = -2$. Thus $h^{-1}(-3) = -2$.

28. $f(x) = 1 + 2x^3$
Let $y = f^{-1}(x)$.
Thus $x = f(y) = 1 + 2y^3$.
$1 = 6y^2 \dfrac{dy}{dx}$ so $(f^{-1})'(x) = \dfrac{dy}{dx} = \dfrac{1}{6y^2} = \dfrac{1}{6[f^{-1}(x)]^2}$

30. If $f(x) = x\sqrt{3 + x^2}$ and $y = f^{-1}(x)$, then
$x = f(y) = y\sqrt{3 + y^2}$, so,

$$1 = y'\sqrt{3 + y^2} + y\frac{2yy'}{2\sqrt{3 + y^2}} \quad \Rightarrow \quad y' = \frac{\sqrt{3 + y^2}}{3 + 2y^2}.$$

Since $f(-1) = -2$ implies that $f^{-1}(-2) = -1$, we have

$$\left(f^{-1}\right)'(-2) = \frac{\sqrt{3 + y^2}}{3 + 2y^2}\bigg|_{y=-1} = \frac{2}{5}.$$

Note: $f(x) = x\sqrt{3 + x^2} = -2 \Rightarrow x^2(3 + x^2) = 4$
$\Rightarrow x^4 + 3x^2 - 4 = 0 \Rightarrow (x^2 + 4)(x^2 - 1) = 0$.
Since $(x^2 + 4) = 0$ has no real solution, therefore
$x^2 - 1 = 0$ and $x = 1$ or -1. Since it is given that
$f(x) = -2$, therefore x must be -1.

32. $g(x) = 2x + \sin x \Rightarrow g'(x) = 2 + \cos x \geq 1$ for
all x. Therefore g is increasing, and so one-to-one and
invertible on the whole real line.

$y = g^{-1}(x) \Leftrightarrow x = g(y) = 2y + \sin y$. For $y = g^{-1}(2)$,
we need to solve $2y + \sin y - 2 = 0$. The root is between
0 and 1; to five decimal places $g^{-1}(2) = y \approx 0.68404$.
Also

$$1 = \frac{dx}{dx} = (2 + \cos y)\frac{dy}{dx}$$

$$(g^{-1})'(2) = \frac{dy}{dx}\bigg|_{x=2} = \frac{1}{2 + \cos y} \approx 0.36036.$$

34. If $y = (f \circ g)^{-1}(x)$, then $x = f \circ g(y) = f(g(y))$. Thus
$g(y) = f^{-1}(x)$ and $y = g^{-1}(f^{-1}(x)) = g^{-1} \circ f^{-1}(x)$.
That is, $(f \circ g)^{-1} = g^{-1} \circ f^{-1}$.

36. Let $f(x)$ be an even function. Then $f(x) = f(-x)$.
Hence, f is not one-to-one and it is not invertible.
Therefore, it cannot be self-inverse.
An odd function $g(x)$ may be self-inverse if its graph is
symmetric about the line $x = y$. Examples are $g(x) = x$
and $g(x) = 1/x$.

38. First we consider the case where the domain of f is a
closed interval. Suppose that f is one-to-one and continuous on $[a, b]$, and that $f(a) < f(b)$. We show that
f must be increasing on $[a, b]$. Suppose not. Then there
are numbers x_1 and x_2 with $a \leq x_1 < x_2 \leq b$ and
$f(x_1) > f(x_2)$. If $f(x_1) > f(a)$, let u be a number
such that $u < f(x_1)$, $f(x_2) < u$, and $f(a) < u$. By
the Intermediate-Value Theorem there exist numbers c_1 in
(a, x_1) and c_2 in (x_1, x_2) such that $f(c_1) = u = f(c_2)$,
contradicting the one-to-oneness of f. A similar contradiction arises if $f(x_1) \leq f(a)$ because, in this case,
$f(x_2) < f(b)$ and we can find c_1 in (x_1, x_2) and c_2 in
(x_2, b) such that $f(c_1) = f(c_2)$. Thus f must be increasing on $[a, b]$.

A similar argument shows that if $f(a) > f(b)$, then
f must be decreasing on $[a, b]$.

Finally, if the interval I where f is defined is not
necessarily closed, the same argument shows that if $[a, b]$
is a subinterval of I on which f is increasing (or decreasing), then f must also be increasing (or decreasing)
on any intervals of either of the forms $[x_1, b]$ or $[a, x_2]$,
where x_1 and x_2 are in I and $x_1 \leq a < b \leq x_2$. So f
must be increasing (or decreasing) on the whole of I.

Section 3.2 Exponential and Logarithmic Functions (page 171)

2. $2^{1/2}8^{1/2} = 2^{1/2}2^{3/2} = 2^2 = 4$

4. $(\frac{1}{2})^x 4^{x/2} = \dfrac{2^x}{2^x} = 1$

6. If $\log_4(\frac{1}{8}) = y$ then $4^y = \frac{1}{8}$, or $2^{2y} = 2^{-3}$. Thus
$2y = -3$ and $\log_4(\frac{1}{8}) = y = -\frac{3}{2}$.

8. $4^{3/2} = 8 \Rightarrow \log_4 8 = \frac{3}{2} \Rightarrow 2^{\log_4 8} = 2^{3/2} = 2\sqrt{2}$

10. Since $\log_a\left(x^{1/(\log_a x)}\right) = \dfrac{1}{\log_a x} \log_a x = 1$, therefore
$x^{1/(\log_a x)} = a^1 = a$.

12. $\log_x\left(x(\log_y y^2)\right) = \log_x(2x) = \log_x x + \log_x 2$

$$= 1 + \log_x 2 = 1 + \frac{1}{\log_2 x}$$

14. $\log_{15} 75 + \log_{15} 3 = \log_{15} 225 = 2$
$$\text{(since } 15^2 = 225\text{)}$$

16. $2\log_3 12 - 4\log_3 6 = \log_3\left(\dfrac{4^2 \cdot 3^2}{2^4 \cdot 3^4}\right)$

$$= \log_3(3^{-2}) = -2$$

18. $\log_\pi(1 - \cos x) + \log_\pi(1 + \cos x) - 2\log_\pi \sin x$

$$= \log_\pi\left[\frac{(1 - \cos x)(1 + \cos x)}{\sin^2 x}\right] = \log_\pi \frac{\sin^2 x}{\sin^2 x}$$

$$= \log_\pi 1 = 0$$

20. $\log_3 5 = (\log_{10} 5)/(\log_{10} 3 \approx 1.46497$

22. $x^{\sqrt{2}} = 3$, $\sqrt{2}\log_{10} x = \log_{10} 3$,
$x = 10^{(\log_{10} 3)/\sqrt{2}} \approx 2.17458$

24. $\log_3 x = 5$, $(\log_{10} x)/(\log_{10} 3) = 5$,
$\log_{10} x = 5\log_{10} 3$, $x = 10^{5\log_{10} 3} = 3^5 = 243$

26. Let $\log_a x = u$, $\log_a y = v$.
Then $x = a^u$, $y = a^v$.
Thus $\dfrac{x}{y} = \dfrac{a^u}{a^v} = a^{u-v}$
and $\log_a\left(\dfrac{x}{y}\right) = u - v = \log_a x - \log_a y$.

28. Let $\log_b x = u$, $\log_b a = v$.
Thus $b^u = x$ and $b^v = a$.
Therefore $x = b^u = b^{v(u/v)} = a^{u/v}$
and $\log_a x = \dfrac{u}{v} = \dfrac{\log_b x}{\log_b a}$.

30. First observe that $\log_9 x = \log_3 x / \log_3 9 = \frac{1}{2}\log_3 x$. Now
$2\log_3 x + \log_9 x = 10$
$\log_3 x^2 + \log_3 x^{1/2} = 10$
$\log_3 x^{5/2} = 10$
$x^{5/2} = 3^{10}$, so $x = (3^{10})^{2/5} = 3^4 = 81$

32. Note that $\log_x(1/2) = -\log_x 2 = -1/\log_2 x$.
Since $\lim_{x \to 0+} \log_2 x = -\infty$, therefore
$\lim_{x \to 0+} \log_x(1/2) = 0$.

34. Note that $\log_x 2 = 1/\log_2 x$.
Since $\lim_{x \to 1-} \log_2 x = 0-$, therefore
$\lim_{x \to 1-} \log_x 2 = -\infty$.

36. $y = f^{-1}(x) \Rightarrow x = f(y) = a^y$
$\Rightarrow 1 = \dfrac{dx}{dx} = ka^y \dfrac{dy}{dx}$
$\Rightarrow \dfrac{dy}{dx} = \dfrac{1}{ka^y} = \dfrac{1}{kx}$.

Thus $(f^{-1})'(x) = 1/(kx)$.

Section 3.3 The Natural Logarithm and Exponential (page 179)

2. $\ln(e^{1/2}e^{2/3}) = \frac{1}{2} + \frac{2}{3} = \frac{7}{6}$

4. $e^{(3\ln 9)/2} = 9^{3/2} = 27$

6. $e^{2\ln\cos x} + \left(\ln e^{\sin x}\right)^2 = \cos^2 x + \sin^2 x = 1$

8. $4\ln\sqrt{x} + 6\ln(x^{1/3}) = 2\ln x + 2\ln x = 4\ln x$

10. $\ln(x^2 + 6x + 9) = \ln[(x+3)^2] = 2\ln(x+3)$

12. $3^x = 9^{1-x} \Rightarrow 3^x = 3^{2(1-x)}$
$\Rightarrow x = 2(1-x) \Rightarrow x = \frac{2}{3}$

14. $2^{x^2-3} = 4^x = 2^{2x} \Rightarrow x^2 - 3 = 2x$
$x^2 - 2x - 3 = 0 \Rightarrow (x-3)(x+1) = 0$
Hence, $x = -1$ or 3.

16. $\ln(x^2 - x - 2) = \ln[(x-2)(x+1)]$ is defined if
$(x-2)(x+1) > 0$, that is, if $x < -1$ or $x > 2$. The
domain is the union $(-\infty, -1) \cup (2, \infty)$.

18. $\ln(x^2 - 2) \le \ln x$ holds if $x^2 > 2$, $x > 0$, and $x^2 - 2 \le x$.
Thus we need $x > \sqrt{2}$ and $x^2 - x - 2 \le 0$. This latter
inequality says that $(x-2)(x+1) \le 0$, so it holds for
$-1 \le x \le 2$. The solution set of the given inequality is
$(\sqrt{2}, 2]$.

20. $y = xe^x - x$, $\qquad y' = e^x + xe^x - 1$

22. $y = x^2 e^{x/2}$, $\qquad y' = 2xe^{x/2} + \frac{1}{2}x^2 e^{x/2}$

24. $y = \ln|3x - 2|$, $\qquad y' = \dfrac{3}{3x-2}$

26. $f(x) = e^{x^2}$, $\qquad f'(x) = (2x)e^{x^2}$

28. $x = e^{3t}\ln t$, $\qquad \dfrac{dx}{dt} = 3e^{3t}\ln t + \dfrac{1}{t}e^{3t}$

30. $y = \dfrac{e^x}{1+e^x} = 1 - \dfrac{1}{1+e^x}$, $\quad y' = \dfrac{e^x}{(1+e^x)^2}$

32. $y = e^{-x}\cos x$, $\qquad y' = -e^{-x}\cos x - e^{-x}\sin x$

34. $y = x\ln x - x$
$y' = \ln x + x\left(\dfrac{1}{x}\right) - 1 = \ln x$

36. $y = \ln|\sin x|$, $\qquad y' = \dfrac{\cos x}{\sin x} = \cot x$

38. $y = 2^{(x^2-3x+8)}$, $\quad y' = (2x-3)(\ln 2)2^{(x^2-3x+8)}$

40. $h(t) = t^x - x^t$, $\qquad h'(t) = xt^{x-1} - x^t\ln x$

42. $g(x) = \log_x(2x+3) = \dfrac{\ln(2x+3)}{\ln x}$
$g'(x) = \dfrac{\ln x\left(\dfrac{2}{2x+3}\right) - [\ln(2x+3)]\left(\dfrac{1}{x}\right)}{(\ln x)^2}$
$= \dfrac{2x\ln x - (2x+3)\ln(2x+3)}{x(2x+3)(\ln x)^2}$

44. Given that $y = \left(\dfrac{1}{x}\right)^{\ln x}$, let $u = \ln x$. Then $x = e^u$ and
$y = \left(\dfrac{1}{e^u}\right)^u = (e^{-u})^u = e^{-u^2}$. Hence,
$\dfrac{dy}{dx} = \dfrac{dy}{du} \cdot \dfrac{du}{dx} = (-2ue^{-u^2})\left(\dfrac{1}{x}\right) = -\dfrac{2\ln x}{x}\left(\dfrac{1}{x}\right)^{\ln x}$.

46. $y = \ln|x + \sqrt{x^2 - a^2}|$
$y' = \dfrac{1 + \dfrac{2x}{2\sqrt{x^2 - a^2}}}{x + \sqrt{x^2 - a^2}} = \dfrac{1}{\sqrt{x^2 - a^2}}$

48. $y = (\cos x)^x - x^{\cos x} = e^{x \ln \cos x} - e^{(\cos x)(\ln x)}$

$y' = e^{x \ln \cos x} \left[\ln \cos x + x \left(\frac{1}{\cos x} \right)(-\sin x) \right]$

$\qquad - e^{(\cos x)(\ln x)} \left[-\sin x \ln x + \frac{1}{x} \cos x \right]$

$\quad = (\cos x)^x (\ln \cos x - x \tan x)$

$\qquad - x^{\cos x} \left(-\sin x \ln x + \frac{1}{x} \cos x \right)$

50. Since

$$\frac{d}{dx}(ax^2 + bx + c)e^x = (2ax + b)e^x + (ax^2 + bx + c)e^x$$

$$= [ax^2 + (2a+b)x + (b+c)]e^x$$

$$= [Ax^2 + Bx + C]e^x.$$

Thus, differentiating $(ax^2 + bx + c)e^x$ produces another function of the same type with different constants. Any number of differentiations will do likewise.

52. $f(x) = \ln(2x+1)$ $\qquad f'(x) = 2(2x+1)^{-1}$

$f''(x) = (-1)2^2(2x+1)^{-2}$ $\qquad f'''(x) = (2)2^3(2x+1)^{-3}$

$f^{(4)}(x) = -(3!)2^4(2x+1)^{-4}$

Thus, if $n = 1, 2, 3, \dots$ we have
$f^{(n)}(x) = (-1)^{n-1}(n-1)!2^n(2x+1)^{-n}$.

54. Given that $x^{x^{x^{\cdot^{\cdot^{\cdot}}}}} = a$ where $a > 0$, then

$$\ln a = x^{x^{x^{\cdot^{\cdot^{\cdot}}}}} \ln x = a \ln x.$$

Thus $\ln x = \frac{1}{a} \ln a = \ln a^{1/a}$, so $x = a^{1/a}$.

56. $F(x) = \dfrac{\sqrt{1+x}(1-x)^{1/3}}{(1+5x)^{4/5}}$

$\ln F(x) = \frac{1}{2} \ln(1+x) + \frac{1}{3} \ln(1-x) - \frac{4}{5} \ln(1+5x)$

$\dfrac{F'(x)}{F(x)} = \dfrac{1}{2(1+x)} - \dfrac{1}{3(1-x)} - \dfrac{4}{(1+5x)}$

$F'(0) = F(0) \left[\frac{1}{2} - \frac{1}{3} - \frac{4}{1} \right] = (1) \left[\frac{1}{2} - \frac{1}{3} - 4 \right] = -\frac{23}{6}$

58. Since $y = x^2 e^{-x^2}$, then

$$y' = 2x e^{-x^2} - 2x^3 e^{-x^2} = 2x(1-x)(1+x)e^{-x^2}.$$

The tangent is horizontal at $(0, 0)$ and $\left(\pm 1, \dfrac{1}{e} \right)$.

60. Since $y = \ln x$ and $y' = \dfrac{1}{x} = 4$ then $x = \frac{1}{4}$ and
$y = \ln \frac{1}{4} = -\ln 4$. The tangent line of slope 4 is
$y = -\ln 4 + 4(x - \frac{1}{4})$, i.e., $y = 4x - 1 - \ln 4$.

62. The slope of $y = \ln x$ at $x = a$ is $y' = \dfrac{1}{x} \bigg|_{x=a} = \dfrac{1}{a}$. The line from $(0, 0)$ to $(a, \ln a)$ is tangent to $y = \ln x$ if

$$\frac{\ln a - 0}{a - 0} = \frac{1}{a}$$

i.e., if $\ln a = 1$, or $a = e$. Thus, the line is $y = \dfrac{x}{e}$.

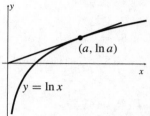

Fig. 3.3.62

64. The tangent line to $y = a^x$ which passes through the origin is tangent at the point (b, a^b) where

$$\frac{a^b - 0}{b - 0} = \frac{d}{dx} a^x \bigg|_{x=b} = a^b \ln a.$$

Thus $\dfrac{1}{b} = \ln a$, so $a^b = a^{1/\ln a} = e$. The line $y = x$ will intersect $y = a^x$ provided the slope of this tangent line does not exceed 1, i.e., provided $\dfrac{e}{b} \le 1$, or $e \ln a \le 1$. Thus we need $a \le e^{1/e}$.

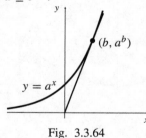

Fig. 3.3.64

66. $xe^y + y - 2x = \ln 2 \Rightarrow e^y + xe^y y' + y' - 2 = 0$.
At $(1, \ln 2)$, $2 + 2y' + y' - 2 = 0 \Rightarrow y' = 0$.
Therefore, the tangent line is $y = \ln 2$.

68. $F_{A,B}(x) = Ae^x \cos x + Be^x \sin x$

$\dfrac{d}{dx} F_{A,B}(x)$

$\quad = Ae^x \cos x - Ae^x \sin x + Be^x \sin x + Be^x \cos x$

$\quad = (A+B)e^x \cos x + (B-A)e^x \sin x = F_{A+B, B-A}(x)$

70. $\dfrac{d}{dx}(Ae^{ax} \cos bx + Be^{ax} \sin bx)$

$\quad = Aae^{ax} \cos bx - Abe^{ax} \sin bx + Bae^{ax} \sin bx$

$\qquad + Bbe^{ax} \cos bx$

$\quad = (Aa + Bb)e^{ax} \cos bx + (Ba - Ab)e^{ax} \sin bx.$

(a) If $Aa + Bb = 1$ and $Ba - Ab = 0$, then $A = \dfrac{a}{a^2 + b^2}$
and $B = \dfrac{b}{a^2 + b^2}$. Thus

$$\int e^{ax} \cos bx \, dx$$
$$= \frac{1}{a^2 + b^2}\left(ae^{ax} \cos bx + be^{ax} \sin bx\right) + C.$$

(b) If $Aa + Bb = 0$ and $Ba - Ab = 1$, then $A = \dfrac{-b}{a^2 + b^2}$
and $B = \dfrac{a}{a^2 + b^2}$. Thus

$$\int e^{ax} \sin bx \, dx$$
$$= \frac{1}{a^2 + b^2}\left(ae^{ax} \sin bx - be^{ax} \cos bx\right) + C.$$

72. $\ln \dfrac{x}{y} = \ln\left(x\,\dfrac{1}{y}\right) = \ln x + \ln\dfrac{1}{y} = \ln x - \ln y.$

74. Let $x > 0$, and $F(x)$ be the area bounded by $y = t^2$, the t-axis, $t = 0$ and $t = x$. For $h > 0$, $F(x + h) - F(x)$ is the shaded area in the following figure.

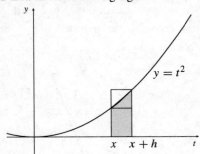

Fig. 3.3.74

Comparing this area with that of the two rectangles, we see that

$$hx^2 < F(x + h) - F(x) < h(x + h)^2.$$

Hence, the Newton quotient for $F(x)$ satisfies

$$x^2 < \frac{F(x + h) - F(x)}{h} < (x + h)^2.$$

Letting h approach 0 from the right (by the Squeeze Theorem applied to one-sided limits)

$$\lim_{h \to 0+} \frac{F(x + h) - F(x)}{h} = x^2.$$

If $h < 0$ and $0 < x + h < x$, then

$$(x + h)^2 < \frac{F(x + h) - F(x)}{h} < x^2,$$

so similarly,

$$\lim_{h \to 0-} \frac{F(x + h) - F(x)}{h} = x^2.$$

Combining these two limits, we obtain

$$\frac{d}{dx} F(x) = \lim_{h \to 0} \frac{F(x + h) - F(x)}{h} = x^2.$$

Therefore $F(x) = \displaystyle\int x^2 \, dx = \frac{1}{3}x^3 + C$. Since $F(0) = C = 0$, therefore $F(x) = \frac{1}{3}x^3$. For $x = 2$, the area of the region is $F(2) = \frac{8}{3}$ square units.

Section 3.4 Growth and Decay (page 187)

2. $\displaystyle\lim_{x \to \infty} x^{-3}e^x = \lim_{x \to \infty} \frac{e^x}{x^3} = \infty$

4. $\displaystyle\lim_{x \to \infty} \frac{x - 2e^{-x}}{x + 3e^{-x}} = \lim_{x \to \infty} \frac{1 - 2/(xe^x)}{1 + 3/(xe^x)} = \frac{1 - 0}{1 + 0} = 1$

6. $\displaystyle\lim_{x \to 0+} \frac{\ln x}{x} = -\infty$

8. $\displaystyle\lim_{x \to \infty} \frac{(\ln x)^3}{\sqrt{x}} = 0$ (power wins)

10. Let $y(t)$ be the number of kg undissolved after t hours. Thus, $y(0) = 50$ and $y(5) = 20$. Since $y'(t) = ky(t)$, therefore $y(t) = y(0)e^{kt} = 50e^{kt}$. Then

$$20 = y(5) = 50e^{5k} \Rightarrow k = \tfrac{1}{5}\ln\tfrac{2}{5}.$$

If 90% of the sugar is dissolved at time T then $5 = y(T) = 50e^{kT}$, so

$$T = \frac{1}{k}\ln\frac{1}{10} = \frac{5\ln(0.1)}{\ln(0.4)} \approx 12.56.$$

Hence, 90% of the sugar will dissolved in about 12.56 hours.

12. Let $P(t)$ be the percentage remaining after t years. Thus $P'(t) = kP(t)$ and $P(t) = P(0)e^{kt} = 100e^{kt}$. Then,

$$50 = P(1690) = 100e^{1690k} \Rightarrow k = \frac{1}{1690}\ln\frac{1}{2} \approx 0.0004101.$$

a) $P(100) = 100e^{100k} \approx 95.98$, i.e., about 95.98% remains after 100 years.

b) $P(1000) = 100e^{1000k} \approx 66.36$, i.e., about 66.36% remains after 1000 years.

14. Let $N(t)$ be the number of bacteria in the culture t days after the culture was set up. Thus $N(3) = 3N(0)$ and $N(7) = 10 \times 10^6$. Since $N(t) = N(0)e^{kt}$, we have

$$3N(0) = N(3) = N(0)e^{3k} \Rightarrow k = \tfrac{1}{3}\ln 3.$$
$$10^7 = N(7) = N(0)e^{7k} \Rightarrow N(0) = 10^7 e^{-(7/3)\ln 3} \approx 770400.$$

There were approximately 770,000 bacteria in the culture initially. (Note that we are approximating a discrete quantity (number of bacteria) by a continuous quantity $N(t)$ in this exercise.)

16. Since

$$I'(t) = kI(t) \Rightarrow I(t) = I(0)e^{kt} = 40e^{kt},$$
$$15 = I(0.01) = 40e^{0.01k} \Rightarrow k = \frac{1}{0.01}\ln\frac{15}{40} = 100\ln\frac{3}{8},$$

thus,

$$I(t) = 40\exp\left(100t\ln\frac{3}{8}\right) = 40\left(\frac{3}{8}\right)^{100t}.$$

18. Let $y(t)$ be the value of the investment after t years. Thus $y(0) = 1000$ and $y(5) = 1500$. Since $y(t) = 1000e^{kt}$ and $1500 = y(5) = 1000e^{5k}$, therefore, $k = \tfrac{1}{5}\ln\tfrac{3}{2}$.

a) Let t be the time such that $y(t) = 2000$, i.e.,

$$1000e^{kt} = 2000$$
$$\Rightarrow \quad t = \frac{1}{k}\ln 2 = \frac{5\ln 2}{\ln(\frac{3}{2})} = 8.55.$$

Hence, the doubling time for the investment is about 8.55 years.

b) Let $r\%$ be the effective annual rate of interest; then

$$1000(1 + \frac{r}{100}) = y(1) = 1000e^k$$
$$\Rightarrow r = 100(e^k - 1) = 100[\exp(\tfrac{1}{5}\ln\tfrac{3}{2}) - 1]$$
$$= 8.447.$$

The effective annual rate of interest is about 8.45%.

20. Let $i\%$ be the effective rate, then an original investment of $\$A$ will grow to $\$A\left(1 + \dfrac{i}{100}\right)$ in one year. Let $r\%$ be the nominal rate per annum compounded n times per year, then an original investment of $\$A$ will grow to

$$\$A\left(1 + \frac{r}{100n}\right)^n$$

in one year, if compounding is performed n times per year. For $i = 9.5$ and $n = 12$, we have

$$\$A\left(1 + \frac{9.5}{100}\right) = \$A\left(1 + \frac{r}{1200}\right)^{12}$$
$$\Rightarrow r = 1200\left(\sqrt[12]{1.095} - 1\right) = 9.1098.$$

The nominal rate of interest is about 9.1098%.

22. Let $N(t)$ be the number of rats on the island t months after the initial population was released and before the first cull. Thus $N(0) = R$ and $N(3) = 2R$. Since $N(t) = Re^{kt}$, we have $e^{3k} = 2$, so $e^k = 2^{1/3}$. Hence $N(5) = Re^{5k} = 2^{5/3}R$. After the first 1,000 rats are killed the number remaining is $2^{5/3}R - 1,000$. If this number is less than R, the number at the end of succeeding 5-year periods will decline. The minimum value of R for which this won't happen must satisfy $2^{5/3}R - 1,000 = R$, that is, $R = 1,000/(2^{5/3} - 1) \approx 459.8$. Thus $R = 460$ rats should be brought to the island initially.

24.

a) The concentration $x(t)$ satisfies $\dfrac{dx}{dt} = a - bx(t)$. This says that $x(t)$ is increasing if it is less than a/b and decreasing if it is greater than a/b. Thus, the limiting concentration is a/b.

b) The differential equation for $x(t)$ resembles that of Exercise 21(b), except that $y(x)$ is replaced by $x(t)$, and b is replaced by $-b$. Using the result of Exercise 21(b), we obtain, since $x(0) = 0$,

$$x(t) = \left(x(0) - \frac{a}{b}\right)e^{-bt} + \frac{a}{b}$$
$$= \frac{a}{b}\left(1 - e^{-bt}\right).$$

c) We will have $x(t) = \tfrac{1}{2}(a/b)$ if $1 - e^{-bt} = \tfrac{1}{2}$, that is, if $e^{-bt} = \tfrac{1}{2}$, or $-bt = \ln(1/2) = -\ln 2$. The time required to attain half the limiting concentration is $t = (\ln 2)/b$.

26. Let $T(t)$ be the temperature of the object t minutes after its temperature was $45°$ C. Thus $T(0) = 45$ and $T(40) = 20$. Also $\dfrac{dT}{dt} = k(T + 5)$. Let $u(t) = T(t) + 5$, so $u(0) = 50$, $u(40) = 25$, and $\dfrac{du}{dt} = \dfrac{dT}{dt} = k(T + 5) = ku$. Thus,

$$u(t) = 50e^{kt},$$
$$25 = u(40) = 50e^{40k},$$
$$\Rightarrow k = \frac{1}{40}\ln\frac{25}{50} = \frac{1}{40}\ln\frac{1}{2}.$$

We wish to know t such that $T(t) = 0$, i.e., $u(t) = 5$, hence

$$5 = u(t) = 50e^{kt}$$

$$t = \frac{40 \ln \left(\frac{5}{50} \right)}{\ln \left(\frac{1}{2} \right)} = 132.88 \text{ min.}$$

Hence, it will take about $(132.88 - 40) = 92.88$ minutes more to cool to $0°$ C.

28. By the solution given for the logistic equation, we have

$$y_1 = \frac{Ly_0}{y_0 + (L - y_0)e^{-k}}, \qquad y_2 = \frac{Ly_0}{y_0 + (L - y_0)e^{-2k}}$$

Thus $y_1(L - y_0)e^{-k} = (L - y_1)y_0$, and
$y_2(L - y_0)e^{-2k} = (L - y_2)y_0$.
Square the first equation and thus eliminate e^{-k}:

$$\left(\frac{(L - y_1)y_0}{y_1(L - y_0)} \right)^2 = \frac{(L - y_2)y_0}{y_2(L - y_0)}$$

Now simplify: $y_0 y_2 (L - y_1)^2 = y_1^2 (L - y_0)(L - y_2)$
$y_0 y_2 L^2 - 2y_1 y_0 y_2 L + y_0 y_1^2 y_2 = y_1^2 L^2 - y_1^2 (y_0 + y_2)L + y_0 y_1^2 y_2$

Assuming $L \neq 0$, $L = \dfrac{y_1^2(y_0 + y_2) - 2y_0 y_1 y_2}{y_1^2 - y_0 y_2}$.
If $y_0 = 3$, $y_1 = 5$, $y_2 = 6$, then
$L = \dfrac{25(9) - 180}{25 - 18} = \dfrac{45}{7} \approx 6.429$.

30. The solution $y = \dfrac{Ly_0}{y_0 + (L - y_0)e^{-kt}}$ is valid on the largest interval containing $t = 0$ on which the denominator does not vanish.
If $y_0 > L$ then $y_0 + (L - y_0)e^{-kt} = 0$ if
$t = t^* = -\dfrac{1}{k} \ln \dfrac{y_0}{y_0 - L}$.
Then the solution is valid on (t^*, ∞).
$\lim_{t \to t^*+} y(t) = \infty$.

32.
$$y(t) = \frac{L}{1 + Me^{-kt}}$$
$$200 = y(0) = \frac{L}{1 + M}$$
$$1,000 = y(1) = \frac{L}{1 + Me^{-k}}$$
$$10,000 = \lim_{t \to \infty} y(t) = L$$

Thus $200(1 + M) = L = 10,000$, so $M = 49$. Also $1,000(1 + 49e^{-k}) = L = 10,000$, so $e^{-k} = 9/49$ and $k = \ln(49/9) \approx 1.695$.

Section 3.5 The Inverse Trigonometric Functions (page 195)

2. $\cos^{-1} \left(-\dfrac{1}{2} \right) = \dfrac{2\pi}{3}$

4. $\sec^{-1} \sqrt{2} = \dfrac{\pi}{4}$

6. $\cos(\sin^{-1} 0.7) = \sqrt{1 - \sin^2(\arcsin 0.7)}$
$\qquad\qquad = \sqrt{1 - 0.49} = \sqrt{0.51}$

8. $\sin^{-1}(\cos 40°) = 90° - \cos^{-1}(\cos 40°) = 50°$

10. $\sin\left(\cos^{-1}(-\tfrac{1}{3})\right) = \sqrt{1 - \cos^2(\arccos(-\tfrac{1}{3}))}$
$\qquad\qquad = \sqrt{1 - \tfrac{1}{9}} = \dfrac{\sqrt{8}}{3} = \dfrac{2\sqrt{2}}{3}$

12. $\tan(\tan^{-1} 200) = 200$

14. $\cos(\sin^{-1} x) = \sqrt{1 - \sin^2(\sin^{-1} x)} = \sqrt{1 - x^2}$

16. $\tan(\arctan x) = x \Rightarrow \sec(\arctan x) = \sqrt{1 + x^2}$
$\qquad\qquad \Rightarrow \cos(\arctan x) = \dfrac{1}{\sqrt{1 + x^2}}$
$\qquad\qquad \Rightarrow \sin(\arctan x) = \dfrac{x}{\sqrt{1 + x^2}}$

18. $\cos(\sec^{-1} x) = \dfrac{1}{x} \Rightarrow \sin(\sec^{-1} x) = \sqrt{1 - \dfrac{1}{x^2}} = \dfrac{\sqrt{x^2 - 1}}{|x|}$
$\qquad\qquad \Rightarrow \tan(\sec^{-1} x) = \sqrt{x^2 - 1} \operatorname{sgn} x$
$\qquad\qquad = \begin{cases} \sqrt{x^2 - 1} & \text{if } x \geq 1 \\ -\sqrt{x^2 - 1} & \text{if } x \leq -1 \end{cases}$

20. $y = \tan^{-1}(ax + b)$, $\qquad y' = \dfrac{a}{1 + (ax + b)^2}$.

22. $f(x) = x \sin^{-1} x$
$\qquad f'(x) = \sin^{-1} x + \dfrac{x}{\sqrt{1 - x^2}}$.

24. $u = z^2 \sec^{-1}(1 + z^2)$
$\dfrac{du}{dz} = 2z \sec^{-1}(1 + z^2) + \dfrac{z^2(2z)}{(1 + z^2)\sqrt{(1 + z^2)^2 - 1}}$
$\qquad = 2z \sec^{-1}(1 + z^2) + \dfrac{2z^2 \operatorname{sgn}(z)}{(1 + z^2)\sqrt{z^2 + 2}}$

26. $y = \sin^{-1} \left(\dfrac{a}{x} \right) \qquad (|x| > |a|)$
$y' = \dfrac{1}{\sqrt{1 - \left(\dfrac{a}{x} \right)^2}} \left[-\dfrac{a}{x^2} \right] = -\dfrac{a}{|x|\sqrt{x^2 - a^2}}$

28. $H(t) = \dfrac{\sin^{-1} t}{\sin t}$

$H'(t) = \dfrac{\sin t \left(\dfrac{1}{\sqrt{1 - t^2}}\right) - \sin^{-1} t \cos t}{\sin^2 t}$

$\quad = \dfrac{1}{(\sin t)\sqrt{1 - t^2}} - \csc t \cot t \sin^{-1} t$

30. $y = \cos^{-1}\left(\dfrac{a}{\sqrt{a^2 + x^2}}\right)$

$y' = -\left(1 - \dfrac{a^2}{a^2 + x^2}\right)^{-1/2}\left[-\dfrac{a}{2}(a^2 + x^2)^{-3/2}(2x)\right]$

$\quad = \dfrac{a\,\mathrm{sgn}\,(x)}{a^2 + x^2}$

32. $y = a\cos^{-1}\left(1 - \dfrac{x}{a}\right) - \sqrt{2ax - x^2} \qquad (a > 0)$

$y' = -a\left[1 - \left(1 - \dfrac{x}{a}\right)^2\right]^{-1/2}\left(-\dfrac{1}{a}\right) - \dfrac{2a - 2x}{2\sqrt{2ax - x^2}}$

$\quad = \dfrac{x}{\sqrt{2ax - x^2}}$

34. If $y = \sin^{-1} x$, then $y' = \dfrac{1}{\sqrt{1 - x^2}}$. If the slope is 2 then $\dfrac{1}{\sqrt{1 - x^2}} = 2$ so that $x = \pm\dfrac{\sqrt{3}}{2}$. Thus the equations of the two tangent lines are

$y = \dfrac{\pi}{3} + 2\left(x - \dfrac{\sqrt{3}}{2}\right)$ and $y = -\dfrac{\pi}{3} + 2\left(x + \dfrac{\sqrt{3}}{2}\right)$.

36. Since the domain of $\sec^{-1}$ consists of two disjoint intervals $(-\infty, -1]$ and $[1, \infty)$, the fact that the derivative of $\sec^{-1}$ is positive wherever defined does not imply that $\sec^{-1}$ is increasing over its whole domain, only that it is increasing on each of those intervals taken independently. In fact, $\sec^{-1}(-1) = \pi > 0 = \sec^{-1}(1)$ even though $-1 < 1$.

38. $\cot^{-1} x = \arctan(1/x)$;

$\dfrac{d}{dx}\cot^{-1} x = \dfrac{1}{1 + \dfrac{1}{x^2}}\dfrac{-1}{x^2} = -\dfrac{1}{1 + x^2}$

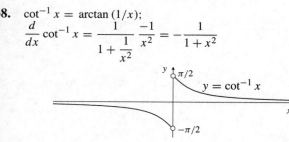

Fig. 3.5.38

Remark: the domain of $\cot^{-1}$ can be extended to include 0 by defining, say, $\cot^{-1} 0 = \pi/2$. This will make $\cot^{-1}$ right-continuous (but not continuous) at $x = 0$. It is also possible to define $\cot^{-1}$ in such a way that it is continuous on the whole real line, but we would then lose the identity $\cot^{-1} x = \tan^{-1}(1/x)$, which we prefer to maintain for calculation purposes.

40. If $g(x) = \tan(\tan^{-1} x)$ then

$g'(x) = \dfrac{\sec^2(\tan^{-1} x)}{1 + x^2}$

$\quad = \dfrac{1 + [\tan(\tan^{-1} x)]^2}{1 + x^2} = \dfrac{1 + x^2}{1 + x^2} = 1.$

If $h(x) = \tan^{-1}(\tan x)$ then h is periodic with period π, and

$h'(x) = \dfrac{\sec^2 x}{1 + \tan^2 x} = 1$

provided that $x \neq (k + \frac{1}{2})\pi$ where k is an integer. $h(x)$ is not defined at odd multiples of $\dfrac{\pi}{2}$.

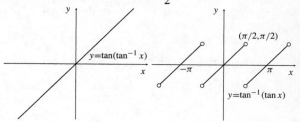

Fig. 3.5.40(a) Fig. 3.5.40(b)

42. $\dfrac{d}{dx}\sin^{-1}(\cos x) = \dfrac{1}{\sqrt{1 - \cos^2 x}}(-\sin x)$

$\quad = \begin{cases} -1 & \text{if } \sin x > 0 \\ 1 & \text{if } \sin x < 0 \end{cases}$

$\sin^{-1}(\cos x)$ is continuous everywhere and differentiable everywhere except at $x = n\pi$ for integers n.

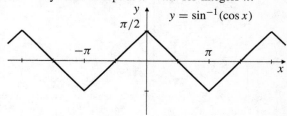

Fig. 3.5.42

44. $\dfrac{d}{dx}\tan^{-1}(\cot x) = \dfrac{1}{1 + \cot^2 x}(-\csc^2 x) = -1$ except at integer multiples of π.

$\tan^{-1}(\cot x)$ is continuous and differentiable everywhere except at $x = n\pi$ for integers n. It is not defined at those points.

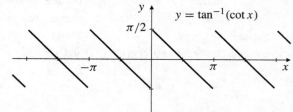

Fig. 3.5.44

46. If $x \geq 1$ and $y = \tan^{-1} \sqrt{x^2 - 1}$, then $\tan y = \sqrt{x^2 - 1}$ and $\sec y = x$, so that $y = \sec^{-1} x$.
If $x \leq -1$ and $y = \pi - \tan^{-1} \sqrt{x^2 - 1}$, then $\frac{\pi}{2} < y < \frac{3\pi}{2}$, so $\sec y < 0$. Therefore

$$\tan y = \tan(\pi - \tan^{-1} \sqrt{x^2 - 1}) = -\sqrt{x^2 - 1}$$
$$\sec^2 y = 1 + (x^2 - 1) = x^2$$
$$\sec y = x,$$

because both x and $\sec y$ are negative. Thus $y = \sec^{-1} x$ in this case also.

48. If $x \geq 1$ and $y = \sin^{-1} \dfrac{\sqrt{x^2 - 1}}{x}$, then $0 \leq y < \frac{\pi}{2}$ and

$$\sin y = \frac{\sqrt{x^2 - 1}}{x}$$
$$\cos^2 y = 1 - \frac{x^2 - 1}{x^2} = \frac{1}{x^2}$$
$$\sec^2 y = x^2.$$

Thus $\sec y = x$ and $y = \sec^{-1} x$.
If $x \leq -1$ and $y = \pi - \sin^{-1} \dfrac{\sqrt{x^2 - 1}}{x}$, then $\frac{\pi}{2} \leq y < \frac{3\pi}{2}$ and $\sec y < 0$. Therefore

$$\sin y = \sin\left(\pi - \sin^{-1} \frac{\sqrt{x^2 - 1}}{x}\right) = \frac{\sqrt{x^2 - 1}}{x}$$
$$\cos^2 y = 1 - \frac{x^2 - 1}{x^2} = \frac{1}{x^2}$$
$$\sec^2 y = x^2$$
$$\sec y = x,$$

because both x and $\sec y$ are negative. Thus $y = \sec^{-1} x$ in this case also.

50. Since $f(x) = x - \tan^{-1}(\tan x)$ then

$$f'(x) = 1 - \frac{\sec^2 x}{1 + \tan^2 x} = 1 - 1 = 0$$

if $x \neq -(k + \frac{1}{2})\pi$ where k is an integer. Thus, f is constant on intervals not containing odd multiples of $\dfrac{\pi}{2}$. $f(0) = 0$ but $f(\pi) = \pi - 0 = \pi$. There is no contradiction here because $f'\left(\dfrac{\pi}{2}\right)$ is not defined, so f is not constant on the interval containing 0 and π.

52. $y' = \dfrac{1}{1 + x^2} \Rightarrow y = \tan^{-1} x + C$
$y(0) = C = 1$
Thus, $y = \tan^{-1} x + 1$.

54. $y' = \dfrac{1}{\sqrt{1 - x^2}} \Rightarrow y = \sin^{-1} x + C$
$y(\frac{1}{2}) = \sin^{-1}(\frac{1}{2}) + C = 1$
$\Rightarrow \dfrac{\pi}{6} + C = 1 \Rightarrow C = 1 - \dfrac{\pi}{6}$.
Thus, $y = \sin^{-1} x + 1 - \dfrac{\pi}{6}$.

Section 3.6   Hyperbolic Functions (page 200)

2. $\cosh x \cosh y + \sinh x \sinh y$
$= \frac{1}{4}[(e^x + e^{-x})(e^y + e^{-y}) + (e^x - e^{-x})(e^y - e^{-y})]$
$= \frac{1}{4}(2e^{x+y} + 2e^{-x-y}) = \frac{1}{2}(e^{x+y} + e^{-(x+y)})$
$= \cosh(x + y)$.
$\sinh x \cosh y + \cosh x \sinh y$
$= \frac{1}{4}[(e^x - e^{-x})(e^y + e^{-y}) + (e^x + e^{-x})(e^y - e^{-y})]$
$= \frac{1}{2}(e^{x+y} - e^{-(x+y)}) = \sinh(x + y)$.
$\cosh(x - y) = \cosh[x + (-y)]$
$= \cosh x \cosh(-y) + \sinh x \sinh(-y)$
$= \cosh x \cosh y - \sinh x \sinh y$.
$\sinh(x - y) = \sinh[x + (-y)]$
$= \sinh x \cosh(-y) + \cosh x \sinh(-y)$
$= \sinh x \cosh y - \cosh x \sinh y$.

4. $y = \coth x = \dfrac{e^x + e^{-x}}{e^x - e^{-x}}$ $y = \text{sech}\, x = \dfrac{2}{e^x + e^{-x}}$

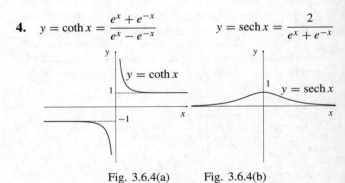

Fig. 3.6.4(a)   Fig. 3.6.4(b)

$y = \text{csch}\, x = \dfrac{2}{e^x - e^{-x}}$

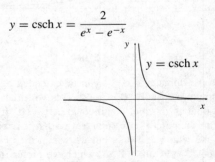

Fig. 3.6.4

6. Let $y = \sinh^{-1}\left(\dfrac{x}{a}\right) \Leftrightarrow x = a \sinh y \Rightarrow 1 = a(\cosh y)\dfrac{dy}{dx}$.
Thus,

$$\frac{d}{dx}\sinh^{-1}\left(\frac{x}{a}\right) = \frac{1}{a\cosh y}$$

$$= \frac{1}{a\sqrt{1 + \sinh^2 y}} = \frac{1}{\sqrt{a^2 + x^2}}$$

$$\int \frac{dx}{\sqrt{a^2 + x^2}} = \sinh^{-1}\frac{x}{a} + C. \qquad (a > 0)$$

Let $y = \cosh^{-1}\dfrac{x}{a} \Leftrightarrow x = a\operatorname{Cosh} y = a\cosh y$

for $y \geq 0$, $x \geq a$. We have $1 = a(\sinh y)\dfrac{dy}{dx}$. Thus,

$$\frac{d}{dx}\cosh^{-1}\frac{x}{a} = \frac{1}{a\sinh y}$$

$$= \frac{1}{a\sqrt{\cosh^2 y - 1}} = \frac{1}{\sqrt{x^2 - a^2}}$$

$$\int \frac{dx}{\sqrt{x^2 - a^2}} = \cosh^{-1}\frac{x}{a} + C. \qquad (a > 0,\; x \geq a)$$

Let $y = \tanh^{-1}\dfrac{x}{a} \Leftrightarrow x = a\tanh y \Rightarrow 1 = a(\operatorname{sech}^2 y)\dfrac{dy}{dx}$.
Thus,

$$\frac{d}{dx}\tanh^{-1}\frac{x}{a} = \frac{1}{a\operatorname{sech}^2 y}$$

$$= \frac{a}{a^2 - a^2\tanh^2 x} = \frac{a}{a^2 - x^2}$$

$$\int \frac{dx}{a^2 - x^2} = \frac{1}{a}\tanh^{-1}\frac{x}{a} + C.$$

8. $\operatorname{csch}^{-1} x = \sinh^{-1}(1/x) = \ln\left(\dfrac{1}{x} + \sqrt{\dfrac{1}{x^2} + 1}\right)$ has

domain and range consisting of all real numbers x except $x = 0$. We have

$$\frac{d}{dx}\operatorname{csch}^{-1} x = \frac{d}{dx}\sinh^{-1}\frac{1}{x}$$

$$= \frac{1}{\sqrt{1 + \left(\dfrac{1}{x}\right)^2}}\left(\frac{-1}{x^2}\right) = \frac{-1}{|x|\sqrt{x^2 + 1}}.$$

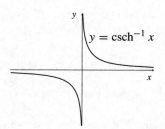

Fig. 3.6.8

10. Let $y = \operatorname{Sech}^{-1} x$ where $\operatorname{Sech} x = \operatorname{sech} x$ for $x \geq 0$.
Hence, for $y \geq 0$,

$$x = \operatorname{sech} y \Leftrightarrow \frac{1}{x} = \cosh y$$

$$\Leftrightarrow \frac{1}{x} = \operatorname{Cosh} y \Leftrightarrow y = \operatorname{Cosh}^{-1}\frac{1}{x}.$$

Thus,

$$\operatorname{Sech}^{-1} x = \operatorname{Cosh}^{-1}\frac{1}{x}$$

$$\mathcal{D}(\operatorname{Sech}^{-1}) = \mathcal{R}(\operatorname{sech}) = (0, 1]$$

$$\mathcal{R}(\operatorname{Sech}^{-1}) = \mathcal{D}(\operatorname{sech}) = [0, \infty).$$

Also,

$$\frac{d}{dx}\operatorname{Sech}^{-1} x = \frac{d}{dx}\operatorname{Cosh}^{-1}\frac{1}{x}$$

$$= \frac{1}{\sqrt{\left(\dfrac{1}{x}\right)^2 - 1}}\left(\frac{-1}{x^2}\right) = \frac{-1}{x\sqrt{1 - x^2}}.$$

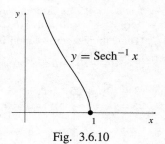

Fig. 3.6.10

12. Since

$$h_{L,M}(x) = L\cosh k(x - a) + M\sinh k(x - a)$$

$$h''_{L,M}(x) = Lk^2\cosh k(x - a) + Mk^2\sinh k(x - a)$$

$$= k^2 h_{L,M}(x)$$

hence, $h_{L,M}(x)$ is a solution of $y'' - k^2 y = 0$ and

$$h_{L,M}(x)$$

$$= \frac{L}{2}\left(e^{kx - ka} + e^{-kx + ka}\right) + \frac{M}{2}\left(e^{kx - ka} - e^{-kx + ka}\right)$$

$$= \left(\frac{L}{2}e^{-ka} + \frac{M}{2}e^{-ka}\right)e^{kx} + \left(\frac{L}{2}e^{ka} - \frac{M}{2}e^{ka}\right)e^{-kx}$$

$$= Ae^{kx} + Be^{-kx} = f_{A,B}(x)$$

where $A = \frac{1}{2}e^{-ka}(L + M)$ and $B = \frac{1}{2}e^{ka}(L - M)$.

Section 3.7 Second-Order Linear DEs with Constant Coefficients (page 206)

2.
$$y'' - 2y' - 3y = 0$$
auxiliary eqn $r^2 - 2r - 3 = 0 \Rightarrow r = -1, \ r = 3$
$$y = Ae^{-t} + Be^{3t}$$

4. $4y'' - 4y' - 3y = 0$
$4r^2 - 4r - 3 = 0 \Rightarrow (2r+1)(2r-3) = 0$
Thus, $r_1 = -\frac{1}{2}$, $r_2 = \frac{3}{2}$, and $y = Ae^{-(1/2)t} + Be^{(3/2)t}$.

6. $y'' - 2y' + y = 0$
$r^2 - 2r + 1 = 0 \Rightarrow (r-1)^2 = 0$
Thus, $r = 1, \ 1$, and $y = Ae^t + Bte^t$.

8. $9y'' + 6y' + y = 0$
$9r^2 + 6r + 1 = 0 \Rightarrow (3r+1)^2 = 0$
Thus, $r = -\frac{1}{3}, \ -\frac{1}{3}$, and $y = Ae^{-(1/3)t} + Bte^{-(1/3)t}$.

10. For $y'' - 4y' + 5y = 0$ the auxiliary equation is $r^2 - 4r + 5 = 0$, which has roots $r = 2 \pm i$. Thus, the general solution of the DE is $y = Ae^{2t} \cos t + Be^{2t} \sin t$.

12. Given that $y'' + y' + y = 0$, hence $r^2 + r + 1 = 0$. Since $a = 1$, $b = 1$ and $c = 1$, the discriminant is $D = b^2 - 4ac = -3 < 0$ and $-(b/2a) = -\frac{1}{2}$ and $\omega = \sqrt{3}/2$. Thus, the general solution is
$$y = Ae^{-(1/2)t} \cos\left(\frac{\sqrt{3}}{2}t\right) + Be^{-(1/2)t} \sin\left(\frac{\sqrt{3}}{2}t\right).$$

14. Given that $y'' + 10y' + 25y = 0$, hence $r^2 + 10r + 25 = 0 \Rightarrow (r+5)^2 = 0 \Rightarrow r = -5$. Thus,
$$y = Ae^{-5t} + Bte^{-5t}$$
$$y' = -5e^{-5t}(A + Bt) + Be^{-5t}.$$
Since
$$0 = y(1) = Ae^{-5} + Be^{-5}$$
$$2 = y'(1) = -5e^{-5}(A+B) + Be^{-5},$$
we have $A = -2e^5$ and $B = 2e^5$.
Thus, $y = -2e^5 e^{-5t} + 2te^5 e^{-5t} = 2(t-1)e^{-5(t-1)}$.

16. The auxiliary equation $r^2 - (2+\epsilon)r + (1+\epsilon)$ factors to $(r - 1 - \epsilon)(r - 1) = 0$ and so has roots $r = 1 + \epsilon$ and $r = 1$. Thus the DE $y'' - (2+\epsilon)y' + (1+\epsilon)y = 0$ has general solution $y = Ae^{(1+\epsilon)t} + Be^t$. The function $y_\epsilon(t) = \dfrac{e^{(1+\epsilon)t} - e^t}{\epsilon}$ is of this form with $A = -B = 1/\epsilon$. We have, substituting $\epsilon = h/t$,
$$\lim_{\epsilon \to 0} y_\epsilon(t) = \lim_{\epsilon \to 0} \frac{e^{(1+\epsilon)t} - e^t}{\epsilon}$$
$$= t \lim_{h \to 0} \frac{e^{t+h} - e^t}{h}$$
$$= t \left(\frac{d}{dt} e^t\right) = t e^t$$

which is, along with e^t, a solution of the CASE II DE $y'' - 2y' + y = 0$.

18. The auxiliary equation $ar^2 + br + c = 0$ has roots
$$r_1 = \frac{-b - \sqrt{D}}{2a}, \quad r_2 = \frac{-b + \sqrt{D}}{2a},$$
where $D = b^2 - 4ac$. Note that $a(r_2 - r_1) = \sqrt{D} = -(2ar_1 + b)$. If $y = e^{r_1 t}u$, then $y' = e^{r_1 t}(u' + r_1 u)$, and $y'' = e^{r_1 t}(u'' + 2r_1 u' + r_1^2 u)$. Substituting these expressions into the DE $ay'' + by' + cy = 0$, and simplifying, we obtain
$$e^{r_1 t}(au'' + 2ar_1 u' + bu') = 0,$$
or, more simply, $u'' - (r_2 - r_1)u' = 0$. Putting $v = u'$ reduces this equation to first order:
$$v' = (r_2 - r_1)v,$$
which has general solution $v = Ce^{(r_2 - r_1)t}$. Hence
$$u = \int Ce^{(r_2 - r_1)t}\, dt = Be^{(r_2 - r_1)t} + A,$$
and $y = e^{r_1 t}u = Ae^{r_1 t} + Be^{r_2 t}$.

19. If $y = A \cos \omega t + B \sin \omega t$ then
$$y'' + \omega^2 y = -A\omega^2 \cos \omega t - B\omega^2 \sin \omega t$$
$$+ \omega^2(A \cos \omega t + B \sin \omega t) = 0$$
for all t. So y is a solution of (†).

20. If $f(t)$ is any solution of (†) then $f''(t) = -\omega^2 f(t)$ for all t. Thus,
$$\frac{d}{dt}\left[\omega^2 \Big(f(t)\Big)^2 + \Big(f'(t)\Big)^2\right]$$
$$= 2\omega^2 f(t)f'(t) + 2f'(t)f''(t)$$
$$= 2\omega^2 f(t)f'(t) - 2\omega^2 f(t)f'(t) = 0$$
for all t. Thus, $\omega^2 \Big(f(t)\Big)^2 + \Big(f'(t)\Big)^2$ is constant. (This can be interpreted as a conservation of energy statement.)

21. If $g(t)$ satisfies (†) and also $g(0) = g'(0) = 0$, then by Exercise 20,
$$\omega^2 \Big(g(t)\Big)^2 + \Big(g'(t)\Big)^2$$
$$= \omega^2 \Big(g(0)\Big)^2 + \Big(g'(0)\Big)^2 = 0.$$
Since a sum of squares cannot vanish unless each term vanishes, $g(t) = 0$ for all t.

22. If $f(t)$ is any solution of (†), let
$g(t) = f(t) - A\cos\omega t - B\sin\omega t$ where $A = f(0)$
and $B\omega = f'(0)$. Then g is also solution of (†). Also
$g(0) = f(0) - A = 0$ and $g'(0) = f'(0) - B\omega = 0$.
Thus, $g(t) = 0$ for all t by Exercise 24, and therefore
$f(x) = A\cos\omega t + B\sin\omega t$. Thus, it is proved that every
solution of (†) is of this form.

24. Because $y'' + 4y = 0$, therefore $y = A\cos 2t + B\sin 2t$.
Now
$$y(0) = 2 \Rightarrow A = 2,$$
$$y'(0) = -5 \Rightarrow B = -\tfrac{5}{2}.$$

Thus, $y = 2\cos 2t - \frac{5}{2}\sin 2t$.
circular frequency $= \omega = 2$, frequency $=$
$\dfrac{\omega}{2\pi} = \dfrac{1}{\pi} \approx 0.318$
period $= \dfrac{2\pi}{\omega} = \pi \approx 3.14$
amplitude $= \sqrt{(2)^2 + (-\tfrac{5}{2})^2} \simeq 3.20$

26. $y = \mathcal{A}\cos\big(\omega(t-c)\big) + \mathcal{B}\sin\big(\omega(t-c)\big)$
(easy to calculate $y'' + \omega^2 y = 0$)
$y = \mathcal{A}\big(\cos(\omega t)\cos(\omega c) + \sin(\omega t)\sin(\omega c)\big)$
$\qquad + \mathcal{B}\big(\sin(\omega t)\cos(\omega c) - \cos(\omega t)\sin(\omega c)\big)$
$\quad = \big(\mathcal{A}\cos(\omega c) - \mathcal{B}\sin(\omega c)\big)\cos\omega t$
$\qquad + \big(\mathcal{A}\sin(\omega c) + \mathcal{B}\cos(\omega c)\big)\sin\omega t$
$\quad = A\cos\omega t + B\sin\omega t$
where $A = \mathcal{A}\cos(\omega c) - \mathcal{B}\sin(\omega c)$ and
$B = \mathcal{A}\sin(\omega c) + \mathcal{B}\cos(\omega c)$

28. $\begin{cases} y'' + \omega^2 y = 0 \\ y(a) = A \\ y'(a) = B \end{cases}$
$y = A\cos\big(\omega(t-a)\big) + \dfrac{B}{\omega}\sin\big(\omega(t-a)\big)$

30. Frequency $= \dfrac{\omega}{2\pi}$, $\omega^2 = \dfrac{k}{m}$ (k = spring const, m = mass)
Since the spring does not change, $\omega^2 m = k$ (constant)
For $m = 400$ gm, $\omega = 2\pi(24)$ (frequency = 24 Hz)
If $m = 900$ gm, then $\omega^2 = \dfrac{4\pi^2(24)^2(400)}{900}$
so $\omega = \dfrac{2\pi \times 24 \times 2}{3} = 32\pi$.
Thus frequency $= \dfrac{32\pi}{2\pi} = 16$ Hz
For $m = 100$ gm, $\omega = \dfrac{4\pi^2(24)^2 400}{100}$
so $\omega = 96\pi$ and frequency $= \dfrac{\omega}{2\pi} = 48$ Hz.

32. Expanding the hyperbolic functions in terms of exponentials,
$$y = e^{kt}[A\cosh\omega(t-t_0)B\sinh\omega(t-t_0)]$$
$$= e^{kt}\left[\frac{A}{2}e^{\omega(t-t_0)} + \frac{A}{2}e^{-\omega(t-t_0)}\right.$$
$$\left. + \frac{B}{2}e^{\omega(t-t_0)} - \frac{B}{2}e^{-\omega(t-t_0)}\right]$$
$$= A_1 e^{(k+\omega)t} + B_1 e^{(k-\omega)t}$$

where $A_1 = (A/2)e^{-\omega t_0} + (B/2)e^{-\omega t_0}$ and
$B_1 = (A/2)e^{\omega t_0} - (B/2)e^{\omega t_0}$. Under the conditions of
this problem we know that $Rr = k \pm \omega$ are the two real
roots of the auxiliary equation $ar^2 + br + c = 0$, so $e^{(k\pm\omega)t}$
are independent solutions of $ay'' + by' + cy = 0$, and our
function y must also be a solution. Since it involves two
arbitrary constants, it is a general solution.

34. $\begin{cases} y'' + 4y' + 3y = 0 \\ y(3) = 1 \\ y'(3) = 0 \end{cases}$

The DE has auxiliary equation $r^2 + 4r + 3 = 0$ with roots
$r = -2 + 1 = -1$ and $r = -2 - 1 = -3$ (i.e. $k \pm \omega$,
where $k = -2$ and $\omega = 1$). By the second previous
problem, a general solution can be expressed in the form
$y = e^{-2t}[A\cosh(t-3) + B\sinh(t-3)]$ for which
$$y' = -2e^{-2t}[A\cosh(t-3) + B\sinh(t-3)]$$
$$+ e^{-2t}[A\sinh(t-3) + B\cosh(t-3)].$$

The initial conditions give
$$1 = y(3) = e^{-6}A$$
$$0 = y'(3) = -e^{-6}(-2A + B)$$

Thus $A = e^6$ and $B = 2A = 2e^6$. The IVP has solution
$$y = e^{6-2t}[\cosh(t-3) + 2\sinh(t-3)].$$

36. Since $x'(0) = 0$ and $x(0) = 1 > 1/5$, the motion will be
governed by $x'' = -x + (1/5)$ until such time $t > 0$ when
$x'(t) = 0$ again.

Let $u = x - (1/5)$. Then $u'' = x'' = -(x - 1/5) = -u$,
$u(0) = 4/5$, and $u'(0) = x'(0) = 0$. This simple harmonic motion initial-value problem has solution
$u(t) = (4/5)\cos t$. Thus $x(t) = (4/5)\cos t + (1/4)$ and
$x'(t) = u'(t) = -(4/5)\sin t$. These formulas remain
valid until $t = \pi$ when $x'(t)$ becomes 0 again. Note that
$x(\pi) = -(4/5) + (1/5) = -(3/5)$.

Since $x(\pi) < -(1/5)$, the motion for $t > \pi$ will be
governed by $x'' = -x - (1/5)$ until such time $t > \pi$
when $x'(t) = 0$ again.

Let $v = x + (1/5)$. Then $v'' = x'' = -(x + 1/5) = -v$, $v(\pi) = -(3/5) + (1/5) = -(2/5)$, and $v'(\pi) = x'(\pi) = 0$. Thius initial-value problem has solution $v(t) = -(2/5)\cos(t - \pi) = (2/5)\cos t$, so that $x(t) = (2/5)\cos t - (1/5)$ and $x'(t) = -(2/5)\sin t$. These formulas remain valid for $t \geq \pi$ until $t = 2\pi$ when x' becomes 0 again. We have $x(2\pi) = (2/5) - (1/5) = 1/5$ and $x'(2\pi) = 0$.

The conditions for stopping the motion are met at $t = 2\pi$; the mass remains at rest thereafter. Thus

$$x(t) = \begin{cases} \frac{4}{5}\cos t + \frac{1}{5} & \text{if } 0 \leq t \leq \pi \\ \frac{2}{5}\cos t - \frac{1}{5} & \text{if } \pi < t \leq 2\pi \\ \frac{1}{5} & \text{if } t > 2\pi \end{cases}$$

Review Exercises 3   (page 208)

2. $f(x) = \sec^2 x \tan x \Rightarrow f'(x) = 2\sec^2 x \tan^2 x + \sec^4 x > 0$ for x in $(-\pi/2, \pi/2)$, so f is increasing and therefore one-to-one and invertible there. The domain of f^{-1} is $(-\infty, \infty)$, the range of f. Since $f(\pi/4) = 2$, therefore $f^{-1}(2) = \pi/4$, and

$$(f^{-1})'(2) = \frac{1}{f'(f^{-1}(2))} = \frac{1}{f'(\pi/4)} = \frac{1}{8}.$$

4. Observe $f'(x) = e^{-x^2}(1 - 2x^2)$ is positive if $x^2 < 1/2$ and is negative if $x^2 > 1/2$. Thus f is increasing on $(-1/\sqrt{2}, 1/\sqrt{2})$ and is decreasing on $(-\infty, -1/\sqrt{2})$ and on $(1/\sqrt{2}, \infty)$.

6. $y = e^{-x}\sin x$, $(0 \leq x \leq 2\pi)$ has a horizontal tangent where

$$0 = \frac{dy}{dx} = e^{-x}(\cos x - \sin x).$$

This occurs if $\tan x = 1$, so $x = \pi/4$ or $x = 5\pi/4$. The points are $(\pi/4, e^{-\pi/4}/\sqrt{2})$ and $(5\pi/4, -e^{-5\pi/4}/\sqrt{2})$.

8. Let the length, radius, and volume of the clay cylinder at time t be ℓ, r, and V, respectively. Then $V = \pi r^2 \ell$, and

$$\frac{dV}{dt} = 2\pi r\ell\frac{dr}{dt} + \pi r^2 \frac{d\ell}{dt}.$$

Since $dV/dt = 0$ and $d\ell/dt = k\ell$ for some constant $k > 0$, we have

$$2\pi r\ell\frac{dr}{dt} = -k\pi r^2\ell, \quad \Rightarrow \quad \frac{dr}{dt} = -\frac{kr}{2}.$$

That is, r is decreasing at a rate proportional to itself.

10. a) $\displaystyle\lim_{h\to 0}\frac{a^h - 1}{h} = \lim_{h\to 0}\frac{a^{0+h} - a^0}{h} = \frac{d}{dx}a^x\bigg|_{x=0} = \ln a.$

Putting $h = 1/n$, we get $\displaystyle\lim_{n\to\infty} n\left(a^{1/n} - 1\right) = \ln a.$

b) Using the technique described in the exercise, we calculate

$$2^{10}\left(2^{1/2^{10}} - 1\right) \approx 0.69338183$$

$$2^{11}\left(2^{1/2^{11}} - 1\right) \approx 0.69326449$$

Thus $\ln 2 \approx 0.693$.

12. If $f(x) = (\ln x)/x$, then $f'(x) = (1 - \ln x)/x^2$. Thus $f'(x) > 0$ if $\ln x < 1$ (i.e., $x < e$) and $f'(x) < 0$ if $\ln x > 1$ (i.e., $x > e$). Since f is increasing to the left of e and decreasing to the right, it has a maximum value $f(e) = 1/e$ at $x = e$. Thus, if $x > 0$ and $x \neq e$, then

$$\frac{\ln x}{x} < \frac{1}{e}.$$

Putting $x = \pi$ we obtain $(\ln\pi)/\pi < 1/e$. Thus

$$\ln(\pi^e) = e\ln\pi < \pi = \pi\ln e = \ln e^\pi,$$

and $\pi^e < e^\pi$ follows because $\ln$ is increasing.

14. a) $\dfrac{\ln x}{x} = \dfrac{\ln 2}{2}$ is satisfied if $x = 2$ or $x = 4$ (because $\ln 4 = 2\ln 2$).

b) The line $y = mx$ through the origin intersects the curve $y = \ln x$ at $(b, \ln b)$ if $m = (\ln b)/b$. The same line intersects $y = \ln x$ at a different point $(x, \ln x)$ if $(\ln x)/x = m = (\ln b)/b$. This equation will have only one solution $x = b$ if the line $y = mx$ intersects the curve $y = \ln x$ only once, at $x = b$, that is, if the line is tangent to the curve at $x = b$. In this case m is the slope of $y = \ln x$ at $x = b$, so

$$\frac{1}{b} = m = \frac{\ln b}{b}.$$

Thus $\ln b = 1$, and $b = e$.

16. If $y = \cos^{-1} x$, then $x = \cos y$ and $0 \leq y \leq \pi$. Thus

$$\tan y = \text{sgn}\, x\sqrt{\sec^2 y - 1} = \text{sgn}\, x\sqrt{\frac{1}{x^2} - 1} = \frac{\sqrt{1-x^2}}{x}.$$

Thus $\cos^{-1} x = \tan^{-1}((\sqrt{1-x^2})/x)$.

Since $\cot x = 1/\tan x$, $\cot^{-1} x = \tan^{-1}(1/x)$.

$$\csc^{-1} x = \sin^{-1}\frac{1}{x} = \frac{\pi}{2} - \cos^{-1}\frac{1}{x}$$

$$= \frac{\pi}{2} - \tan^{-1}\frac{\sqrt{1 - (1/x)^2}}{1/x}$$

$$= \frac{\pi}{2} - \text{sgn}\, x\tan^{-1}\sqrt{x^2 - 1}.$$

18. Let $T(t)$ be the temperature of the milk t minutes after it is removed from the refrigerator. Let $U(t) = T(t) - 20$. By Newton's law,

$$U'(t) = kU(t) \quad \Rightarrow \quad U(t) = U(0)e^{kt}.$$

Now $T(0) = 5 \Rightarrow U(0) = -15$ and $T(12) = 12 \Rightarrow U(12) = -8$. Thus

$$-8 = U(12) = U(0)e^{12k} = -15e^{12k}$$
$$e^{12k} = 8/15, \qquad k = \tfrac{1}{12}\ln(8/15).$$

If $T(s) = 18$, then $U(s) = -2$, so $-2 = -15e^{sk}$. Thus $sk = \ln(2/15)$, and

$$s = \frac{\ln(2/15)}{k} = 12\frac{\ln(2/15)}{\ln(8/15)} \approx 38.46.$$

It will take another $38.46 - 12 = 26.46$ min for the milk to warm up to $18°$.

20. Let $f(x) = e^x - 1 - x$. Then $f(0) = 0$ and by the MVT,

$$\frac{f(x)}{x} = \frac{f(x) - f(0)}{x - 0} = f'(c) = e^c - 1$$

for some c between 0 and x. If $x > 0$, then $c > 0$, and $f'(c) > 0$. If $x < 0$, then $c < 0$, and $f'(c) < 0$. In either case $f(x) = xf'(c) > 0$, which is what we were asked to show.

Challenging Problems 3　(page 209)

2. $\dfrac{dv}{dt} = -g - kv.$

a) Let $u(t) = -g - kv(t)$. Then $\dfrac{du}{dt} = -k\dfrac{dv}{dt} = -ku$, and

$$u(t) = u(0)e^{-kt} = -(g + kv_0)e^{-kt}$$
$$v(t) = -\frac{1}{k}\Big(g + u(t)\Big) = -\frac{1}{k}\Big(g - (g + kv_0)e^{-kt}\Big).$$

b) $\lim_{t\to\infty} v(t) = -g/k$

c) $\dfrac{dy}{dt} = v(t) = -\dfrac{g}{k} + \dfrac{g + kv_0}{k}e^{-kt}, \quad y(0) = y_0$

$$y(t) = -\frac{gt}{k} - \frac{g + kv_0}{k^2}e^{-kt} + C$$

$$y_0 = -0 - \frac{g + kv_0}{k^2} + C \Rightarrow C = y_0 + \frac{g + kv_0}{k^2}$$

$$y(t) = y_0 - \frac{gt}{k} + \frac{g + kv_0}{k^2}\Big(1 - e^{-kt}\Big)$$

4. If $p = e^{-bt}y$, then $\dfrac{dp}{dt} = e^{-bt}\left(\dfrac{dy}{dt} - by\right)$.

The DE $\dfrac{dp}{dt} = kp\left(1 - \dfrac{p}{e^{-bt}M}\right)$ therefore transforms to

$$\frac{dy}{dt} = by + kpe^{bt}\left(1 - \frac{p}{e^{-bt}M}\right)$$
$$= (b + k)y - \frac{ky^2}{M} = Ky\left(1 - \frac{y}{L}\right),$$

where $K = b + k$ and $L = \dfrac{b + k}{k}M$. This is a standard Logistic equation with solution (as obtained in Section 3.4) given by

$$y = \frac{Ly_0}{y_0 + (L - y_0)e^{-Kt}},$$

where $y_0 = y(0) = p(0) = p_0$. Converting this solution back in terms of the function $p(t)$, we obtain

$$p(t) = \frac{Lp_0e^{-bt}}{p_0 + (L - p_0)e^{-(b+k)t}}$$
$$= \frac{(b + k)Mp_0}{p_0ke^{bt} + \big((b + k)M - kp_0\big)e^{-kt}}.$$

Since p represents a percentage, we must have $(b + k)M/k < 100$.

If $k = 10$, $b = 1$, $M = 90$, and $p_0 = 1$, then $\dfrac{b + k}{k}M = 99 < 100$. The numerator of the final expression for $p(t)$ given above is a constant. Therefore $p(t)$ will be largest when the derivative of the denominator,

$$f(t) = p_0ke^{bt} + \big((b + k)M - kp_0\big)e^{-kt} = 10e^t + 980e^{-10t}$$

is zero. Since $f'(t) = 10e^t - 9,800e^{-10t}$, this will happen at $t = \ln(980)/11$. The value of p at this t is approximately 48.1. Thus the maximum percentage of potential clients who will adopt the technology is about 48.1%.

CHAPTER 4. SOME APPLICATIONS OF DERIVATIVES

Section 4.1 Related Rates (page 214)

2. As in Exercise 1, $dA/dt = 2x\,dx/dt$. If $dA/dt = -2$ ft^2/s and $x = 8$ ft, then $dx/dt = -2/(16)$. The side length is decreasing at 1/8 ft/s.

4. Let A and r denote the area and radius of the circle. Then

$$A = \pi r^2 \Rightarrow r = \sqrt{\frac{A}{\pi}}$$

$$\Rightarrow \frac{dr}{dt} = \left(\frac{1}{2\sqrt{A\pi}}\right)\frac{dA}{dt}.$$

When $\dfrac{dA}{dt} = -2$, and $A = 100$, $\dfrac{dr}{dt} = -\dfrac{1}{10\sqrt{\pi}}$. The radius is decreasing at the rate $\dfrac{1}{10\sqrt{\pi}}$ cm/min when the area is 100 cm^2.

6. Let the length, width, and area be l, w, and A at time t. Thus $A = lw$.

$$\frac{dA}{dt} = l\frac{dw}{dt} + w\frac{dl}{dt}.$$

When $l = 16$, $w = 12$, $\dfrac{dw}{dt} = 3$, $\dfrac{dA}{dt} = 0$, we have

$$0 = 16 \times 3 + 12\frac{dl}{dt} \Rightarrow \frac{dl}{dt} = -\frac{48}{12} = -4.$$

The length is decreasing at 4 m/s.

8. The volume V of the ball is given by

$$V = \frac{4}{3}\pi r^3 = \frac{4\pi}{3}\left(\frac{D}{2}\right)^3 = \frac{\pi}{6}D^3,$$

where $D = 2r$ is the diameter of the ball. We have

$$\frac{dV}{dt} = \frac{\pi}{2}D^2\frac{dD}{dt}.$$

When $D = 6$ cm, $dD/dt = -.5$ cm/h. At that time

$$\frac{dV}{dt} = \frac{\pi}{2}(36)(-0.5) = -9\pi \approx -28.3.$$

The volume is decreasing at about 28.3 cm^3/h.

10. Let V, r and h denote the volume, radius and height of the cylinder at time t. Thus, $V = \pi r^2 h$ and

$$\frac{dV}{dt} = 2\pi r h\frac{dr}{dt} + \pi r^2\frac{dh}{dt}.$$

If $V = 60$, $\dfrac{dV}{dt} = 2$, $r = 5$, $\dfrac{dr}{dt} = 1$, then

$$h = \frac{V}{\pi r^2} = \frac{60}{25\pi} = \frac{12}{5\pi}$$

$$\frac{dh}{dt} = \frac{1}{\pi r^2}\left(\frac{dV}{dt} - 2\pi r h\frac{dr}{dt}\right)$$

$$= \frac{1}{25\pi}\left(2 - 10\pi\frac{12}{5\pi}\right) = -\frac{22}{25\pi}.$$

The height is decreasing at the rate $\dfrac{22}{25\pi}$ cm/min.

12. Let the length, width and area at time t be x, y and A respectively. Thus $A = xy$ and

$$\frac{dA}{dt} = x\frac{dy}{dt} + y\frac{dx}{dt}.$$

If $\dfrac{dA}{dt} = 5$, $\dfrac{dx}{dt} = 10$, $x = 20$, $y = 16$, then

$$5 = 20\frac{dy}{dt} + 16(10) \Rightarrow \frac{dy}{dt} = -\frac{31}{4}.$$

Thus, the width is decreasing at $\dfrac{31}{4}$ m/s.

14. Since $x^2 y^3 = 72$, then

$$2xy^3\frac{dx}{dt} + 3x^2 y^2\frac{dy}{dt} = 0 \Rightarrow \frac{dy}{dt} = -\frac{2y}{3x}\frac{dx}{dt}.$$

If $x = 3$, $y = 2$, $\dfrac{dx}{dt} = 2$, then $\dfrac{dy}{dt} = -\dfrac{8}{9}$. Hence, the vertical velocity is $-\dfrac{8}{9}$ units/s.

16. From the figure, $x^2 + k^2 = s^2$. Thus

$$x\frac{dx}{dt} = s\frac{ds}{dt}.$$

When angle $PCA = 45°$, $x = k$ and $s = \sqrt{2}k$. The radar gun indicates that $ds/dt = 100$ km/h. Thus $dx/dt = 100\sqrt{2}k/k \approx 141$. The car is travelling at about 141 km/h.

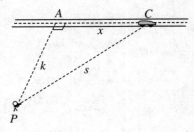

Fig. 4.1.16

18. Let the distances x and y be as shown at time t. Thus
$x^2 + y^2 = 25$ and $2x\dfrac{dx}{dt} + 2y\dfrac{dy}{dt} = 0$.
If $\dfrac{dx}{dt} = \dfrac{1}{3}$ and $y = 3$, then $x = 4$ and $\dfrac{4}{3} + 3\dfrac{dy}{dt} = 0$ so
$\dfrac{dy}{dt} = -\dfrac{4}{9}$.
The top of the ladder is slipping down at a rate of $\dfrac{4}{9}$ m/s.

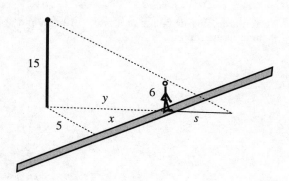

Fig. 4.1.18

20.

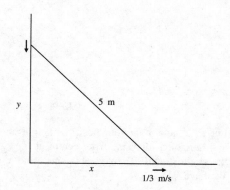

Fig. 4.1.20

Refer to the figure. s, y, and x are, respectively, the length of the woman's shadow, the distances from the woman to the lamppost, and the distances from the woman to the point on the path nearest the lamppost. From one of triangles in the figure we have

$$y^2 = x^2 + 25.$$

If $x = 12$, then $y = 13$. Moreover,

$$2y\frac{dy}{dt} = 2x\frac{dx}{dt}.$$

We are given that $dx/dt = 2$ ft/s, so $dy/dt = 24/13$ ft/s when $x = 12$ ft. Now the similar triangles in the figure show that

$$\frac{s}{6} = \frac{s+y}{15},$$

so that $s = 2y/3$. Hence $ds/dt = 48/39$. The woman's shadow is changing at rate 48/39 ft/s when she is 12 ft from the point on the path nearest the lamppost.

22. Let x, y be distances travelled by A and B from their positions at 1:00 pm in t hours.
Thus $\dfrac{dx}{dt} = 16$ km/h, $\dfrac{dy}{dt} = 20$ km/h.
Let s be the distance between A and B at time t.
Thus $s^2 = x^2 + (25 + y)^2$

$$2s\frac{ds}{dt} = 2x\frac{dx}{dt} + 2(25 + y)\frac{dy}{dt}$$

At 1:30 $\left(t = \tfrac{1}{2}\right)$ we have $x = 8$, $y = 10$,
$s = \sqrt{8^2 + 35^2} = \sqrt{1289}$ so

$$\sqrt{1289}\frac{ds}{dt} = 8 \times 16 + 35 \times 20 = 828$$

and $\dfrac{ds}{dt} = \dfrac{828}{\sqrt{1289}} \approx 23.06$. At 1:30, the ships are separating at about 23.06 km/h.

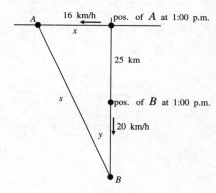

Fig. 4.1.22

24. Let y be the height of balloon t seconds after release.
Then $y = 5t$ m.
Let θ be angle of elevation at B of balloon at time t.
Then $\tan\theta = y/100$. Thus

$$\sec^2\theta\frac{d\theta}{dt} = \frac{1}{100}\frac{dy}{dt} = \frac{5}{100} = \frac{1}{20}$$

$$\left(1 + \tan^2\theta\right)\frac{d\theta}{dt} = \frac{1}{20}$$

$$\left[1 + \left(\frac{y}{100}\right)^2\right]\frac{d\theta}{dt} = \frac{1}{20}.$$

When $y = 200$ we have $5\dfrac{d\theta}{dt} = \dfrac{1}{20}$ so $\dfrac{d\theta}{dt} = \dfrac{1}{100}$.

The angle of elevation of balloon at B is increasing at a rate of $\dfrac{1}{100}$ rad/s.

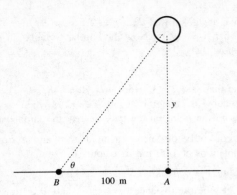

Fig. 4.1.24

26. Let r, h, and V be the top radius, depth, and volume of the water in the tank at time t. Then $\dfrac{r}{h} = \dfrac{10}{8}$ and

$$V = \frac{1}{3}\pi r^2 h = \frac{\pi}{3}\frac{25}{16}h^3. \text{ We have}$$

$$\frac{1}{10} = \frac{\pi}{3}\frac{25}{16}3h^2\frac{dh}{dt} \Rightarrow \frac{dh}{dt} = \frac{16}{250\pi h^2}.$$

When $h = 4$ m, we have $\dfrac{dh}{dt} = \dfrac{1}{250\pi}$.

The water level is rising at a rate of $\dfrac{1}{250\pi}$ m/min when depth is 4 m.

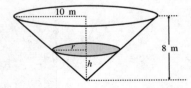

Fig. 4.1.26

28. Let r, h, and V be the top radius, depth, and volume of the water in the tank at time t. Then

$$\frac{r}{h} = \frac{3}{9} = \frac{1}{3}$$
$$V = \frac{1}{3}\pi r^2 h = \frac{\pi}{27}h^3$$
$$\frac{dV}{dt} = \frac{\pi}{9}h^2\frac{dh}{dt}.$$

If $\dfrac{dh}{dt} = 20$ cm/h $= \dfrac{2}{10}$ m/h when $h = 6$ m, then

$$\frac{dV}{dt} = \frac{\pi}{9} \times 36 \times \frac{2}{10} = \frac{4\pi}{5} \approx 2.51 \text{ m}^3/\text{h}.$$

Since water is coming in at a rate of 10 m³/h, it must be leaking out at a rate of $10 - 2.51 \approx 7.49$ m³/h.

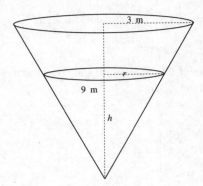

Fig. 4.1.28

30. Let P, x, and y be your position, height above centre, and horizontal distance from centre at time t. Let θ be the angle shown. Then $y = 10\sin\theta$, and $x = 10\cos\theta$. We have

$$\frac{dy}{dt} = 10\cos\theta\frac{d\theta}{dt}, \qquad \frac{d\theta}{dt} = 1 \text{ rpm} = 2\pi \text{ rad/min}.$$

When $x = 6$, then $\cos\theta = \dfrac{6}{10}$, so $\dfrac{dy}{dt} = 10 \times \dfrac{6}{10} \times 12\pi$. You are rising or falling at a rate of 12π m/min at the time in question.

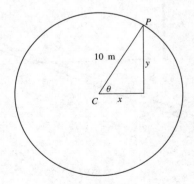

Fig. 4.1.30

32.
$$P = \frac{1}{3}x^{0.6}y^{0.4}$$
$$\frac{dP}{dt} = \frac{0.6}{3}x^{-0.4}y^{0.4}\frac{dx}{dt} + \frac{0.4}{3}x^{0.6}y^{-0.6}\frac{dy}{dt}.$$

If $dP/dt = 0$, $x = 40$, $dx/dt = 1$, and $y = 10,000$, then

$$\frac{dy}{dt} = -\frac{6y^{0.4}}{x^{0.4}}\frac{y^{0.6}}{4x^{0.6}}\frac{dx}{dt} = -\frac{6y}{4x}\frac{dx}{dt} = -375.$$

The daily expenses are decreasing at \$375 per day.

34. Let x and y be the distances travelled from the intersection point by the boat and car respectively in t minutes. Then

$$\frac{dx}{dt} = 20 \times \frac{1000}{60} = \frac{1000}{3}\text{m/min}$$

$$\frac{dy}{dt} = 80 \times \frac{1000}{60} = \frac{4000}{3}\text{m/min}$$

The distance s between the boat and car satisfy

$$s^2 = x^2 + y^2 + 20^2, \qquad s\frac{ds}{dt} = x\frac{dx}{dt} + y\frac{dy}{dt}.$$

After one minute, $x = \frac{1000}{3}$, $y = \frac{4000}{3}$ so $s \approx 1374.$ m. Thus

$$1374.5\frac{ds}{dt} = \frac{1000}{3}\frac{1000}{3} + \frac{4000}{3}\frac{4000}{3} \approx 1,888,889.$$

Hence $\frac{ds}{dt} \approx 1374.2$ m/min ≈ 82.45 km/h after 1 minute.

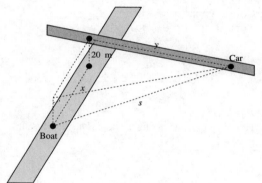

Fig. 4.1.34

36. Let V and h be the volume and depth of water in the pool at time t. If $h \le 2$, then

$$\frac{x}{h} = \frac{20}{2} = 10, \quad \text{so } V = \frac{1}{2}xh8 = 40h^2.$$

If $2 \le h \le 3$, then $V = 160 + 160(h - 2)$.

a) If $h = 2.5$m, then $-1 = \frac{dV}{dt} = 160\frac{dh}{dt}$.
So surface of water is dropping at a rate of $\frac{1}{160}$ m/min.

b) If $h = 1$m, then $-1 = \frac{dV}{dt} = 80h\frac{dh}{dt} = 80\frac{dh}{dt}$.
So surface of water is dropping at a rate of $\frac{1}{80}$m/min.

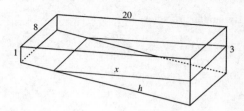

Fig. 4.1.36

38. Let x, y, and s be distances shown at time t. Then

$$s^2 = x^2 + 16, \qquad (15 - s)^2 = y^2 + 16$$

$$s\frac{ds}{dt} = x\frac{dx}{dt}, \qquad -(15 - s)\frac{ds}{dt} = y\frac{dy}{dt}.$$

When $x = 3$ and $\frac{dx}{dt} = \frac{1}{2}$, then $s = 5$ and $y = \sqrt{10^2 - 4^2} = \sqrt{84}$.
Also $\frac{ds}{dt} = \frac{3}{5}\left(\frac{1}{2}\right) = \frac{3}{10}$ so

$$\frac{dy}{dt} = -\frac{10}{\sqrt{84}}\frac{3}{10} = -\frac{3}{\sqrt{84}} \approx 0.327.$$

Crate B is moving toward Q at a rate of 0.327 m/s.

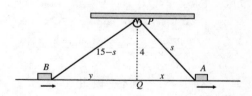

Fig. 4.1.38

40. Let y be height of ball t seconds after it drops. Thus $\frac{d^2y}{dt^2} = -9.8$, $\frac{dy}{dt}|_{t=0} = 0$, $y|_{t=0} = 20$, and

$$y = -4.9t^2 + 20, \qquad \frac{dy}{dt} = -9.8t.$$

Let s be distance of shadow of ball from base of pole.
By similar triangles, $\dfrac{s-10}{y} = \dfrac{s}{20}$.

$$20s - 200 = sy, \quad s = \dfrac{200}{20-y}$$

$$20\dfrac{ds}{dt} = y\dfrac{ds}{dt} + s\dfrac{dy}{dt}.$$

a) At $t = 1$, we have $\dfrac{dy}{dt} = -9.8$, $y = 15.1$,

$4.9\dfrac{ds}{dt} = \dfrac{200}{4.9}(-9.8)$.

The shadow is moving at a rate of 81.63 m/s after one second.

b) As the ball hits the ground, $y = 0$, $s = 10$,

$t = \sqrt{\dfrac{20}{4.9}}$, and $\dfrac{dy}{dt} = -9.8\sqrt{\dfrac{20}{4.9}}$, so

$20\dfrac{ds}{dt} = 0 + 10\dfrac{dy}{dt}.$

Now $y = 0$ implies that $t = \sqrt{\dfrac{20}{4.9}}$. Thus

$$\dfrac{ds}{dt} = -\dfrac{1}{2}(9.8)\sqrt{\dfrac{20}{4.9}} \approx -9.90.$$

The shadow is moving at about 9.90 m/s when the ball hits the ground.

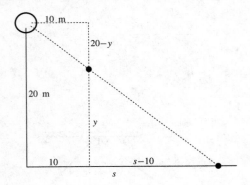

Fig. 4.1.40

Section 4.2 Extreme Values (page 222)

2. $f(x) = x + 2$ on $(-\infty, 0]$
abs max 2 at $x = 0$, no min.

4. $f(x) = x^2 - 1$
no max, abs min -1 at $x = 0$.

6. $f(x) = x^2 - 1$ on $(2, 3)$
no max or min values.

8. $f(x) = x^3 + x - 4$ on (a, b)
Since $f'(x) = 3x^2 + 1 > 0$ for all x, therefore f is increasing. Since (a, b) is open, f has no max or min values.

10. $f(x) = \dfrac{1}{x-1}$. Since $f'(x) = \dfrac{-1}{(x-1)^2} < 0$ for all x in the domain of f, therefore f has no max or min values.

12. $f(x) = \dfrac{1}{x-1}$ on $[2, 3]$
abs min $\frac{1}{2}$ at $x = 3$, abs max 1 at $x = 2$.

14. Let $f(x) = |x^2 - x - 2| = |(x - 2)(x + 1)|$ on $[-3, 3]$:
$f(-3) = 10$, $f(3) = 4$.
$f'(x) = (2x - 1)\operatorname{sgn}(x^2 - x - 2)$.
CP $x = 1/2$; SP $x = -1$, and $x = 2$. $f(1/2) = 9/4$,
$f(-1) = 0$, $f(2) = 0$.
Max value of f is 10 at $x = -3$; min value is 0 at $x = -1$ or $x = 2$.

16. $f(x) = (x + 2)^{(2/3)}$
no max, abs min 0 at $x = -2$.

18. $f(x) = x^2 + 2x$, $f'(x) = 2x + 2 = 2(x + 1)$
Critical point: $x = -1$.
$f(x) \to \infty$ as $x \to \pm\infty$.

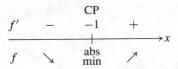

Hence, $f(x)$ has no max value, and the abs min is -1 at $x = -1$.

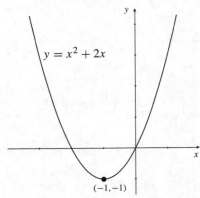

$y = x^2 + 2x$

$(-1, -1)$

Fig. 4.2.18

20. $f(x) = (x^2 - 4)^2$, $f'(x) = 4x(x^2 - 4) = 4x(x + 2)(x - 2)$
Critical points: $x = 0, \pm 2$.
$f(x) \to \infty$ as $x \to \pm\infty$.

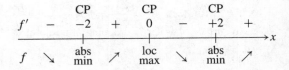

Hence, $f(x)$ has abs min 0 at $x = \pm 2$ and loc max 16 at $x = 0$.

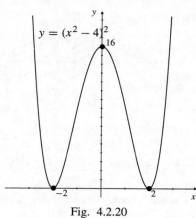

Fig. 4.2.20

22. $f(x) = x^2(x-1)^2$,
$f'(x) = 2x(x-1)^2 + 2x^2(x-1) = 2x(2x-1)(x-1)$
Critical points: $x = 0, \frac{1}{2}$ and 1.
$f(x) \to \infty$ as $x \to \pm\infty$.

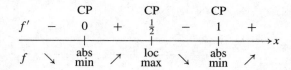

Hence, $f(x)$ has loc max $\frac{1}{16}$ at $x = \frac{1}{2}$ and abs min 0 at $x = 0$ and $x = 1$.

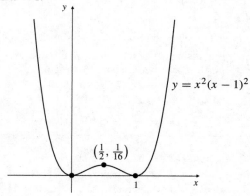

Fig. 4.2.22

24. $f(x) = \dfrac{x}{x^2 + 1}$, $f'(x) = \dfrac{1 - x^2}{(x^2 + 1)^2}$
Critical point: $x = \pm 1$.
$f(x) \to 0$ as $x \to \pm\infty$.

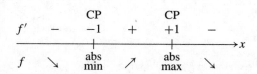

Hence, f has abs max $\frac{1}{2}$ at $x = 1$ and abs min $-\frac{1}{2}$ at $x = -1$.

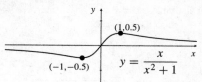

Fig. 4.2.24

26. $f(x) = \dfrac{x}{\sqrt{x^4 + 1}}$, $f'(x) = \dfrac{1 - x^4}{(x^4 + 1)^{3/2}}$
Critical points: $x = \pm 1$.
$f(x) \to 0$ as $x \to \pm\infty$.

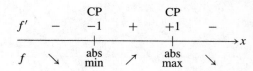

Hence, f has abs max $\frac{1}{\sqrt{2}}$ at $x = 1$ and abs min $-\frac{1}{\sqrt{2}}$ at $x = -1$.

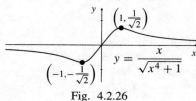

Fig. 4.2.26

28. $f(x) = x + \sin x$, $f'(x) = 1 + \cos x \geq 0$
$f'(x) = 0$ at $x = \pm\pi, \pm 3\pi, \ldots$
$f(x) \to \pm\infty$ as $x \to \pm\infty$.
Hence, f has no max or min values.

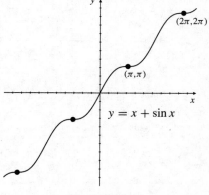

Fig. 4.2.28

61

30. $f(x) = x - 2\tan^{-1} x$, $f'(x) = 1 - \dfrac{2}{1+x^2} = \dfrac{x^2-1}{x^2+1}$

Critical points: $x = \pm 1$.

$f(x) \to \pm\infty$ as $x \to \pm\infty$.

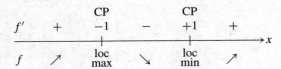

Hence, f has loc max $-1 + \dfrac{\pi}{2}$ at $x = -1$ and loc min $1 - \dfrac{\pi}{2}$ at $x = 1$.

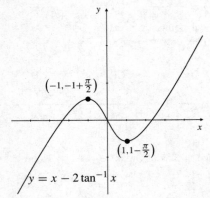

Fig. 4.2.30

32. $f(x) = e^{-x^2/2}$, $f'(x) = -xe^{-x^2/2}$

Critical point: $x = 0$.

$f(x) \to 0$ as $x \to \pm\infty$.

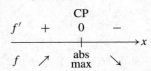

Hence, f has abs max 1 at $x = 0$ and no min value.

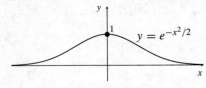

Fig. 4.2.32

34. $f(x) = x^2 e^{-x^2}$, $f'(x) = 2xe^{-x^2}(1-x^2)$

Critical points: $x = 0, \pm 1$.

$f(x) \to 0$ as $x \to \pm\infty$.

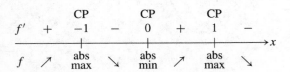

Hence, f has abs max $1/e$ at $x = \pm 1$ and abs min 0 at $x = 0$.

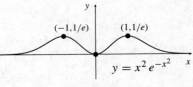

Fig. 4.2.34

36. Since $f(x) = |x+1|$,

$$f'(x) = \operatorname{sgn}(x+1) = \begin{cases} 1, & \text{if } x > -1; \\ -1, & \text{if } x < -1. \end{cases}$$

-1 is a singular point; f has no max but has abs min 0 at $x = -1$.

$f(x) \to \infty$ as $x \to \pm\infty$.

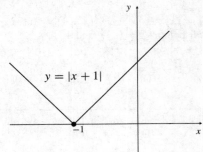

Fig. 4.2.36

38. $f(x) = \sin|x|$

$f'(x) = \operatorname{sgn}(x)\cos|x| = 0$ at $x = \pm\dfrac{\pi}{2}$, $\pm\dfrac{3\pi}{2}$, $\pm\dfrac{5\pi}{2}$, ...

0 is a singular point. Since $f(x)$ is an even function, its graph is symmetric about the origin.

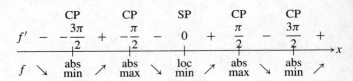

Hence, f has abs max 1 at $x = \pm(4k+1)\dfrac{\pi}{2}$ and abs min -1 at $x = \pm(4k+3)\dfrac{\pi}{2}$ where $k = 0, 1, 2, \ldots$ and loc min 0 at $x = 0$.

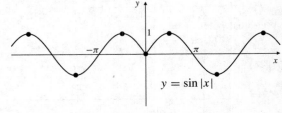

Fig. 4.2.38

40. $f(x) = (x-1)^{2/3} - (x+1)^{2/3}$
$f'(x) = \frac{2}{3}(x-1)^{-1/3} - \frac{2}{3}(x+1)^{-1/3}$
Singular point at $x = \pm 1$. For critical points:
$(x-1)^{-1/3} = (x+1)^{-1/3} \Rightarrow x-1 = x+1 \Rightarrow 2 = 0$, so
there are no critical points.

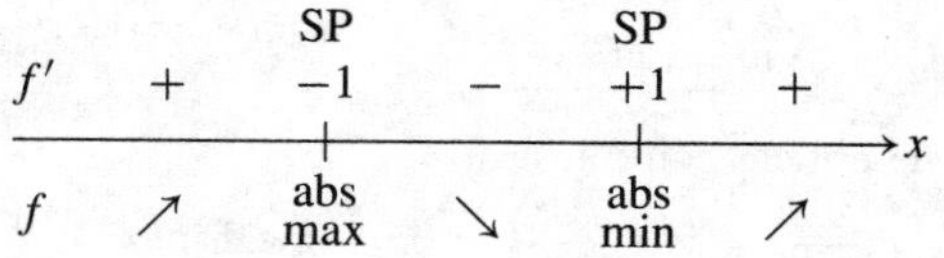

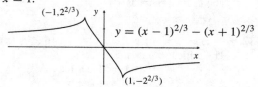

Hence, f has abs max $2^{2/3}$ at $x = -1$ and abs min
$-2^{2/3}$ at $x = 1$.

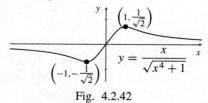

$y = (x-1)^{2/3} - (x+1)^{2/3}$

Fig. 4.2.40

42. $f(x) = x/\sqrt{x^4+1}$. f is continuous on $\mathbb{R}$, and
$\lim_{x \to \pm\infty} f(x) = 0$. Since $f(1) > 0$ and $f(-1) < 0$,
f must have both maximum and minimum values.

$$f'(x) = \frac{\sqrt{x^4+1} - x\dfrac{4x^3}{2\sqrt{x^4+1}}}{x^4+1} = \frac{1-x^4}{(x^4+1)^{3/2}}.$$

CP $x = \pm 1$. $f(\pm 1) = \pm 1/\sqrt{2}$. f has max value $1/\sqrt{2}$
and min value $-1/\sqrt{2}$.

$\left(1, \frac{1}{\sqrt{2}}\right)$

$y = \dfrac{x}{\sqrt{x^4+1}}$

$\left(-1, -\dfrac{1}{\sqrt{2}}\right)$

Fig. 4.2.42

44. $f(x) = x^2/\sqrt{4-x^2}$ is continuous on $(-2, 2)$, and
$\lim_{x \to -2+} f(x) = \lim_{x \to 2-} f(x) = \infty$. Thus f can
have no maximum value, but will have a minimum value.

$$f'(x) = \frac{2x\sqrt{4-x^2} - x^2 \dfrac{-2x}{2\sqrt{4-x^2}}}{4-x^2} = \frac{8x - x^3}{(4-x^2)^{3/2}}.$$

CP $x = 0$, $x = \pm\sqrt{8}$. $f(0) = 0$, and $\pm\sqrt{8}$ is not in the
domain of f. f has minimum value 0 at $x = 0$.

46. $f(x) = (\sin x)/x$ is continuous and differentiable on $\mathbb{R}$
except at $x = 0$ where it is undefined.
Since $\lim_{x \to 0} f(x) = 1$ (Theorem 8 of Section 2.5), and
$|f(x)| < 1$ for all $x \ne 0$ (because $|\sin x| < |x|$), f cannot
have a maximum value.
Since $\lim_{x \to \pm\infty} f(x) = 0$ and since $f(x) < 0$ at some
points, f must have a minimum value occurring at a crit-
ical point. In fact, since $|f(x)| \le 1/|x|$ for $x \ne 0$ and f
is even, the minimum value will occur at the two critical
points closest to $x = 0$. (See Figure 2.20 in Section 2.5
of the text.)

48. No. $f(x) = -x^2$ has abs max value 0, but
$g(x) = |f(x)| = x^2$ has no abs max value.

Section 4.3 Concavity and Inflections (page 227)

2. $f(x) = 2x - x^2$, $f'(x) = 2 - 2x$, $f''(x) = -2 < 0$.
Thus, f is concave down on $(-\infty, \infty)$.

4. $f(x) = x - x^3$, $f'(x) = 1 - 3x^2$,
$f''(x) = -6x$.

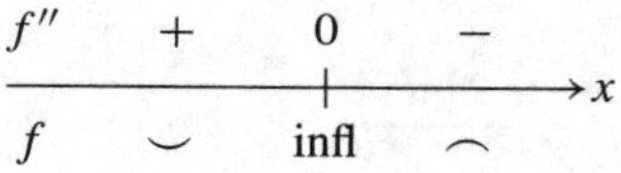

f''	$+$	0	$-$	
f	$\smile$	infl	$\frown$	$\to x$

6. $f(x) = 10x^3 + 3x^5$, $f'(x) = 30x^2 + 15x^4$,
$f''(x) = 60x + 60x^3 = 60x(1 + x^2)$.

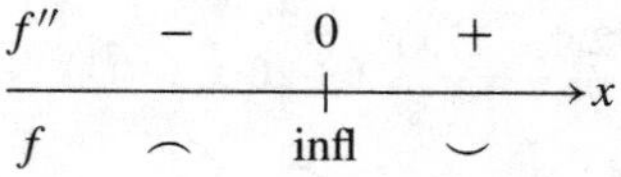

f''	$-$	0	$+$	
f	$\frown$	infl	$\smile$	$\to x$

8. $f(x) = (2 + 2x - x^2)^2$, $f'(x) = 2(2 + 2x - x^2)(2 - 2x)$,
$f''(x) = 2(2 - 2x)^2 + 2(2 + 2x - x^2)(-2)$
$= 12x(x - 2)$.

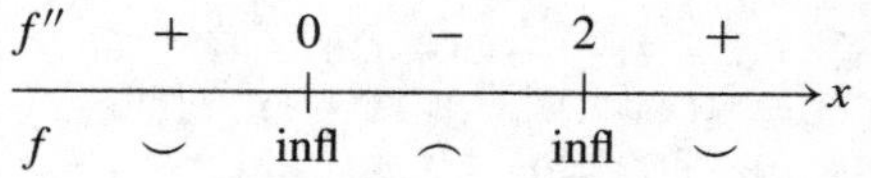

f''	$+$	0	$-$	2	$+$	
f	$\smile$	infl	$\frown$	infl	$\smile$	$\to x$

10. $f(x) = \dfrac{x}{x^2 + 3}$, $f'(x) = \dfrac{3 - x^2}{(x^2 + 3)^2}$,
$f''(x) = \dfrac{2x(x^2 - 9)}{(x^2 + 3)^3}$.

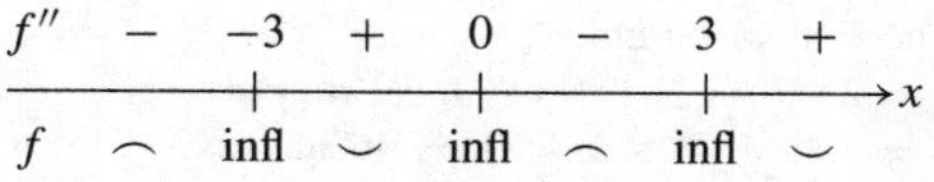

f''	$-$	-3	$+$	0	$-$	3	$+$	
f	$\frown$	infl	$\smile$	infl	$\frown$	infl	$\smile$	$\to x$

12. $f(x) = \cos 3x$, $f'(x) = -3\sin 3x$, $f''(x) = -9\cos 3x$.
Inflection points: $x = \left(n + \frac{1}{2}\right)\frac{\pi}{3}$ for $n = 0, \pm 1, \pm 2, \dots$.

f is concave up on $\left(\frac{4n+1}{6}\pi, \frac{4n+3}{6}\pi\right)$ and concave

down on $\left(\frac{4n+3}{6}\pi, \frac{4n+5}{6}\pi\right)$.

14. $f(x) = x - 2\sin x$, $f'(x) = 1 - 2\cos x$, $f''(x) = 2\sin x$.
Inflection points: $x = n\pi$ for $n = 0, \pm 1, \pm 2, \dots$.
f is concave down on $\left((2n+1)\pi, (2n+2)\pi\right)$ and concave

up on $\left((2n)\pi, (2n+1)\pi\right)$.

16. $f(x) = xe^x$, $f'(x) = e^x(1+x)$,
$f''(x) = e^x(2+x)$.

18. $f(x) = \dfrac{\ln(x^2)}{x}$, $f'(x) = \dfrac{2 - \ln(x^2)}{x^2}$,

$f''(x) = \dfrac{-6 + 2\ln(x^2)}{x^3}$.

f has inflection point at $x = \pm e^{3/2}$ and f is undefined at $x = 0$. f is concave up on $(-e^{3/2}, 0)$ and $(e^{3/2}, \infty)$; and concave down on $(-\infty, -e^{3/2})$ and $(0, e^{3/2})$.

20. $f(x) = (\ln x)^2$, $f'(x) = \dfrac{2}{x}\ln x$,

$f''(x) = \dfrac{2(1 - \ln x)}{x^2}$ for all $x > 0$.

22. $f(x) = (x-1)^{1/3} + (x+1)^{1/3}$,
$f'(x) = \frac{1}{3}[(x-1)^{-2/3} + (x+1)^{-2/3}]$,
$f''(x) = -\frac{2}{9}[(x-1)^{-5/3} + (x+1)^{-5/3}]$.
$f(x) = 0 \Leftrightarrow x - 1 = -(x+1) \Leftrightarrow x = 0$.
Thus, f has inflection point at $x = 0$. $f''(x)$ is undefined at $x = \pm 1$. f is defined at ± 1 and $x = \pm 1$ are also inflection points. f is concave up on $(-\infty, -1)$ and $(0, 1)$; and down on $(-1, 0)$ and $(1, \infty)$.

24. $f(x) = 3x^3 - 36x - 3$, $f'(x) = 9(x^2 - 4)$, $f''(x) = 18x$.
The critical points are
$x = 2$, $f''(2) > 0 \Rightarrow$ local min;
$x = -2$, $f''(-2) < 0 \Rightarrow$ local max.

26. $f(x) = x + \dfrac{4}{x}$, $f'(x) = 1 - \dfrac{4}{x^2}$, $f''(x) = 8x^{-3}$.
The critical points are
$x = 2$, $f''(2) > 0 \Rightarrow$ local min;
$x = -2$, $f''(-2) < 0 \Rightarrow$ local max.

28. $f(x) = \dfrac{x}{2^x}$, $f'(x) = \dfrac{1 - x\ln 2}{2^x}$,

$f''(x) = \dfrac{\ln 2(x\ln 2 - 2)}{2^x}$.

The critical point is
$x = \dfrac{1}{\ln 2}$, $f''\left(\dfrac{1}{\ln 2}\right) < 0 \Rightarrow$ local max.

30. $f(x) = xe^x$, $f'(x) = e^x(1+x)$, $f''(x) = e^x(2+x)$.
The critical point is $x = -1$.
$f''(-1) > 0, \Rightarrow$ local min.

32. $f(x) = (x^2 - 4)^2$, $f'(x) = 4x^3 - 16x$, $f''(x) = 12x^2 - 16$.
The critical points are
$x = 0$, $f''(0) < 0 \Rightarrow$ local max;
$x = 2$, $f''(2) > 0 \Rightarrow$ local min;
$x = -2$, $f''(-2) > 0 \Rightarrow$ local min.

34. $f(x) = (x^2 - 3)e^x$,
$f'(x) = (x^2 + 2x - 3)e^x = (x+3)(x-1)e^x$,
$f''(x) = (x^2 + 4x - 1)e^x$.
The critical points are
$x = -3$, $f''(-3) < 0 \Rightarrow$ local max;
$x = 1$, $f''(1) > 0 \Rightarrow$ local min.

36. Since
$$f(x) = \begin{cases} x^2 & \text{if } x \geq 0 \\ -x^2 & \text{if } x < 0, \end{cases}$$
we have
$$f'(x) = \begin{cases} 2x & \text{if } x \geq 0 \\ -2x & \text{if } x < 0 \end{cases} = 2|x|$$
$$f''(x) = \begin{cases} 2 & \text{if } x > 0 \\ -2 & \text{if } x < 0 \end{cases} = 2\operatorname{sgn} x.$$

$f'(x) = 0$ if $x = 0$. Thus, $x = 0$ is a critical point of f. It is also an inflection point since the conditions of Definition 3 are satisfied. $f''(0)$ does not exist. If a the graph of a function has a tangent line, vertical or not, at x_0, and has opposite concavity on opposite sides of x_0, the x_0 is an inflection point of f, whether or not $f''(x_0)$ even exists.

38. Suppose that f has an inflection point at x_0. To be specific, suppose that $f''(x) < 0$ on (a, x_0) and $f''(x) > 0$ on (x_0, b) for some numbers a and b satisfying $a < x_0 < b$.
If the graph of f has a non-vertical tangent line at x_0, then $f'(x_0)$ exists. Let
$$F(x) = f(x) - f(x_0) - f'(x_0)(x - x_0).$$

$F(x)$ represents the signed vertical distance between the graph of f and its tangent line at x_0. To show that the graph of f crosses its tangent line at x_0, it is sufficient to show that $F(x)$ has opposite signs on opposite sides of x_0.

Observe that $F(x_0) = 0$, and $F'(x) = f'(x) - f'(x_0)$, so that $F'(x_0) = 0$ also. Since $F''(x) = f''(x)$, the assumptions above show that F' has a local minimum value at x_0 (by the First Derivative Test). Hence $F(x) > 0$ if $a < x < x_0$ or $x_0 < x < b$. It follows (by Theorem 6) that $F(x) < 0$ if $a < x < x_0$, and $F(x) > 0$ if $x_0 < x < b$. This completes the proof for the case of a nonvertical tangent.

If f has a vertical tangent at x_0, then its graph necessarily crosses the tangent (the line $x = x_0$) at x_0, since the graph of a function must cross any vertical line through a point of its domain that is not an endpoint.

40. Let there be a function f such that

$$f'(x_0) = f''(x_0) = ... = f^{(k-1)}(x_0) = 0,$$
$$f^{(k)}(x_0) \neq 0 \qquad \text{for some } k \geq 2.$$

If k is even, then f has a local min value at $x = x_0$ when $f^{(k)}(x_0) > 0$, and f has a local max value at $x = x_0$ when $f^{(k)}(x_0) < 0$.

If k is odd, then f has an inflection point at $x = x_0$.

42. We are given that

$$f(x) = \begin{cases} x^2 \sin \dfrac{1}{x}, & \text{if } x \neq 0; \\ 0, & \text{if } x = 0. \end{cases}$$

If $x \neq 0$, then

$$f'(x) = 2x \sin \frac{1}{x} - \cos \frac{1}{x}$$
$$f''(x) = 2 \sin \frac{1}{x} - \frac{2}{x} \cos \frac{1}{x} - \frac{1}{x^2} \sin \frac{1}{x}.$$

If $x = 0$, then

$$f'(x) = \lim_{h \to 0} \frac{h^2 \sin \dfrac{1}{h} - 0}{h} = 0.$$

Thus 0 is a critical point of f. There are points x arbitrarily close to 0 where $f(x) > 0$, for example $x = \dfrac{2}{(4n + 1)\pi}$, and other such points where $f(x) < 0$, for example $x = \dfrac{2}{(4n + 3)\pi}$. Therefore f does not have a local max or min at $x = 0$. Also, there are points arbitrarily close to 0 where $f''(x) > 0$, for example $x = \dfrac{1}{(2n + 1)\pi}$, and other such points where $f''(x) < 0$, for instance $x = \dfrac{1}{2n\pi}$. Therefore f does not have constant concavity on any interval $(0, a)$ where $a > 0$, so 0 is not an inflection point of f either.

Section 4.4   Sketching the Graph of a Function   (page 236)

2.

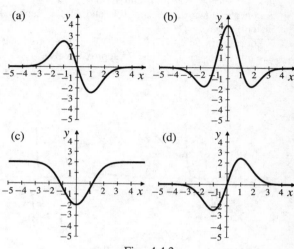

Fig. 4.4.2

The function graphed in Fig. 4.2(a):
is odd, is asymptotic to $y = 0$ at $\pm\infty$,
is increasing on $(-\infty, -1)$ and $(1, \infty)$,
is decreasing on $(-1, 1)$,
has CPs at $x = -1$ (max) and 1 (min),
is concave up on $(-\infty, -2)$ and $(0, 2)$ (approximately),
is concave down on $(-2, 0)$ and $(2, \infty)$ (approximately),
has inflections at $x = \pm 2$ (approximately).

The function graphed in Fig. 4.2(b):
is even, is asymptotic to $y = 0$ at $\pm\infty$,
is increasing on $(-1.7, 0)$ and $(1.7, \infty)$ (approximately),
is decreasing on $(-\infty, -1.7)$ and $(0, 1.7)$ (approximately),
has CPs at $x = 0$ (max) and ± 1.7 (min) (approximately),
is concave up on $(-2.5, -1)$ and $(1, 2.5)$ (approximately),
is concave down on $(-\infty, -2.5)$, $(-1, 1)$, and $(2.5, \infty)$ (approximately),
has inflections at ± 2.5 and ± 1 (approximately).

The function graphed in Fig. 4.2(c):
is even, is asymptotic to $y = 2$ at $\pm\infty$,
is increasing on $(0, \infty)$,
is decreasing on $(-\infty, 0)$,
has a CP at $x = 0$ (min),
is concave up on $(-1, 1)$ (approximately),
is concave down on $(-\infty, -1)$ and $(1, \infty)$ (approximately),
has inflections at $x = \pm 1$ (approximately).

The function graphed in Fig. 4.2(d):
is odd, is asymptotic to $y = 0$ at $\pm\infty$,
is increasing on $(-1, 1)$,
is decreasing on $(-\infty, -1)$ and $(1, \infty)$,
has CPs at $x = -1$ (min) and 1 (max),
is concave down on $(-\infty, -1.7)$ and $(0, 1.7)$ (approximately),
is concave up on $(-1.7, 0)$ and $(1.7, \infty)$ (approximately),
has inflections at 0 and ± 1.7 (approximately).

The function graphed in Fig. 4.4(d):
is odd, is asymptotic to $y = \pm 2$,
is increasing on $(-\infty, -0.7)$ and $(0.7, \infty)$ (approximately),
is decreasing on $(-0.7, 0.7)$ (approximately),
has CPs at $x = \pm 0.7$ (approximately),
is concave up on $(-\infty, -1)$ and $(0, 1)$ (approximately),
is concave down on $(-1, 0)$ and $(1, \infty)$ (approximately),
has an inflection at $x = 0$ and $x = \pm 1$ (approximately).

4.

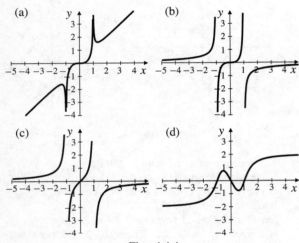

Fig. 4.4.4

6. According to the given properties:
Oblique asymptote: $y = x - 1$.
Critical points: $x = 0, 2$. Singular point: $x = -1$.
Local max 2 at $x = 0$; local min 0 at $x = 2$.

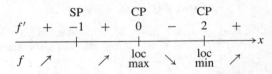

Inflection points: $x = -1, 1, 3$.

The function graphed in Fig. 4.4(a):
is odd, is asymptotic to $x = \pm 1$ and $y = x$,
is increasing on $(-\infty, -1.5)$, $(-1, 1)$, and $(1.5, \infty)$ (approximately),
is decreasing on $(-1.5, -1)$ and $(1, 1.5)$ (approximately),
has CPs at $x = -1.5$, $x = 0$, and $x = 1.5$,
is concave up on $(0, 1)$ and $(1, \infty)$,
is concave down on $(-\infty, -1)$ and $(-1, 0)$,
has an inflection at $x = 0$.

The function graphed in Fig. 4.4(b):
is odd, is asymptotic to $x = \pm 1$ and $y = 0$,
is increasing on $(-\infty, -1)$, $(-1, 1)$, and $(1, \infty)$,
has a CP at $x = 0$,
is concave up on $(-\infty, -1)$ and $(0, 1)$,
is concave down on $(-1, 0)$ and $(1, \infty)$,
has an inflection at $x = 0$.

The function graphed in Fig. 4.4(c):
is odd, is asymptotic to $x = \pm 1$ and $y = 0$,
is increasing on $(-\infty, -1)$, $(-1, 1)$, and $(1, \infty)$,
has no CP,
is concave up on $(-\infty, -1)$ and $(0, 1)$,
is concave down on $(-1, 0)$ and $(1, \infty)$,
has an inflection at $x = 0$.

Since $\lim\limits_{x \to \pm\infty} \left(f(x) + 1 - x \right) = 0$, the line $y = x - 1$ is an oblique asymptote.

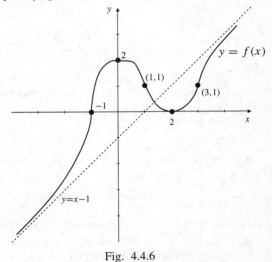

Fig. 4.4.6

8. $y = x(x^2 - 1)^2$, $y' = (x^2 - 1)(5x^2 - 1)$, $y'' = 4x(5x^2 - 3)$.
From y: Intercepts: $(0, 0)$, $(1, 0)$. Symmetry: odd (i.e., about the origin).

From y': Critical point: $x = \pm 1$, $\pm\dfrac{1}{\sqrt{5}}$.

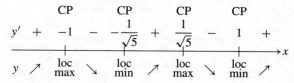

From y'': Inflection points at
$x = 0$, $\pm\sqrt{\dfrac{3}{5}}$.

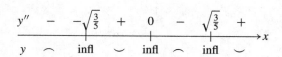

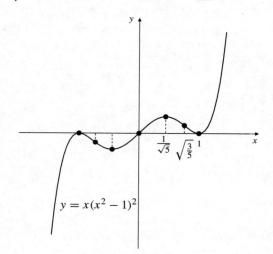

Fig. 4.4.8

10. $y = \dfrac{x-1}{x+1} = 1 - \dfrac{2}{x+1}$, $y' = \dfrac{2}{(x+1)^2}$, $y'' = \dfrac{-4}{(x+1)^3}$.
From y: Intercepts: $(0, -1)$, $(1, 0)$. Asymptotes: $y = 1$ (horizontal), $x = -1$ (vertical). No obvious symmetry.
Other points: $(-2, 3)$.
From y': No critical point.

		ASY	
y'	$+$	-1	$+$
y	$\nearrow$		$\nearrow$

From y'': No inflection point.

		ASY	
y''	$+$	-1	$-$
y	$\smile$		$\frown$

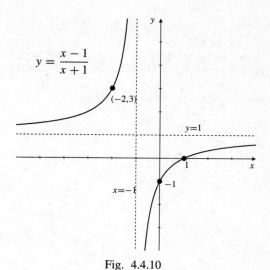

Fig. 4.4.10

12. $y = \dfrac{1}{4 + x^2}$, $y' = \dfrac{-2x}{(4+x^2)^2}$, $y'' = \dfrac{6x^2 - 8}{(4+x^2)^3}$.
From y: Intercept: $(0, \frac{1}{4})$. Asymptotes: $y = 0$ (horizontal). Symmetry: even (about y-axis).
From y': Critical point: $x = 0$.

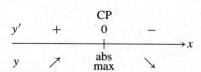

From y'': $y'' = 0$ at $x = \pm\dfrac{2}{\sqrt{3}}$.

		$-\frac{2}{\sqrt{3}}$		$\frac{2}{\sqrt{3}}$	
y''	$+$		$-$		$+$
y	$\smile$	infl	$\frown$	infl	$\smile$

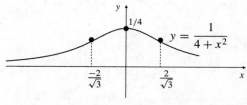

Fig. 4.4.12

14. $y = \dfrac{x}{x^2-1}$, $y' = -\dfrac{x^2+1}{(x^2-1)^2}$, $y'' = \dfrac{2x(x^2+3)}{(x^2-1)^3}$.
From y: Intercept: $(0,0)$. Asymptotes: $y=0$ (horizontal), $x=\pm1$ (vertical). Symmetry: odd. Other points: $(2,\frac{2}{3})$, $(-2,-\frac{2}{3})$.
From y': No critical or singular points.

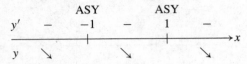

From y'': $y''=0$ at $x=0$.

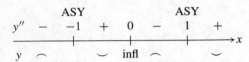

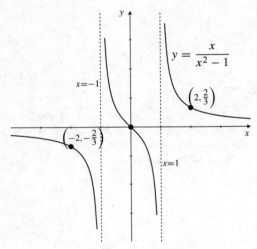

Fig. 4.4.14

16. $y = \dfrac{x^3}{x^2-1}$, $y' = \dfrac{x^2(x^2-3)}{(x^2-1)^2}$, $y'' = \dfrac{2x(x^2+3)}{(x^2-1)^3}$.
From y: Intercept: $(0,0)$. Asymptotes: $x=\pm1$ (vertical), $y=x$ (oblique). Symmetry: odd. Other points: $\left(\pm\sqrt{3},\pm\dfrac{3\sqrt{3}}{2}\right)$.
From y': Critical point: $x=0,\ \pm\sqrt{3}$.

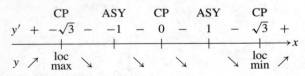

From y'': $y''=0$ at $x=0$.

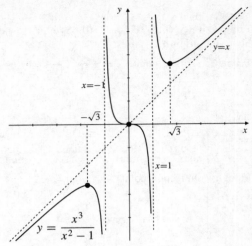

Fig. 4.4.16

18. $y = \dfrac{x^2}{x^2+1}$, $y' = \dfrac{2x}{(x^2+1)^2}$, $y'' = \dfrac{2(1-3x^2)}{(x^2+1)^3}$.
From y: Intercept: $(0,0)$. Asymptotes: $y=1$ (horizontal). Symmetry: even.
From y': Critical point: $x=0$.

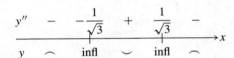

From y'': $y''=0$ at $x=\pm\dfrac{1}{\sqrt{3}}$.

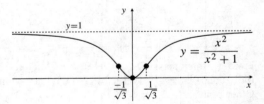

Fig. 4.4.18

20. $y = \dfrac{x^2-2}{x^2-1}$, $y' = \dfrac{2x}{(x^2-1)^2}$, $y'' = \dfrac{-2(3x^2+1)}{(x^2-1)^3}$.
From y: Intercept: $(0,2)$, $(\pm\sqrt{2},0)$. Asymptotes: $y=1$ (horizontal), $x=\pm1$ (vertical). Symmetry: even.
From y': Critical point: $x=0$.

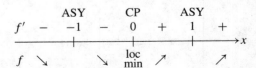

From y'': $y'' = 0$ nowhere.

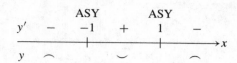

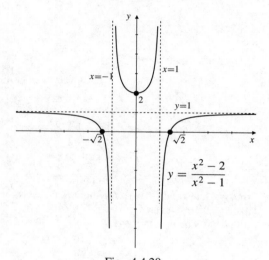

Fig. 4.4.20

22. $y = \dfrac{x^2 - 1}{x^2} = 1 - \dfrac{1}{x^2}$, $y' = \dfrac{2}{x^3}$, $y'' = -\dfrac{6}{x^4}$.
From y: Intercepts: $(\pm 1, 0)$. Asymptotes: $y = 1$ (horizontal), $x = 0$ (vertical). Symmetry: even.
From y': No critical points.

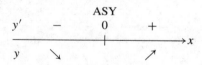

From y'': y'' is negative for all x.

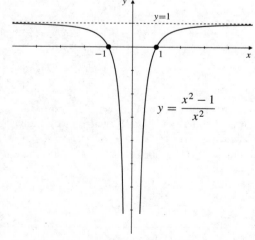

Fig. 4.4.22

24. $y = \dfrac{(2 - x)^2}{x^3}$, $y' = -\dfrac{(x - 2)(x - 6)}{x^4}$,
$y'' = \dfrac{2(x^2 - 12x + 24)}{x^5} = \dfrac{2(x - 6 + 2\sqrt{3})(x - 6 - 2\sqrt{3})}{x^5}$.

From y: Intercept: $(2, 0)$. Asymptotes: $y = 0$ (horizontal), $x = 0$ (vertical). Symmetry: none obvious. Other points: $(-2, -2)$, $(-10, -0.144)$.
From y': Critical points: $x = 2$, 6.

From y'': $y'' = 0$ at $x = 6 \pm 2\sqrt{3}$.

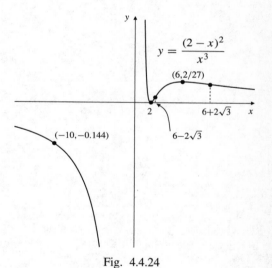

Fig. 4.4.24

26. $y = \dfrac{x}{x^2 + x - 2} = \dfrac{x}{(2 + x)(x - 1)}$,
$y' = \dfrac{-(x^2 + 2)}{(x + 2)^2(x - 1)^2}$, $y'' = \dfrac{2(x^3 + 6x + 2)}{(x + 2)^3(x - 1)^3}$.
From y: Intercepts: $(0, 0)$. Asymptotes: $y = 0$ (horizontal), $x = 1$, $x = -2$ (vertical). Other points: $(-3, -\frac{3}{4})$, $(2, \frac{1}{2})$.
From y': No critical point.

From y'': $y'' = 0$ if $f(x) = x^3 + 6x + 2 = 0$. Since $f'(x) = 3x^2 + 6 \geq 6$, f is increasing and can only have one root. Since $f(0) = 2$ and $f(-1) = -5$, that root must be between -1 and 0. Let the root be r.

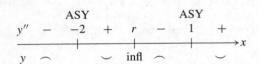

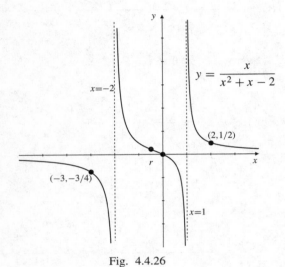

Fig. 4.4.26

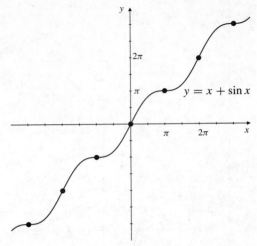

Fig. 4.4.28

30. $y = e^{-x^2}$, $y' = -2xe^{-x^2}$, $y'' = (4x^2 - 2)e^{-x^2}$.
From y: Intercept: $(0, 1)$. Asymptotes: $y = 0$ (horizontal). Symmetry: even.
From y': Critical point: $x = 0$.

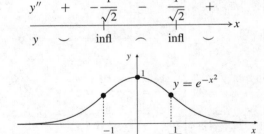

From y'': $y'' = 0$ at $x = \pm \dfrac{1}{\sqrt{2}}$.

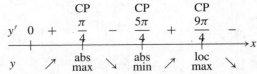

Fig. 4.4.30

28. $y = x + \sin x$, $y' = 1 + \cos x$, $y'' = -\sin x$.
From y: Intercept: $(0, 0)$. Other points: $(k\pi, k\pi)$, where k is an integer. Symmetry: odd.
From y': Critical point: $x = (2k + 1)\pi$, where k is an integer.

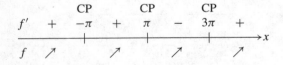

From y'': $y'' = 0$ at $x = k\pi$, where k is an integer.

32. $y = e^{-x} \sin x$ $(x \geq 0)$,
$y' = e^{-x}(\cos x - \sin x)$, $y'' = -2e^{-x} \cos x$.
From y: Intercept: $(k\pi, 0)$, where k is an integer. Asymptotes: $y = 0$ as $x \to \infty$.
From y': Critical points: $x = \dfrac{\pi}{4} + k\pi$, where k is an integer.

From y'': $y'' = 0$ at $x = (k + \frac{1}{2})\pi$, where k is an integer.

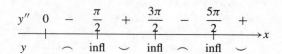

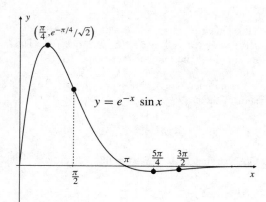

Fig. 4.4.32

34. $y = x^2 e^x$, $y' = (2x + x^2)e^x = x(2 + x)e^x$,
$y'' = (x^2 + 4x + 2)e^x = (x + 2 - \sqrt{2})(x + 2 + \sqrt{2})e^x$.
From y: Intercept: $(0, 0)$.
Asymptotes: $y = 0$ as $x \to -\infty$.
From y': Critical point: $x = 0$, $x = -2$.

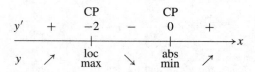

From y'': $y'' = 0$ at $x = -2 \pm \sqrt{2}$.

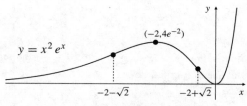

Fig. 4.4.34

36. $y = \dfrac{\ln x}{x^2}$ $(x > 0)$,
$y' = \dfrac{1 - 2\ln x}{x^3}$, $y'' = \dfrac{6\ln x - 5}{x^4}$.
From y: Intercepts: $(1, 0)$. Asymptotes: $y = 0$, since $\lim\limits_{x \to \infty} \dfrac{\ln x}{x^2} = 0$, and $x = 0$, since $\lim\limits_{x \to 0+} \dfrac{\ln x}{x^2} = -\infty$.
From y': Critical point: $x = e^{1/2}$.

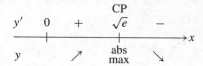

From y'': $y'' = 0$ at $x = e^{5/6}$.

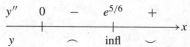

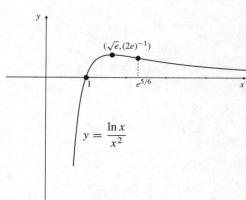

Fig. 4.4.36

38. $y = \dfrac{x}{\sqrt{x^2 + 1}}$, $y' = (x^2 + 1)^{-3/2}$, $y'' = -3x(x^2 + 1)^{-5/2}$.
From y: Intercept: $(0, 0)$. Asymptotes: $y = 1$ as $x \to \infty$, and $y = -1$ as $x \to -\infty$. Symmetry: odd.
From y': No critical point. $y' > 0$ and y is increasing for all x.
From y'': $y'' = 0$ at $x = 0$.

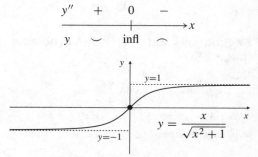

Fig. 4.4.38

40. According to Theorem 5 of Section 4.4,

$$\lim_{x \to 0+} x \ln x = 0.$$

Thus,

$$\lim_{x \to 0} x \ln |x| = \lim_{x \to 0+} x \ln x = 0.$$

If $f(x) = x \ln |x|$ for $x \neq 0$, we may define $f(0)$ such that $f(0) = \lim_{x \to 0} x \ln |x| = 0$. Then f is continuous on the whole real line and

$$f'(x) = \ln |x| + 1, \qquad f''(x) = \frac{1}{|x|} \operatorname{sgn}(x).$$

From f: Intercept: $(0, 0)$, $(\pm 1, 0)$. Asymptotes: none. Symmetry: odd.

From f': CP: $x = \pm \dfrac{1}{e}$. SP: $x = 0$.

	CP		SP		CP	
f'	$+$	$-\frac{1}{e}$ $-$	0 $-$		$\frac{1}{e}$ $+$	$\to x$
f	$\nearrow$	loc max $\searrow$	$\searrow$		loc min $\nearrow$	

From f'': f'' is undefined at $x = 0$.

	$-$	0	$+$	
f''				$\to x$
f	$\frown$	infl	$\smile$	

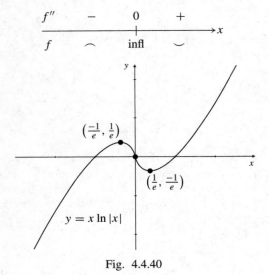

Fig. 4.4.40

Section 4.5 Extreme-Value Problems (page 242)

2. Let the numbers be x and $\dfrac{8}{x}$ where $x > 0$. Their sum is $S = x + \dfrac{8}{x}$. Since $S \to \infty$ as $x \to \infty$ or $x \to 0+$, the minimum sum must occur at a critical point:

$$0 = \frac{dS}{dx} = 1 - \frac{8}{x^2} \Rightarrow x = 2\sqrt{2}.$$

Thus, the smallest possible sum is $2\sqrt{2} + \dfrac{8}{2\sqrt{2}} = 4\sqrt{2}$.

4. Let the numbers be x and $16 - x$. Let $P(x) = x^3(16 - x)^5$. Since $P(x) \to -\infty$ as $x \to \pm\infty$, so the maximum must occur at a critical point:

$$0 = P'(x) = 3x^2(16 - x)^5 - 5x^3(16 - x)^4$$
$$= x^2(16 - x)^4(48 - 8x).$$

The critical points are 0, 6 and 16. Clearly, $P(0) = P(16) = 0$, and $P(6) = 216 \times 10^5$. Thus, $P(x)$ is maximum if the numbers are 6 and 10.

6. If the numbers are x and $n - x$, then $0 \leq x \leq n$ and the sum of their squares is

$$S(x) = x^2 + (n - x)^2.$$

Observe that $S(0) = S(n) = n^2$. For critical points:

$$0 = S'(x) = 2x - 2(n - x) = 2(2x - n) \Rightarrow x = n/2.$$

Since $S(n/2) = n^2/2$, this is the smallest value of the sum of squares.

8. Let the width and the length of a rectangle of given perimeter $2P$ be x and $P - x$. Then the area of the rectangle is

$$A(x) = x(P - x) = Px - x^2.$$

Since $A(x) \to -\infty$ as $x \to \pm\infty$ the maximum must occur at a critical point:

$$0 = \frac{dA}{dx} = P - 2x \Rightarrow x = \frac{P}{2}$$

Hence, the width and the length are $\dfrac{P}{2}$ and $\left(P - \dfrac{P}{2}\right) = \dfrac{P}{2}$. Since the width equals the length, it is a square.

10. Let the various dimensions be as shown in the figure. Since $h = 10 \sin \theta$ and $b = 20 \cos \theta$, the area of the triangle is

$$A(\theta) = \tfrac{1}{2} bh = 100 \sin \theta \cos \theta$$
$$= 50 \sin 2\theta \qquad \text{for } 0 < \theta < \frac{\pi}{2}.$$

Since $A(\theta) \to 0$ as $\theta \to 0$ and $\theta \to \dfrac{\pi}{2}$, the maximum must be at a critial point:

$$0 = A'(\theta) = 100 \cos 2\theta \Rightarrow 2\theta = \frac{\pi}{2} \Rightarrow \theta = \frac{\pi}{4}.$$

Hence, the largest possible area is

$$A(\pi/4) = 50 \sin\left[2\left(\frac{\pi}{4}\right)\right] = 50 \, \text{m}^2.$$

(Remark: alternatively, we may simply observe that the largest value of $\sin 2\theta$ is 1; therefore the largest possible area is $50(1) = 50 \, \text{m}^2$.)

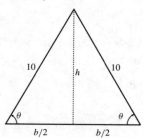

Fig. 4.5.10

12. Let x be as shown in the figure. The perimeter of the rectangle is

$$P(x) = 4x + 2\sqrt{R^2 - x^2} \qquad (0 \le x \le R).$$

For critical points:

$$0 = \frac{dP}{dx} = 4 + \frac{-2x}{\sqrt{R^2 - x^2}}$$

$$\Rightarrow 2\sqrt{R^2 - x^2} = x \Rightarrow x = \frac{2R}{\sqrt{5}}.$$

Since

$$\frac{d^2P}{dx^2} = \frac{-2R^2}{(R^2 - x^2)^{3/2}} < 0,$$

therefore $P(x)$ is concave down on $[0, R]$, so it must have an absolute maximum value at $x = \frac{2R}{\sqrt{5}}$. The largest perimeter is therefore

$$P\left(\frac{2R}{\sqrt{5}}\right) = 4\left(\frac{2R}{\sqrt{5}}\right) + \sqrt{R^2 - \frac{4R^2}{5}} = \frac{10R}{\sqrt{5}} \, \text{units}.$$

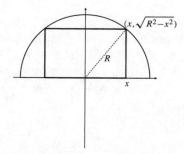

Fig. 4.5.12

14. See the diagrams below.

a) The area of the rectangle is $A = xy$. Since

$$\frac{y}{a - x} = \frac{b}{a} \Rightarrow y = \frac{b(a - x)}{a}.$$

Thus, the area is

$$A = A(x) = \frac{bx}{a}(a - x) \qquad (0 \le x \le a).$$

For critical points:

$$0 = A'(x) = \frac{b}{a}(a - 2x) \Rightarrow x = \frac{a}{2}.$$

Since $A''(x) = -\frac{2b}{a} < 0$, A must have a maximum value at $x = \frac{a}{2}$. Thus, the largest area for the rectangle is

$$\frac{b}{a}\left(\frac{a}{2}\right)\left(a - \frac{a}{2}\right) = \frac{ab}{4} \text{ square units,}$$

that is, half the area of the triangle ABC.

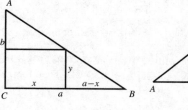

Fig. 4.5.14(a)　　　Fig. 4.5.14(b)

(b) This part has the same answer as part (a). To see this, let $CD \perp AB$, and solve separate problems for the largest rectangles in triangles ACD and BCD as shown. By part (a), both maximizing rectangles have the same height, namely half the length of CD. Thus, their union is a rectangle of area half of that of triangle ABC.

16. NEED FIGURE If the equal sides of the isosceles triangle are 10 cm long and the angles opposite these sides are θ, then the area of the triangle is

$$A(\theta) = \frac{1}{2}(10)(10 \sin \theta) = 50 \sin \theta \text{ cm}^2,$$

which is evidently has maximum value 50 cm^2 when $\theta = \pi/2$, that is, when the triangle is right-angled. This solution requires no calculus, and so is easier than the one given for the previous problem.

18. Let x be the side of the cut-out squares. Then the volume of the box is

$$V(x) = x(70 - 2x)(150 - 2x) \qquad (0 \le x \le 35).$$

Since $V(0) = V(35) = 0$, the maximum value will occur at a critical point:

$$
\begin{aligned}
0 = V'(x) &= 4(2625 - 220x + 3x^2) \\
&= 4(3x - 175)(x - 15) \\
&\Rightarrow x = 15 \text{ or } \frac{175}{3}.
\end{aligned}
$$

The only critical point in $[0, 35]$ is $x = 15$. Thus, the largest possible volume for the box is

$$V(15) = 15(70 - 30)(150 - 30) = 72,000\,\text{cm}^3.$$

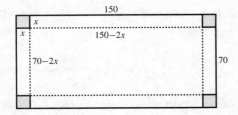

Fig. 4.5.18

20. If the manager charges $\$(40+x)$ per room, then $(80-2x)$ rooms will be rented.
The total income will be $\$(80 - 2x)(40 + x)$ and the total cost will be $\$(80 - 2x)(10) + (2x)(2)$. Therefore, the profit is

$$
\begin{aligned}
P(x) &= (80 - 2x)(40 + x) - [(80 - 2x)(10) + (2x)(2)] \\
&= 2400 + 16x - 2x^2 \qquad \text{for } x > 0.
\end{aligned}
$$

If $P'(x) = 16 - 4x = 0$, then $x = 4$. Since $P''(x) = -4 < 0$, P must have a maximum value at $x = 4$. Therefore, the manager should charge $\$44$ per room.

22. This problem is similar to the previous one except that the 10 in the numerator of the second fraction in the expression for T is replaced with a 4. This has no effect on the critical point of T, namely $x = 5$, which now lies outside the appropriate interval $0 \le x \le 4$. Minimum T must occur at an endpoint. Note that

$$
\begin{aligned}
T(0) &= \frac{12}{15} + \frac{4}{39} = 0.9026 \\
T(4) &= \frac{1}{15}\sqrt{12^2 + 4^2} = 0.8433.
\end{aligned}
$$

The minimum travel time corresponds to $x = 4$, that is, to driving in a straight line to B.

24. Let the dimensions of the rectangle be as shown in the figure. Clearly,

$$
\begin{aligned}
x &= a\sin\theta + b\cos\theta, \\
y &= a\cos\theta + b\sin\theta.
\end{aligned}
$$

Therefore, the area is

$$
\begin{aligned}
A(\theta) &= xy \\
&= (a\sin\theta + b\cos\theta)(a\cos\theta + b\sin\theta) \\
&= ab + (a^2 + b^2)\sin\theta\cos\theta \\
&= ab + \frac{1}{2}(a^2 + b^2)\sin 2\theta \qquad \text{for } 0 \le \theta \le \frac{\pi}{2}.
\end{aligned}
$$

If $A'(\theta) = (a^2 + b^2)\cos 2\theta = 0$, then $\theta = \dfrac{\pi}{4}$. Since $A''(\theta) = -2(a^2 + b^2)\sin 2\theta < 0$ when $0 \le \theta \le \dfrac{\pi}{2}$, therefore $A(\theta)$ must have a maximum value at $\theta = \dfrac{\pi}{4}$. Hence, the area of the largest rectangle is

$$
\begin{aligned}
A\left(\frac{\pi}{4}\right) &= ab + \frac{1}{2}(a^2 + b^2)\sin\left(\frac{\pi}{2}\right) \\
&= ab + \frac{1}{2}(a^2 + b^2) = \frac{1}{2}(a + b)^2 \quad \text{sq. units.}
\end{aligned}
$$

(Note: $x = y = \dfrac{a}{\sqrt{2}} + \dfrac{b}{\sqrt{2}}$ indicates that the rectangle containing the given rectangle with sides a and b, has largest area when it is a square.)

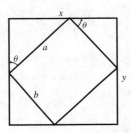

Fig. 4.5.24

26. The longest beam will have length equal to the minimum of $L = x + y$, where x and y are as shown in the figure below:

$$x = \frac{a}{\cos\theta}, \qquad y = \frac{b}{\sin\theta}.$$

Thus,

$$L = L(\theta) = \frac{a}{\cos\theta} + \frac{b}{\sin\theta} \qquad \left(0 < \theta < \frac{\pi}{2}\right).$$

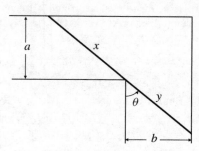

Fig. 4.5.26

If $L'(\theta) = 0$, then

$$\frac{a \sin \theta}{\cos^2 \theta} - \frac{b \cos \theta}{\sin^2 \theta} = 0$$

$$\Leftrightarrow \quad \frac{a \sin^3 \theta - b \cos^3 \theta}{\cos^2 \theta \sin^2 \theta} = 0$$

$$\Leftrightarrow \quad a \sin^3 \theta - b \cos^3 \theta = 0$$

$$\Leftrightarrow \quad \tan^3 \theta = \frac{b}{a}$$

$$\Leftrightarrow \quad \tan \theta = \frac{b^{1/3}}{a^{1/3}}.$$

Clearly, $L(\theta) \to \infty$ as $\theta \to 0+$ or $\theta \to \frac{\pi}{2}-$. Thus, the minimum must occur at $\theta = \tan^{-1}\left(\frac{b^{1/3}}{a^{1/3}}\right)$. Using the triangle above for $\tan \theta = \frac{b^{1/3}}{a^{1/3}}$, it follows that

$$\cos \theta = \frac{a^{1/3}}{\sqrt{a^{2/3} + b^{2/3}}}, \quad \sin \theta = \frac{b^{1/3}}{\sqrt{a^{2/3} + b^{2/3}}}.$$

Hence, the minimum is

$$L(\theta) = \frac{a}{\left(\dfrac{a^{1/3}}{\sqrt{a^{2/3} + b^{2/3}}}\right)} + \frac{b}{\left(\dfrac{b^{1/3}}{\sqrt{a^{2/3} + b^{2/3}}}\right)}$$

$$= \left(a^{2/3} + b^{2/3}\right)^{3/2} \text{ units.}$$

28. Let θ be the angle of inclination of the ladder. The height of the fence is

$$h(\theta) = 6 \sin \theta - 2 \tan \theta \qquad \left(0 < \theta < \frac{\pi}{2}\right).$$

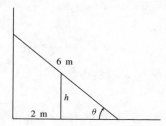

Fig. 4.5.28

For critical points:

$$0 = h'(\theta) = 6 \cos \theta - 2 \sec^2 \theta$$

$$\Rightarrow 3 \cos \theta = \sec^2 \theta \Rightarrow 3 \cos^3 \theta = 1$$

$$\Rightarrow \cos \theta = \left(\tfrac{1}{3}\right)^{1/3}.$$

Since $h''(\theta) = -6 \sin \theta - 4 \sec^2 \theta \tan \theta < 0$ for $0 < \theta < \frac{\pi}{2}$, therefore $h(\theta)$ must be maximum at $\theta = \cos^{-1}\left(\tfrac{1}{3}\right)^{1/3}$. Then

$$\sin \theta = \frac{\sqrt{3^{2/3} - 1}}{3^{1/3}}, \quad \tan \theta = \sqrt{3^{2/3} - 1}.$$

Thus, the maximum height of the fence is

$$h(\theta) = 6 \left(\frac{\sqrt{3^{2/3} - 1}}{3^{1/3}}\right) - 2\sqrt{3^{2/3} - 1}$$

$$= 2(3^{2/3} - 1)^{3/2} \approx 2.24 \, \text{m}.$$

30. The square of the distance from $(8, 1)$ to the curve $y = 1 + x^{3/2}$ is

$$S = (x - 8)^2 + (y - 1)^2$$

$$= (x - 8)^2 + (1 + x^{3/2} - 1)^2$$

$$= x^3 + x^2 - 16x + 64.$$

Note that y, and therefore also S, is only defined for $x \geq 0$. If $x = 0$ then $S = 64$. Also, $S \to \infty$ if $x \to \infty$. For critical points:

$$0 = \frac{dS}{dx} = 3x^2 + 2x - 16 = (3x + 8)(x - 2)$$

$$\Rightarrow x = -\tfrac{8}{3} \text{ or } 2.$$

Only $x = 2$ is feasible. At $x = 2$ we have $S = 44 < 64$. Therefore the minimum distance is $\sqrt{44} = 2\sqrt{11}$ units.

32. Let the radius and the height of the circular cylinder be r and h. By similar triangles,

$$\frac{h}{R - r} = \frac{H}{R} \Rightarrow h = \frac{H(R - r)}{R}.$$

Hence, the volume of the circular cylinder is

$$V(r) = \pi r^2 h = \frac{\pi r^2 H(R - r)}{R}$$

$$= \pi H\left(r^2 - \frac{r^3}{R}\right) \qquad \text{for } 0 \le r \le R.$$

Since $V(0) = V(R) = 0$, the maximum value of V must be at a critical point. If $\dfrac{dV}{dr} = \pi H\left(2r - \dfrac{3r^2}{R}\right) = 0$,

then $r = \dfrac{2R}{3}$. Therefore the cylinder has maximum

volume if its radius is $r = \dfrac{2R}{3}$ units, and its height is

$$h = \frac{H\left(R - \dfrac{2R}{3}\right)}{R} = \frac{H}{3} \text{ units.}$$

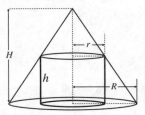

Fig. 4.5.32

34.

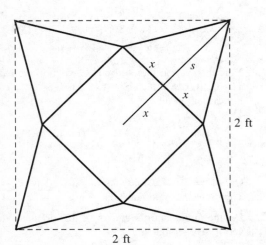

2 ft

2 ft

Fig. 4.5.34

From the figure, if the side of the square base of the pyramid is $2x$, then the slant height of triangular walls of the pyramid is $s = \sqrt{2} - x$. The vertical height of the pyramid is

$$h = \sqrt{s^2 - x^2} = \sqrt{2 - 2\sqrt{2}x + x^2 - x^2} = \sqrt{2}\sqrt{1 - \sqrt{2}x}.$$

Thus the volume of the pyramid is

$$V = \frac{4\sqrt{2}}{3} x^2 \sqrt{1 - \sqrt{2}x},$$

for $0 \le x \le 1/\sqrt{2}$. $V = 0$ at both endpoints, so the maximum will occur at an interior critical point. For CP:

$$0 = \frac{dV}{dx} = \frac{4\sqrt{2}}{3}\left[2x\sqrt{1 - \sqrt{2}x} - \frac{\sqrt{2}x^2}{2\sqrt{1 - \sqrt{2}x}}\right]$$

$$4x(1 - \sqrt{2}x) = \sqrt{2}x^2$$

$$4x = 5\sqrt{2}x^2 \quad , x = 4/(5\sqrt{2}).$$

$V(4/(5\sqrt{2})) = 32\sqrt{2}/(75\sqrt{5})$. The largest volume of such a pyramid is $32\sqrt{2}/(75\sqrt{5})$ ft^3.

36. Let h and r be the length and radius of the cylindrical part of the tank. The volume of the tank is

$$V = \pi r^2 h + \tfrac{4}{3}\pi r^3.$$

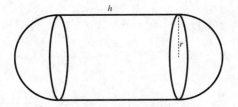

Fig. 4.5.36

If the cylindrical wall costs \$$k$ per unit area and the hemispherical wall \$$2k$ per unit area, then the total cost of the tank wall is

$$C = 2\pi r h k + 8\pi r^2 k$$

$$= 2\pi r k \frac{V - \tfrac{4}{3}\pi r^3}{\pi r^2} + 8\pi r^2 k$$

$$= \frac{2Vk}{r} + \frac{16}{3}\pi r^2 k \qquad (0 < r < \infty).$$

Since $C \to \infty$ as $r \to 0+$ or $r \to \infty$, the minimum cost must occur at a critical point. For critical points,

$$0 = \frac{dC}{dr} = -2Vkr^{-2} + \frac{32}{3}\pi r k \qquad \Leftrightarrow \qquad r = \left(\frac{3V}{16\pi}\right)^{1/3}.$$

Since $V = \pi r^2 h + \frac{4}{3}\pi r^3$,

$$r^3 = \frac{3}{16\pi}\left(\pi r^2 h + \frac{4}{3}\pi r^3\right) \Rightarrow r = \frac{1}{4}h$$

$$\Rightarrow h = 4r = 4\left(\frac{3V}{16\pi}\right)^{1/3}.$$

Hence, in order to minimize the cost, the radius and length of the cylindrical part of the tank should be $\left(\frac{3V}{16\pi}\right)^{1/3}$ and $4\left(\frac{3V}{16\pi}\right)^{1/3}$ units respectively.

38. If the path of the light ray is as shown in the figure then the time of travel from A to B is

$$T = T(x) = \frac{\sqrt{a^2 + x^2}}{v_1} + \frac{\sqrt{b^2 + (c-x)^2}}{v_2}.$$

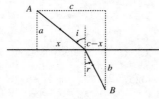

Fig. 4.5.38

To minimize T, we look for a critical point:

$$0 = \frac{dT}{dx} = \frac{1}{v_1}\frac{x}{\sqrt{a^2 + x^2}} - \frac{1}{v_2}\frac{c-x}{\sqrt{b^2 + (c-x)^2}}$$

$$= \frac{1}{v_1}\sin i - \frac{1}{v_2}\sin r.$$

Thus,

$$\frac{\sin i}{\sin r} = \frac{v_1}{v_2}.$$

40. The curve $y = 1 + 2x - x^3$ has slope $m = y' = 2 - 3x^2$. Evidently m is greatest for $x = 0$, in which case $y = 1$ and $m = 2$. Thus the tangent line with maximal slope has equation $y = 1 + 2x$.

42. Let h and r be the height and base radius of the cone and R be the radius of the sphere. From similar triangles,

$$\frac{r}{\sqrt{h^2 + r^2}} = \frac{R}{h - R}$$

$$\Rightarrow \quad h = \frac{2r^2 R}{r^2 - R^2} \qquad (r > R).$$

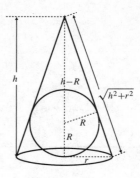

Fig. 4.5.42

Then the volume of the cone is

$$V = \frac{1}{3}\pi r^2 h = \frac{2}{3}\pi R\frac{r^4}{r^2 - R^2} \qquad (R < r < \infty).$$

Clearly $V \to \infty$ if $r \to \infty$ or $r \to R+$. Therefore to minimize V, we look for a critical point:

$$0 = \frac{dV}{dr} = \frac{2}{3}\pi R\left[\frac{(r^2 - R^2)(4r^3) - r^4(2r)}{(r^2 - R^2)^2}\right]$$

$$\Leftrightarrow \quad 4r^5 - 4r^3 R^2 - 2r^5 = 0$$

$$\Leftrightarrow \quad r = \sqrt{2}R.$$

Hence, the smallest possible volume of a right circular cone which can contain sphere of radius R is

$$V = \frac{2}{3}\pi R\left(\frac{4R^4}{2R^2 - R^2}\right) = \frac{8}{3}\pi R^3 \text{ cubic units.}$$

44. Let distances and angles be as shown. Then $\tan\alpha = \frac{2}{x}$, $\tan(\theta + \alpha) = \frac{12}{x}$

$$\frac{12}{x} = \frac{\tan\theta + \tan\alpha}{1 - \tan\theta\tan\alpha} = \frac{\tan\theta + \frac{2}{x}}{1 - \frac{2}{x}\tan\theta}$$

$$\frac{12}{x} - \frac{24}{x^2}\tan\theta = \tan\theta + \frac{2}{x}$$

$$\tan\theta\left(1 + \frac{24}{x^2}\right) = \frac{10}{x}, \quad \text{so } \tan\theta = \frac{10x}{x^2 + 24} = f(x).$$

To maximize θ (i.e., to get the best view of the mural), we can maximize $\tan\theta = f(x)$.
Since $f(0) = 0$ and $f(x) \to 0$ as $x \to \infty$, we look for a critical point.

$$0 = f'(x) = 10\left[\frac{x^2 + 24 - 2x^2}{(x^2 + 24)^2}\right] \Rightarrow x^2 = 24$$

$$\Rightarrow x = 2\sqrt{6}$$

Stand back $2\sqrt{6}$ ft (≈ 4.9 ft) to see the mural best.

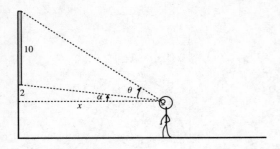

Fig. 4.5.44

46. Let the cone have radius r and height h.
Let sector of angle θ from disk be used.

Then $2\pi r = R\theta$ so $r = \dfrac{R}{2\pi}\theta$.

Also $h = \sqrt{R^2 - r^2} = \sqrt{R^2 - \dfrac{R^2\theta^2}{4\pi^2}} = \dfrac{R}{2\pi}\sqrt{4\pi^2 - \theta^2}$

The cone has volume

$$V = \frac{\pi r^2 h}{3} = \frac{\pi}{3}\frac{R^2}{4\pi^2}\theta^2\frac{R}{2\pi}\sqrt{4\pi^2 - \theta}$$

$$= \frac{R^3}{24\pi^2}f(\theta) \quad \text{where } f(\theta) = \theta^2\sqrt{4\pi^2 - \theta^2} \quad (0 \le \theta \le 2\pi)$$

$V(0) = V(2\pi) = 0$ so maximum V must occur at a critical point. For CP:

$$0 = f'(\theta) = 2\theta\sqrt{4\pi^2 - \theta^2} - \frac{\theta^3}{\sqrt{4\pi^2 - \theta^2}}$$

$$\Rightarrow 2(4\pi^2 - \theta^2) = \theta^2 \qquad \Rightarrow \theta^2 = \frac{8}{3}\pi^2.$$

The largest cone has volume $V\left(\pi\sqrt{\dfrac{8}{3}}\right) = \dfrac{2\pi R^3}{9\sqrt{3}}$

cu. units.

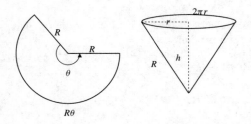

Fig. 4.5.46

Section 4.6 Finding Roots of Equations (page 251)

2. $f(x) = x^2 - 3$, $f'(x) = 2x$.
Newton's formula $x_{n+1} = g(x_n)$, where

$$g(x) = x - \frac{x^2 - 3}{2x} = \frac{x^2 + 3}{2x}.$$

Starting with $x_0 = 1.5$, get $x_4 = x_5 = 1.73205080757$.

4. $f(x) = x^3 + 2x^2 - 2$, $f'(x) = 3x^2 + 4x$.
Newton's formula $x_{n+1} = g(x_n)$, where

$$g(x) = x - \frac{x^3 + 2x^2 - 2}{3x^2 + 4x} = \frac{2x^3 + 2x^2 + 2}{3x^2 + 4x}.$$

Starting with $x_0 = 1.5$, get $x_5 = x_6 = 0.839286755214$.

6. $f(x) = x^3 + 3x^2 - 1$, $f'(x) = 3x^2 + 6x$.
Newton's formula $x_{n+1} = g(x_n)$, where

$$g(x) = x - \frac{x^3 + 3x^2 - 1}{3x^2 + 6x} = \frac{2x^3 + 3x^2 + 1}{3x^2 + 6x}.$$

Because $f(-3) = -1$, $f(-2) = 3$, $f(-1) = 1$, $f(0) = -1$, $f(1) = 3$, there are roots between -3 and -2, between -1 and 0, and between 0 and 1.
Starting with $x_0 = -2.5$, get $x_5 = x_6 = -2.87938524157$.
Starting with $x_0 = -0.5$, get $x_4 = x_5 = -0.652703644666$.
Starting with $x_0 = 0.5$, get $x_4 = x_5 = 0.532088886328$.

8. $f(x) = x^2 - \cos x$, $f'(x) = 2x + \sin x$.
Newton's formula is $x_{n+1} = g(x_n)$, where

$$g(x) = x - \frac{x^2 - \cos x}{2x + \sin x}.$$

The graphs of $\cos x$ and x^2, suggest a root near $x = \pm 0.8$. Starting with $x_0 = 0.8$, get $x_3 = x_4 = 0.824132312303$. The other root is the negative of this one, because $\cos x$ and x^2 are both even functions.

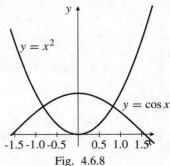

Fig. 4.6.8

10. A graphing calculator shows that the equation

$$(1 + x^2)\sqrt{x} - 1 = 0$$

has a root near $x = 0.6$. Use of a solve routine or Newton's Method gives $x = 0.56984029099806$.

12. Let $f(x) = \dfrac{\sin x}{1 + x^2}$. Since $|f(x)| \leq 1/(1 + x^2) \to 0$ as $x \to \pm\infty$ and $f(0) = 0$, the maximum and minimum values of f will occur at the two critical points of f that are closest to the origin on the right and left, respectively. For CP:

$$0 = f'(x) = \frac{(1 + x^2)\cos x - 2x \sin x}{(1 + x^2)^2}$$
$$0 = (1 + x^2)\cos x - 2x \sin x$$

with $0 < x < \pi$ for the maximum and $-\pi < x < 0$ for the minimum. Solving this equation using a solve routine or Newton's Method starting, say, with $x_0 = 1.5$, we get $x = \pm 0.79801699184239$. The corresponding max and min values of f are ± 0.437414158279.

14. For $x^2 = 0$ we have $x_{n+1} = x_n - (x_n^2/(2x_n)) = x_n/2$. If $x_0 = 1$, then $x_1 = 1/2$, $x_2 = 1/4$, $x_3 = 1/8$.

   a) $x_n = 1/2^n$, by induction.

   b) x_n approximates the root $x = 0$ to within 0.0001 provided $2^n > 10,000$. We need $n \geq 14$ to ensure this.

   c) To ensure that x_n^2 is within 0.0001 of 0 we need $(1/2^n)^2 < 0.0001$, that is, $2^{2n} > 10,000$. We need $n \geq 7$.

   d) Convergence of Newton approximations to the root $x = 0$ of $x^2 = 0$ is slower than usual because the derivative $2x$ of x^2 is zero at the root.

16. Newton's Method formula for $f(x) = x^{1/3}$ is

$$x_{n+1} = x_n - \frac{x_n^{1/3}}{(1/3)x_n^{-2/3}} = x_n - 3x_n = -2x_n.$$

If $x_0 = 1$, then $x_1 = -2$, $x_2 = 4$, $x_3 = -8$, $x_4 = 16$, and, in general, $x_n = (-2)^n$. The successive "approximations" oscillate ever more widely, diverging from the root at $x = 0$.

18. To solve $1 + \frac{1}{4}\sin x = x$, start with $x_0 = 1$ and iterate $x_{n+1} = 1 + \frac{1}{4}\sin x_n$. x_5 and x_6 round to 1.23613.

20. To solve $(x + 9)^{1/3} = x$, start with $x_0 = 2$ and iterate $x_{n+1} = (x_n + 9)^{1/3}$. x_4 and x_5 round to 2.24004.

22. To solve $x^3 + 10x - 10 = 0$, start with $x_0 = 1$ and iterate $x_{n+1} = 1 - \frac{1}{10}x_n^3$. x_7 and x_8 round to 0.92170.

24. Let $g(x) = f(x) - x$ for $a \leq x \leq b$. g is continuous (because f is), and since $a \leq f(x) \leq b$ whenever $a \leq x \leq b$ (by condition (i)), we know that $g(a) \geq 0$ and $g(b) \leq 0$. By the Intermediate-Value Theorem there exists r in $[a, b]$ such that $g(r) = 0$, that is, such that $f(r) = r$.

Section 4.7 Linear Approximations (page 256)

2. $f(x) = x^{-3}$, $f'(x) = -3x^{-4}$, $f(2) = 1/8$, $f'(2) = -3/16$.
Linearization at $x = 2$: $L(x) = \frac{1}{8} - \frac{3}{16}(x - 2)$.

4. $f(x) = \sqrt{3 + x^2}$, $f'(x) = x/\sqrt{3 + x^2}$, $f(1) = 2$, $f'(1) = 1/2$.
Linearization at $x = 1$: $L(x) = 2 + \frac{1}{2}(x - 1)$.

6. $f(x) = x^{-1/2}$, $f'(x) = (-1/2)x^{-3/2}$, $f(4) = 1/2$, $f'(4) = -1/16$.
Linearization at $x = 4$: $L(x) = \frac{1}{2} - \frac{1}{16}(x - 4)$.

8. $f(x) = \cos(2x)$, $f'(x) = -2\sin(2x)$, $f(\pi/3) = -1/2$, $f'(\pi/3) = -\sqrt{3}$.
Linearization at $x = \pi/3$: $L(x) = -\frac{1}{2} - \sqrt{3}\left(x - \frac{\pi}{3}\right)$.

10. $f(x) = \tan x$, $f'(x) = \sec^2 x$, $f(\pi/4) = 1$, $f'(\pi/4) = 2$.
Linearization at $x = \pi/4$: $L(x) = 1 + 2\left(x - \frac{\pi}{4}\right)$.

12. If V and x are the volume and side length of the cube, then $V = x^3$. If $x = 20$ cm and $\Delta V = -12$ cm^3, then

$$-12 = \Delta V \approx \frac{dV}{dx}\Delta x = 3x^2\,\Delta x = 1,200\,\Delta x,$$

so that $\Delta x = -1/100$. The edge length must decrease by about 0.01 cm in to decrease the volume by 12 cm^3.

14. $a = g[R/(R + h)]^2$ implies that

$$\Delta a \approx \frac{da}{dh}\Delta h = gR^2\frac{-2}{(R + h)^3}\Delta h.$$

If $h = 0$ and $\Delta h = 10$ mi, then

$$\Delta a \approx -\frac{20g}{R} = -\frac{20 \times 32}{3960} \approx 0.16 \text{ ft/s}^2.$$

16. Let $f(x) = \sqrt{x}$, then $f'(x) = \frac{1}{2}x^{-1/2}$ and $f''(x) = -\frac{1}{4}x^{-3/2}$. Hence,

$$\sqrt{47} = f(47) \approx f(49) + f'(49)(47 - 49)$$
$$= 7 + \left(\frac{1}{14}\right)(-2) = \frac{48}{7} \approx 6.8571429.$$

Clearly, if $x \geq 36$, then

$$|f''(x)| \leq \frac{1}{4 \times 6^3} = \frac{1}{864} = K.$$

Since $f''(x) < 0$, f is concave down. Therefore, the error $E = \sqrt{47} - \dfrac{48}{7} < 0$ and

$$|E| < \frac{K}{2}(47 - 49)^2 = \frac{1}{432}.$$

Thus,

$$\frac{48}{7} - \frac{1}{432} < \sqrt{47} < \frac{48}{7}$$
$$6.8548 < \sqrt{47} < 6.8572.$$

18. Let $f(x) = \dfrac{1}{x}$, then $f'(x) = -\dfrac{1}{x^2}$ and $f''(x) = \dfrac{2}{x^3}$. Hence,

$$\frac{1}{2.003} = f(2.003) \approx f(2) + f'(2)(0.003)$$
$$= \frac{1}{2} + \left(-\frac{1}{4}\right)(0.003) = 0.49925.$$

If $x \ge 2$, then $|f''(x)| \le \dfrac{2}{8} = \dfrac{1}{4}$. Since $f''(x) > 0$ for $x > 0$, f is concave up. Therefore, the error

$$E = \frac{1}{2.003} - 0.49925 > 0$$

and

$$|E| < \frac{1}{8}(0.003)^2 = 0.000001125.$$

Thus,

$$0.49925 < \frac{1}{2.003} < 0.49925 + 0.000001125$$
$$0.49925 < \frac{1}{2.003} < 0.499251125.$$

20. Let $f(x) = \sin x$, then $f'(x) = \cos x$ and $f''(x) = -\sin x$. Hence,

$$\sin\left(\frac{\pi}{5}\right) = f\left(\frac{\pi}{6} + \frac{\pi}{30}\right) \approx f\left(\frac{\pi}{6}\right) + f'\left(\frac{\pi}{6}\right)\left(\frac{\pi}{30}\right)$$
$$= \frac{1}{2} + \frac{\sqrt{3}}{2}\left(\frac{\pi}{30}\right) \approx 0.5906900.$$

If $x \le \dfrac{\pi}{4}$, then $|f''(x)| \le \dfrac{1}{\sqrt{2}}$. Since $f''(x) < 0$ on $0 < x \le 90°$, f is concave down. Therefore, the error E is negative and

$$|E| < \frac{1}{2\sqrt{2}}\left(\frac{\pi}{30}\right)^2 = 0.0038772.$$

Thus,

$$0.5906900 - 0.0038772 < \sin\left(\frac{\pi}{5}\right) < 0.5906900$$
$$0.5868128 < \sin\left(\frac{\pi}{5}\right) < 0.5906900.$$

22. Let $f(x) = \sin x$, then $f'(x) = \cos x$ and $f''(x) = -\sin x$. The linearization at $x = 30° = \pi/6$ gives

$$\sin(33°) = \sin\left(\frac{\pi}{6} + \frac{\pi}{60}\right)$$
$$\approx \sin\frac{\pi}{6} + \cos\frac{\pi}{6}\left(\frac{\pi}{60}\right)$$
$$= \frac{1}{2} + \frac{\sqrt{3}}{2}\left(\frac{\pi}{60}\right) \approx 0.545345.$$

Since $f''(x) < 0$ between $30°$ and $33°$, the error E in the above approximation is negative: $\sin(33°) < 0.545345$. For $30° \le t \le 33°$, we have

$$|f''(t)| = \sin t \le \sin(33°) < 0.545345.$$

Thus the error satisfies

$$|E| \le \frac{0.545345}{2}\left(\frac{\pi}{60}\right)^2 < 0.000747.$$

Therefore

$$0.545345 - 0.000747 < \sin(33°) < 0.545345$$
$$0.544598 < \sin(33°) < 0.545345.$$

24. From the solution to Exercise 16, the linearization to $f(x) = x^{1/2}$ at $x = 49$ has value at $x = 47$ given by

$$L(47) = f(49) + f'(49)(47 - 49) \approx 6.8571429.$$

Also, $6.8548 \le \sqrt{47} \le 6.8572$, and, since $f''(x) = -1/(4(\sqrt{x})^3)$,

$$\frac{-1}{4(6.8548)^3} \le \frac{-1}{4(\sqrt{47})^3} \le f''(x) \le \frac{-1}{4(7)^3}$$

for $47 \le x \le 49$. Thus, on that interval, $M \le f''(x) \le N$, where $M = -0.000776$ and $N = -0.000729$. By Corollary C,

$$L(47) + \frac{M}{2}(47 - 49)^2 \le f(47) \le L(47) + \frac{N}{2}(47 - 49)^2$$
$$6.855591 \le \sqrt{47} \le 6.855685.$$

Using the midpoint of this interval as a new approximation for $\sqrt{47}$ ensures that the error is no greater than half the length of the interval:

$$\sqrt{47} \approx 6.855638, \quad |\text{error}| \le 0.000047.$$

26. From the solution to Exercise 22, the linearization to $f(x) = \sin x$ at $x = 30° = \pi/6$ has value at $x = 33° = \pi/6 + \pi/60$ given by

$$L(33°) = f(\pi/6) + f'(\pi/6)(\pi/60) \approx 0.545345.$$

Also, $0.544597 \leq \sin(33°) \leq 5.545345$, and, since $f''(x) = -\sin x$,

$$-\sin(33°) \leq f''(x) \leq -\sin(30°)$$

for $30° \leq x \leq 33°$. Thus, on that interval, $M \leq f''(x) \leq N$, where $M = -0.545345$ and $N = -0.5$. By Corollary C,

$$L(33°) + \frac{M}{2}(\pi/60)^2 \leq \sin(33°) \leq L(33°) + \frac{N}{2}(\pi/60)^2$$
$$0.544597 \leq \sin(33°) \leq 0.544660.$$

Using the midpoint of this interval as a new approximation for $\sin(33°)$ ensures that the error is no greater than half the length of the interval:

$$\sin(33°) \approx 0.544629, \quad |\text{error}| \leq 0.000031.$$

28. The linearization of $f(x)$ at $x = 2$ is

$$L(x) = f(2) + f'(2)(x - 2) = 4 - (x - 2).$$

Thus $L(3) = 3$. Also, since $1/(2x) \leq f''(x) \leq 1/x$ for $x > 0$, we have for $2 \leq x \leq 3$, $(1/6) \leq f''(x) \leq (1/2)$. Thus

$$3 + \frac{1}{2}\left(\frac{1}{6}\right)(3 - 2)^2 \leq f(3) \leq 3 + \frac{1}{2}\left(\frac{1}{2}\right)(3 - 2)^2.$$

The best approximation for $f(3)$ is the midpoint of this interval: $f(3) \approx 3\frac{1}{6}$.

30. If $f(\theta) = \sin\theta$, then $f'(\theta) = \cos\theta$ and $f''(\theta) = -\sin\theta$. Since $f(0) = 0$ and $f'(0) = 1$, the linearization of f at $\theta = 0$ is $L(\theta) = 0 + 1(\theta - 0) = \theta$.
If $0 \leq t \leq \theta$, then $f''(t) \leq 0$, so $0 \leq \sin\theta \leq \theta$.
If $0 \geq t \geq \theta$, then $f''(t) \geq 0$, so $0 \geq \sin\theta \geq \theta$.
In either case, $|\sin t| \leq |\sin\theta| \leq |\theta|$ if t is between 0 and θ. Thus the error $E(\theta)$ in the approximation $\sin\theta \approx \theta$ satisfies

$$|E(\theta)| \leq \frac{|\theta|}{2}|\theta|^2 = \frac{|\theta|^3}{2}.$$

If $|\theta| \leq 17° = 17\pi/180$, then

$$\frac{|E(\theta)|}{|\theta|} \leq \frac{1}{2}\left(\frac{17\pi}{180}\right)^2 \approx 0.044.$$

Thus the percentage error is less than 5%.

Section 4.8   Taylor Polynomials   (page 264)

2. If $f(x) = \cos x$, then $f'(x) = -\sin x$, $f''(x) = -\cos x$, and $f'''(x) = \sin x$. In particular, $f(\pi/4) = f'''(\pi/4) = 1/\sqrt{2}$ and $f'(\pi/4) = f''(\pi/4) = -1/\sqrt{2}$. Thus

$$P_3(x) = \frac{1}{\sqrt{2}}\left[1 - \left(x - \frac{\pi}{4}\right) - \frac{1}{2}\left(x - \frac{\pi}{4}\right)^2 + \frac{1}{6}\left(x - \frac{\pi}{4}\right)^3\right].$$

4.
$$\begin{aligned}
f(x) &= \sec x & f(0) &= 1 \\
f'(x) &= \sec x \tan x & f'(0) &= 0 \\
f''(x) &= 2\sec^3 x - \sec x & f''(0) &= 1 \\
f'''(x) &= (6\sec^2 x - 1)\sec x \tan x & f'''(0) &= 0
\end{aligned}$$

Thus $P_3(x) = 1 + (x^2/2)$.

6.
$$\begin{aligned}
f(x) &= (1 - x)^{-1} & f(0) &= 1 \\
f'(x) &= (1 - x)^{-2} & f'(0) &= 1 \\
f''(x) &= 2(1 - x)^{-3} & f''(0) &= 2 \\
f'''(x) &= 3!(1 - x)^{-4} & f'''(0) &= 3! \\
&\vdots & &\vdots \\
f^{(n)}(x) &= n!(1 - x)^{-(n+1)} & f^{(n)}(0) &= n!
\end{aligned}$$

Thus
$$P_n(x) = 1 + x + x^2 + x^3 + \cdots + x^n.$$

8.
$$\begin{aligned}
f(x) &= \sin(2x) & f(\pi/2) &= 0 \\
f'(x) &= 2\cos(2x) & f'(\pi/2) &= -2 \\
f''(x) &= -2^2\sin(2x) & f''(\pi/2) &= 0 \\
f'''(x) &= -2^3\cos(2x) & f'''(\pi/2) &= 2^3 \\
f^{(4)}(x) &= 2^4\sin(2x) = 2^4 f(x) & f^{(4)}(\pi/2) &= 0 \\
f^{(5)}(x) &= 2^4 f'(x) & f^{(5)}(\pi/2) &= -2^5 \\
&\vdots & &\vdots
\end{aligned}$$

Evidently $f^{(2n)}(\pi/2) = 0$ and $f^{(2n-1)}(\pi/2) = (-1)^n 2^{2n-1}$. Thus

$$P_{2n-1}(x) = -2\left(x - \frac{\pi}{2}\right) + \frac{2^3}{3!}\left(x - \frac{\pi}{2}\right)^3 - \frac{2^5}{5!}\left(x - \frac{\pi}{2}\right)^5 + \cdots + (-1$$

10. Since $f(x) = \sqrt{x}$, then $f'(x) = \frac{1}{2}x^{-1/2}$, $f''(x) = -\frac{1}{4}x^{-3/2}$ and $f'''(x) = \frac{3}{8}x^{-5/2}$. Hence,

$$\sqrt{61} \approx f(64) + f'(64)(61 - 64) + \frac{1}{2}f''(64)(61 - 64)^2$$

$$= 8 + \frac{1}{16}(-3) - \frac{1}{2}\left(\frac{1}{2048}\right)(-3)^2 \approx 7.8103027.$$

The error is $R_2 = R_2(f; 64, 61) = \dfrac{f'''(c)}{3!}(61 - 64)^3$ for some c between 61 and 64. Clearly $R_2 < 0$. If $t \geq 49$, and in particular $61 \leq t \leq 64$, then

$$|f'''(t)| \leq \tfrac{3}{8}(49)^{-5/2} = 0.0000223 = K.$$

Hence,

$$|R_2| \leq \frac{K}{3!}|61 - 64|^3 = 0.0001004.$$

Since $R_2 < 0$, therefore,

$$7.8103027 - 0.0001004 < \sqrt{61} < 7.8103027$$
$$7.8102023 < \sqrt{61} < 7.8103027.$$

12. Since $f(x) = \tan^{-1} x$, then

$$f'(x) = \frac{1}{1 + x^2}, \quad f''(x) = \frac{-2x}{(1 + x^2)^2}, \quad f'''(x) = \frac{-2 + 6x^2}{(1 + x^2)^3}.$$

Hence,

$$\tan^{-1}(0.97) \approx f(1) + f'(1)(0.97 - 1) + \tfrac{1}{2}f''(1)(0.97 - 1)^2$$
$$= \frac{\pi}{4} + \frac{1}{2}(-0.03) + \left(-\frac{1}{4}\right)(-0.03)^2$$
$$= 0.7701731.$$

The error is $R_2 = \dfrac{f'''(c)}{3!}(-0.03)^3$ for some c between 0.97 and 1. Note that $R_2 < 0$. If $0.97 \leq t \leq 1$, then

$$|f'''(t)| \leq f'''(1) = \frac{-2 + 6}{(1.97)^3} < 0.5232 = K.$$

Hence,

$$|R_2| \leq \frac{K}{3!}|0.97 - 1|^3 < 0.0000024.$$

Since $R_2 < 0$,

$$0.7701731 - 0.0000024 < \tan^{-1}(0.97) < 0.7701731$$
$$0.7701707 < \tan^{-1}(0.97) < 0.7701731.$$

14. Since $f(x) = \sin x$, then $f'(x) = \cos x$, $f''(x) = -\sin x$ and $f'''(x) = -\cos x$. Hence,

$$\sin(47°) = f\left(\frac{\pi}{4} + \frac{\pi}{90}\right)$$
$$\approx f\left(\frac{\pi}{4}\right) + f'\left(\frac{\pi}{4}\right)\left(\frac{\pi}{90}\right) + \frac{1}{2}f''\left(\frac{\pi}{4}\right)\left(\frac{\pi}{90}\right)^2$$
$$= \frac{1}{\sqrt{2}} + \frac{1}{\sqrt{2}}\left(\frac{\pi}{90}\right) - \frac{1}{2\sqrt{2}}\left(\frac{\pi}{90}\right)^2$$
$$\approx 0.7313587.$$

The error is $R_2 = \dfrac{f'''(c)}{3!}\left(\dfrac{\pi}{90}\right)^3$ for some c between 45° and 47°. Observe that $R_2 < 0$. If $45° \leq t \leq 47°$, then

$$|f'''(t)| \leq |-\cos 45°| = \frac{1}{\sqrt{2}} = K.$$

Hence,

$$|R_2| \leq \frac{K}{3!}\left(\frac{\pi}{90}\right)^3 < 0.0000051.$$

Since $R_2 < 0$, therefore

$$0.7313587 - 0.0000051 < \sin(47°) < 0.7313587$$
$$0.7313536 < \sin(47°) < 0.7313587.$$

16. For $f(x) = \cos x$ we have

$$f'(x) = -\sin x \quad f''(x) = -\cos x \quad f'''(x) = \sin x$$
$$f^{(4)}(x) = \cos x \quad f^{(5)}(x) = -\sin x \quad f^{(6)}(x) = -\cos x.$$

The Taylor's Formula for f with $a = 0$ and $n = 6$ is

$$\cos x = 1 - \frac{x^2}{2!} + \frac{x^4}{4!} - \frac{x^6}{6!} + R_6(f; 0, x)$$

where the Lagrange remainder R_6 is given by

$$R_6 = R_6(f; 0, x) = \frac{f^{(7)}(c)}{7!}x^7 = \frac{\sin c}{7!}x^7,$$

for some c between 0 and x.

18. Given that $f(x) = \dfrac{1}{1 - x}$, then

$$f'(x) = \frac{1}{(1 - x)^2}, \quad f''(x) = \frac{2}{(1 - x)^3}.$$

In general,

$$f^{(n)}(x) = \frac{n!}{(1 - x)^{(n+1)}}.$$

Since $a = 0$, $f^{(n)}(0) = n!$. Hence, for $n = 6$, the Taylor's Formula is

$$\frac{1}{1 - x} = f(0) + \sum_{n=1}^{6} \frac{f^{(n)}(0)}{n!}x^n + R_6(f; 0, x)$$
$$= 1 + x + x^2 + x^3 + x^4 + x^5 + x^6 + R_6(f; 0, x).$$

The Langrange remainder is

$$R_6(f; 0, x) = \frac{f^{(7)}(c)}{7!}x^7 = \frac{x^7}{(1 - c)^8}$$

for some c between 0 and x.

20. Given that $f(x) = \tan x$, then

$$f'(x) = \sec^2 x$$
$$f''(x) = 2 \sec^2 x \tan x$$
$$f^{(3)}(x) = 6 \sec^4 x - 4 \sec^2 x$$
$$f^{(4)}(x) = 8 \tan x (3 \sec^4 x - \sec^2 x).$$

Given that $a = 0$ and $n = 3$, the Taylor's Formula is

$$\tan x = f(0) + f'(0)x + \frac{f''(0)}{2!}x^2 + \frac{f'''(0)}{3!}x^3 + R_3(f; 0, x)$$

$$= x + \frac{2}{3!}x^3 + R_3(f; 0, x)$$

$$= x + \frac{1}{3}x^3 + \frac{2}{15}x^5.$$

The Lagrange remainder is

$$R_3(f; 0, x) = \frac{f^{(4)}(c)}{4!}x^4 = \frac{\tan c(3 \sec^4 X - \sec^2 C)}{3}x^4$$

for some c between 0 and x.

22. For e^u, $P_4(u) = 1 + u + \frac{u^2}{2!} + \frac{u^3}{3!} + \frac{u^4}{4!}$. Let $u = -x^2$. Then for e^{-x^2}:

$$P_8(x) = 1 - x^2 + \frac{x^4}{2!} - \frac{x^6}{3!} + \frac{x^8}{4!}.$$

24. $\sin x = \sin\left(\pi + (x - \pi)\right) = -\sin(x - \pi)$

$$P_5(x) = -(x - \pi) + \frac{(x - \pi)^3}{3!} - \frac{(x - \pi)^5}{5!}$$

26. $\cos(3x - \pi) = -\cos(3x)$

$$P_8(x) = -1 + \frac{3^2 x^2}{2!} - \frac{3^4 x^4}{4!} + \frac{3^6 x^6}{6!} - \frac{3^8 x^8}{8!}.$$

28. Let $t = x - 1$ so that

$$x^3 = (1 + t)^3 = 1 + 3t + 3t^2 + t^3$$
$$= 1_3(x - 1) + 3(x - 1)^2 + (x - 1)^3.$$

Thus the Taylor polynomials for x^3 at $x = 1$ are

$$P_0(x) = 1$$
$$P_1(x) = 1 + 3(x - 1)$$
$$P_1(x) = 1 + 3(x - 1) + 3(x - 1)^2$$
$$P_n(x) = 1 + 3(x - 1) + 3(x - 1)^2 + (x - 1)^3 \quad \text{if } n \geq 3.$$

30. For $\ln(1 + x)$ at $x = 0$ we have

$$P_{2n+1}(x) = x - \frac{x^2}{2} + \frac{x^3}{3} - \cdots + \frac{x^{2n+1}}{2n + 1}.$$

For $\ln(1 - x)$ at $x = 0$ we have

$$P_{2n+1}(x) = -x - \frac{x^2}{2} - \frac{x^3}{3} - \cdots - \frac{x^{2n+1}}{2n + 1}.$$

For $\tanh^{-1} x = \frac{1}{2}\ln(1 + x) - \frac{1}{2}\ln(1 - x)$,

$$P_{2n+1}(x) = x + \frac{x^3}{3} + \frac{x^5}{5} + \cdots + \frac{x^{2n+1}}{2n + 1}.$$

32. In Taylor's Formulas for $f(x) = \sin x$ with $a = 0$, only odd powers of x have nonzero coefficients. Accordingly we can take terms up to order x^{2n+1} but use the remainder after the next term $0x^{2n+2}$. The formula is

$$\sin x = x - \frac{x^3}{3!} + \frac{x^5}{5!} - \cdots + (-1)^n \frac{x^{2n+1}}{(2n + 1)!} + R_{2n+2},$$

where

$$R_{2n+2}(f; 0, x) = (-1)^{n+1}\frac{\cos c}{(2n + 3)!}x^{2n+3}$$

for some c between 0 and x.
In order to use the formula to approximate $\sin(1)$ correctly to 5 decimal places, we need $|R_{2n+2}(f; 0, 1)| < 0.000005$. Since $|\cos c| \leq 1$, it is sufficient to have $1/(2n + 3)! < 0.000005$. $n = 3$ will do since $1/9! \approx 0.000003$. Thus

$$\sin(1) \approx 1 - \frac{1}{3!} + \frac{1}{5!} - \frac{1}{7!} \approx 0.84147$$

correct to five decimal places.

34. $1 - x^{n+1} = (1 - x)(1 + x + x^2 + x^3 + \cdots + x^n)$. Thus

$$\frac{1}{1 - x} = 1 + x + x^2 + x^3 + \cdots + x^n + \frac{x^{n+1}}{1 - x}.$$

If $|x| \leq K < 1$, then $|1 - x| \geq 1 - K > 0$, so

$$\left|\frac{x^{n+1}}{1 - x}\right| \leq \frac{1}{1 - K}|x^{n+1}| = O(x^{n+1})$$

as $x \to 0$. By Theorem 11, the nth-order Maclaurin polynomial for $1/(1 - x)$ must be $P_n(x) = 1 + x + x^2 + x^3 + \cdots + x^n$.

Section 4.9 Indeterminate Forms (page 269)

2. $\displaystyle\lim_{x \to 2} \frac{\ln(2x - 3)}{x^2 - 4} \quad \left[\frac{0}{0}\right]$

$$= \frac{\left(\dfrac{2}{2x - 3}\right)}{2x} = \frac{1}{2}.$$

4. $\displaystyle\lim_{x\to 0}\frac{1-\cos ax}{1-\cos bx}\quad\left[\frac{0}{0}\right]$

$\displaystyle=\lim_{x\to 0}\frac{a\sin ax}{b\sin bx}\quad\left[\frac{0}{0}\right]$

$\displaystyle=\lim_{x\to 0}\frac{a^2\cos ax}{b^2\cos bx}=\frac{a^2}{b^2}.$

6. $\displaystyle\lim_{x\to 1}\frac{x^{1/3}-1}{x^{2/3}-1}\quad\left[\frac{0}{0}\right]$

$\displaystyle=\lim_{x\to 1}\frac{(\frac{1}{3})x^{-2/3}}{(\frac{2}{3})x^{-1/3}}=\frac{1}{2}.$

8. $\displaystyle\lim_{x\to 0}\frac{1-\cos x}{\ln(1+x^2)}\quad\left[\frac{0}{0}\right]$

$\displaystyle=\lim_{x\to 0}\frac{\sin x}{\left(\dfrac{2x}{1+x^2}\right)}$

$\displaystyle=\lim_{x\to 0}(1+x^2)\lim_{x\to 0}\frac{\sin x}{2x}$

$\displaystyle=\lim_{x\to 0}\frac{\cos x}{2}=\frac{1}{2}.$

10. $\displaystyle\lim_{x\to 0}\frac{10^x-e^x}{x}\quad\left[\frac{0}{0}\right]$

$\displaystyle=\lim_{x\to 0}\frac{10^x\ln 10-e^x}{1}=\ln 10-1.$

12. $\displaystyle\lim_{x\to 1}\frac{\ln(ex)-1}{\sin\pi x}\quad\left[\frac{0}{0}\right]$

$\displaystyle=\lim_{x\to 1}\frac{\dfrac{1}{x}}{\pi\cos(\pi x)}=-\frac{1}{\pi}.$

14. $\displaystyle\lim_{x\to 0}\frac{x-\sin x}{x^3}\quad\left[\frac{0}{0}\right]$

$\displaystyle=\lim_{x\to 0}\frac{1-\cos x}{3x^2}\quad\left[\frac{0}{0}\right]$

$\displaystyle=\lim_{x\to 0}\frac{\sin x}{6x}\quad\left[\frac{0}{0}\right]$

$\displaystyle=\lim_{x\to 0}\frac{\cos x}{6}=\frac{1}{6}.$

16. $\displaystyle\lim_{x\to 0}\frac{2-x^2-2\cos x}{x^4}\quad\left[\frac{0}{0}\right]$

$\displaystyle=\lim_{x\to 0}\frac{-2x+2\sin x}{4x^3}\quad\left[\frac{0}{0}\right]$

$\displaystyle=-\frac{1}{2}\lim_{x\to 0}\frac{x-\sin x}{x^3}$

$\displaystyle=-\frac{1}{2}\left(\frac{1}{6}\right)=-\frac{1}{12}\quad\text{(by Exercise 14).}$

18. $\displaystyle\lim_{r\to\pi/2}\frac{\ln\sin r}{\cos r}\quad\left[\frac{0}{0}\right]$

$\displaystyle=\lim_{r\to\pi/2}\frac{\left(\dfrac{\cos r}{\sin r}\right)}{-\sin r}=0.$

20. $\displaystyle\lim_{x\to 1-}\frac{\cos^{-1}x}{x-1}\quad\left[\frac{0}{0}\right]$

$\displaystyle=\lim_{x\to 1-}\frac{-\left(\dfrac{1}{\sqrt{1-x^2}}\right)}{1}=-\infty.$

22. $\displaystyle\lim_{t\to(\pi/2)-}(\sec t-\tan t)\quad[\infty-\infty]$

$\displaystyle=\lim_{t\to(\pi/2)-}\frac{1-\sin t}{\cos t}\quad\left[\frac{0}{0}\right]$

$\displaystyle=\lim_{t\to(\pi/2)-}\frac{-\cos t}{-\sin t}=0.$

24. Since $\displaystyle\lim_{x\to 0+}\sqrt{x}\ln x=\lim_{x\to 0+}\frac{\ln x}{x^{-1/2}}\quad\left[\frac{0}{0}\right]$

$\displaystyle=\lim_{x\to 0+}\frac{\left(\dfrac{1}{x}\right)}{\left(-\dfrac{1}{2}\right)x^{-3/2}}=0,$

hence $\displaystyle\lim_{x\to 0+}x^{\sqrt{x}}$

$\displaystyle=\lim_{x\to 0+}e^{\sqrt{x}\ln x}=e^0=1.$

26. $\displaystyle\lim_{x\to 1+}\left(\frac{x}{x-1}-\frac{1}{\ln x}\right)\quad[\infty-\infty]$

$\displaystyle=\lim_{x\to 1+}\frac{x\ln x-x+1}{(x-1)(\ln x)}\quad\left[\frac{0}{0}\right]$

$\displaystyle=\lim_{x\to 1+}\frac{\ln x}{\ln x+1-\dfrac{1}{x}}\quad\left[\frac{0}{0}\right]$

$\displaystyle=\lim_{x\to 1+}\frac{\dfrac{1}{x}}{\dfrac{1}{x}+\dfrac{1}{x^2}}$

$\displaystyle=\lim_{x\to 1+}\frac{x}{x+1}=\frac{1}{2}.$

28. Let $y = \left(\dfrac{\sin x}{x}\right)^{1/x^2}$.

$$\lim_{x\to 0} \ln y = \lim_{x\to 0} \frac{\ln\left(\dfrac{\sin x}{x}\right)}{x^2} \quad \left[\frac{0}{0}\right]$$

$$= \lim_{x\to 0} \frac{\left(\dfrac{x}{\sin x}\right)\left(\dfrac{x\cos x - \sin x}{x^2}\right)}{2x}$$

$$= \lim_{x\to 0} \frac{x\cos x - \sin x}{2x^2 \sin x} \quad \left[\frac{0}{0}\right]$$

$$= \lim_{x\to 0} \frac{-x\sin x}{4x\sin x + 2x^2\cos x}$$

$$= \lim_{x\to 0} \frac{-\sin x}{4\sin x + 2x\cos x} \quad \left[\frac{0}{0}\right]$$

$$= \lim_{x\to 0} \frac{-\cos x}{6\cos x - 2x\sin x} = -\frac{1}{6}.$$

Thus, $\displaystyle \lim_{x\to 0}\left(\frac{\sin x}{x}\right)^{1/x^2} = e^{-1/6}$.

30. $\displaystyle \lim_{x\to 0+} \frac{\csc x}{\ln x} \quad \left[-\frac{\infty}{\infty}\right]$

$$= \lim_{x\to 0+} \frac{-\csc x \cot x}{\dfrac{1}{x}} \quad \left[-\frac{\infty}{\infty}\right]$$

$$= \lim_{x\to 0+} \frac{-x\cos x}{\sin^2 x} \quad \left[\frac{0}{0}\right]$$

$$= -\left(\lim_{x\to 0+}\cos x\right)\lim_{x\to 0+}\frac{1}{2\sin x\cos x}$$

$$= -\infty.$$

32. Let $y = (1 + \tan x)^{1/x}$.

$$\lim_{x\to 0}\ln y = \lim_{x\to 0}\frac{\ln(1+\tan x)}{x} \quad \left[\frac{0}{0}\right]$$

$$= \lim_{x\to 0}\frac{\sec^2 x}{1 + \tan x} = 1.$$

Thus, $\displaystyle \lim_{x\to 0}(1+\tan x)^{1/x} = e$.

34. $\displaystyle \lim_{h\to 0}\frac{f(x+3h) - 3f(x+h) + 3f(x-h) - f(x-3h)}{h^3}$

$$= \lim_{h\to 0}\frac{3f'(x+3h) - 3f'(x+h) - 3f'(x-h) + 3f'(x-3h)}{3h^2}$$

$$= \lim_{h\to 0}\frac{3f''(x+3h) - f''(x+h) + f''(x-h) - 3f''(x-3h)}{2h}$$

$$= \lim_{h\to 0}\frac{9f'''(x+3h) - f'''(x+h) - f'''(x-h) + 9f'''(x-3h)}{2}$$

$$= 8f'''(x).$$

Review Exercises 4   (page 270)

2. a) Since F must be continuous at $r = R$, we have

$$\frac{mgR^2}{R^2} = mkR, \quad \text{or} \quad k = \frac{g}{R}.$$

b) The rate of change of F as r decreases from R is

$$\left(-\frac{d}{dr}(mkr)\right)\Big|_{r=R} = -mk = -\frac{mg}{R}.$$

The rate of change of F as r increases from R is

$$\left(-\frac{d}{dr}\frac{mgR^2}{r^2}\right)\Big|_{r=R} = -\frac{2mgR^2}{R^3} = -2\frac{mg}{R}.$$

Thus F decreases as r increases from R at twice the rate at which it decreases as r decreases from R.

4. If $pV = 5.0T$, then

$$\frac{dp}{dt}V + p\frac{dV}{dt} = 5.0\frac{dT}{dt}.$$

a) If $T = 400$ K, $dT/dt = 4$ K/min, and $V = 2.0$ m^3, then $dV/dt = 0$, so $dp/dt = 5.0(4)/2.0 = 10$. The pressure is increasing at 10 kPa/min.

b) If $T = 400$ K, $dT/dt = 0$, $V = 2$ m^3, and $dV/dt = 0.05$ m^3/min, then $p = 5.0(400)/2 = 1,000$ kPa, and $2\,dp/dt + 1,000(0.05) = 0$, so $dp/dt = -25$. The pressure is decreasing at 25 kPa/min.

6. If she charges $\$x$ per bicycle, her total profit is $\$P$, where

$$P = (x - 75)N(x) = 4.5 \times 10^6 \frac{x - 75}{x^2}.$$

Evidently $P \le 0$ if $x \le 75$, and $P \to 0$ as $x \to \infty$. P will therefore have a maximum value at a critical point in $(75, \infty)$. For CP:

$$0 = \frac{dP}{dx} = 4.5 \times 10^6 \frac{x^2 - (x-75)2x}{x^4},$$

from which we obtain $x = 150$. She should charge $\$150$ per bicycle and order $N(150) = 200$ of them from the manufacturer.

8.

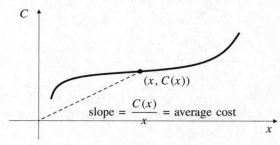

Fig. R-4.8

a) For minimum $C(x)/x$, we need

$$0 = \frac{d}{dx} \frac{C(x)}{x} = \frac{xC'(x) - C(x)}{x^2},$$

so $C'(x) = C(x)/x$; the marginal cost equals the average cost.

b) The line from $(0, 0)$ to $(x, C(x))$ has smallest slope at a value of x which makes it tangent to the graph of $C(x)$. Thus $C'(x) = C(x)/x$, the slope of the line.

c) The line from $(0, 0)$ to $(x, C(x))$ can be tangent to the graph of $C(x)$ at more than one point. Not all such points will provide a minimum value for the average cost. (In the figure, one such line will make the average cost maximum.)

10. If x more trees are planted, the yield of apples will be

$$Y = (60 + x)(800 - 10x)$$
$$= 10(60 + x)(80 - x)$$
$$= 10(4,800 + 20x - x^2).$$

This is a quadratic expression with graph opening downward; its maximum occurs at a CP:

$$0 = \frac{dY}{dx} = 10(20 - 2x) = 20(10 - x).$$

Thus 10 more trees should be planted to maximize the yield.

12. The narrowest hallway in which the table can be turned horizontally through $180°$ has width equal to twice the greatest distance from the origin (the centre of the table) to the curve $x^2 + y^4 = 1/8$ (the edge of the table). We maximize the square of this distance, which we express as a function of y:

$$S(y) = x^2 + y^2 = y^2 + \frac{1}{8} - y^4, \quad (0 \le y \le (1/8)^{1/4}).$$

Note that $S(0) = 1/8$ and $S((1/8)^{1/4}) = 1/\sqrt{8} > S(0)$. For CP:

$$0 = \frac{dS}{dy} = 2y - 4y^3 = 2y(1 - 2y^2).$$

The CPs are given by $y = 0$ (already considered), and $y^2 = 1/2$, where $S(y) = 3/8$. Since $3/8 > 1/\sqrt{8}$, this is the maximum value of S. The hallway must therefore be at least $2\sqrt{3/8} \approx 1.225$ m wide.

14.

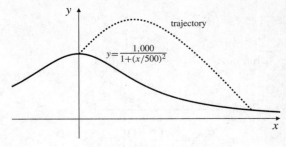

Fig. R-4.14

If the origin is at sea level under the launch point, and $x(t)$ and $y(t)$ are the horizontal and vertical coordinates of the cannon ball's position at time t s after it is fired, then

$$\frac{d^2x}{dt^2} = 0, \qquad \frac{d^2y}{dt^2} = -32.$$

At $t = 0$, we have $dx/dt = dy/dt = 200/\sqrt{2}$, so

$$\frac{dx}{dt} = \frac{200}{\sqrt{2}}, \qquad \frac{dy}{dt} = -32t + \frac{200}{\sqrt{2}}.$$

At $t = 0$, we have $x = 0$ and $y = 1,000$. Thus the position of the ball at time t is given by

$$x = \frac{200t}{\sqrt{2}}, \qquad y = -16t^2 + \frac{200t}{\sqrt{2}} + 1,000.$$

We can obtain the Cartesian equation for the path of the cannon ball by solving the first equation for t and substituting into the second equation:

$$y = -16\frac{2x^2}{200^2} + x + 1,000.$$

The cannon ball strikes the ground when

$$-16\frac{2x^2}{200^2} + x + 1,000 = \frac{1,000}{1 + (x/500)^2}.$$

Graphing both sides of this equation suggests a solution near $x = 1,900$. Newton's Method or a solve routine then gives $x \approx 1,873$. The horizontal range is about 1,873 ft.

16. $$\sin^2 x = \frac{1}{2}\left(1 - \cos(2x)\right)$$
$$= \frac{1}{2}\left[1 - \left(1 - \frac{2^2x^2}{2!} + \frac{2^4x^4}{4!} - \frac{2^6x^6}{6!} + O(x^8)\right)\right]$$
$$= x^2 - \frac{x^4}{3} + \frac{2x^6}{45} + O(x^8)$$
$$\lim_{x \to 0} \frac{3\sin^2 x - 3x^2 + x^4}{x^6}$$
$$= \lim_{x \to 0} \frac{3x^2 - x^4 + \frac{2}{15}x^6 + O(x^8) - 3x^2 - x^4}{x^6}$$
$$= \lim_{x \to 0} \frac{2}{15} + O(x^2) = \frac{2}{15}.$$

18. The second approximation x_1 is the x-intercept of the tangent to $y = f(x)$ at $x = x_0 = 2$; it is the x-intercept of the line $2y = 10x - 19$. Thus $x_1 = 19/10 = 1.9$.

20. The square of the distance from $(2, 0)$ to $(x, \ln x)$ is $S(x) = (x - 2)^2 + (\ln x)^2$, for $x > 0$. Since $S(x) \to \infty$ as $x \to \infty$ or $x \to 0+$, the minimum value of $S(x)$ will occur at a critical point. For CP:

$$0 = S'(x) = 2\left(x - 2 + \frac{\ln x}{x}\right).$$

We solve this equation using a TI-85 solve routine; $x \approx 1.6895797$. The minimum distance from the origin to $y = e^x$ is $\sqrt{S(x)} \approx 0.6094586$.

Challenging Problems 4   (page 272)

2. Let the speed of the tank be v where $v = \dfrac{dy}{dt} = ky$. Thus, $y = Ce^{kt}$. Given that at $t = 0$, $y = 4$, then $4 = y(0) = C$. Also given that at $t = 10$, $y = 2$, thus,

$$2 = y(10) = 4e^{10k} \Rightarrow k = -\tfrac{1}{10}\ln 2.$$

Hence, $y = 4e^{(-\frac{1}{10}\ln 2)t}$ and $v = \dfrac{dy}{dt} = (-\dfrac{1}{10}\ln 2)y$. The slope of the curve $xy = 1$ is $m = \dfrac{dy}{dx} = -\dfrac{1}{x^2}$. Thus, the equation of the tangent line at the point $\left(\dfrac{1}{y_0}, y_0\right)$ is

$$y = y_0 - \frac{1}{\left(\frac{1}{y_0}\right)^2}\left(x - \frac{1}{y_0}\right), \quad \text{i.e.,} \quad y = 2y_0 - xy_0^2.$$

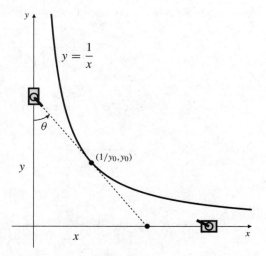

Fig. C-4.2

Hence, the x-intercept is $x = \dfrac{2}{y_0}$ and the y-intercept is $y = 2y_0$. Let θ be the angle between the gun and the y-axis. We have

$$\tan\theta = \frac{x}{y} = \frac{\left(\dfrac{2}{y_0}\right)}{2y_0} = \frac{1}{y_0^2} = \frac{4}{y^2}$$

$$\Rightarrow \quad \sec^2\theta \frac{d\theta}{dt} = \frac{-8}{y^3}\frac{dy}{dt}.$$

Now

$$\sec^2\theta = 1 + \tan^2\theta = 1 + \frac{16}{y^4} = \frac{y^4 + 16}{y^4},$$

so

$$\frac{d\theta}{dt} = -\frac{8y}{y^4 + 16}\frac{dy}{dt} = -\frac{8ky^2}{y^4 + 16}.$$

The maximum value of $\dfrac{y^2}{y^4 + 16}$ occurs at a critical point:

$$0 = \frac{(y^4 + 16)2y - y^2(4y^3)}{(y^4 + 16)^2}$$

$$\Leftrightarrow \quad 2y^5 = 32y,$$

or $y = 2$. Therefore the maximum rate of rotation of the gun turret must be

$$-8k\frac{2^2}{2^4 + 16} = -k = \frac{1}{10}\ln 2 \approx 0.0693 \text{ rad/m},$$

and occurs when your tank is 2 km from the origin.

4. $P = 2\pi\sqrt{L/g} = 2\pi L^{1/2}g^{-1/2}$.

a) If L remains constant, then

$$\Delta P \approx \frac{dP}{dg}\Delta g = -\pi L^{1/2}g^{-3/2}\Delta g$$

$$\frac{\Delta P}{P} \approx \frac{-\pi L^{1/2}g^{-3/2}}{2\pi L^{1/2}g^{-1/2}}\Delta g = -\frac{1}{2}\frac{\Delta g}{g}.$$

If g increases by 1%, then $\Delta g/g = 1/100$, and $\Delta P/P = -1/200$. Thus P decreases by 0.5%.

b) If g remains constant, then

$$\Delta P \approx \frac{dP}{dL}\Delta L = \pi L^{-1/2}g^{-1/2}\Delta L$$

$$\frac{\Delta P}{P} \approx \frac{\pi L^{-1/2}g^{-1/2}}{2\pi L^{1/2}g^{-1/2}}\Delta L = \frac{1}{2}\frac{\Delta L}{L}.$$

If L increases by 2%, then $\Delta L/L = 2/100$, and $\Delta P/P = 1/100$. Thus P increases by 1%.

6. If the depth of liquid in the tank at time t is $y(t)$, then the surface of the liquid has radius $r(t) = Ry(t)/H$, and the volume of liquid in the tank at that time is

$$V(t) = \frac{\pi}{3}\left(\frac{Ry(t)}{H}\right)^2 y(t) = \frac{\pi R^2}{3H^2}\left(y(t)\right)^3.$$

By Torricelli's law, $dV/dt = -k\sqrt{y}$. Thus

$$\frac{\pi R^2}{3H^2}3y^2\frac{dy}{dt} = \frac{dV}{dt} = -k\sqrt{y},$$

or, $dy/dt = -k_1 y^{-3/2}$, where $k_1 = kH^2/(\pi R^2)$.

If $y(t) = y_0\left(1 - \frac{t}{T}\right)^{2/5}$, then $y(0) = y_0$, $y(T) = 0$, and

$$\frac{dy}{dt} = \frac{2}{5}y_0\left(1 - \frac{t}{T}\right)^{-3/5}\left(-\frac{1}{T}\right) = -k_1 y^{-3/2},$$

where $k_1 = 2y_0/(5T)$. Thus this function $y(t)$ satisfies the conditions of the problem.

8. The slope of $y = x^3 + ax^2 + bx + c$ is

$$y' = 3x^2 + 2ax + b,$$

which $\to \infty$ as $x \to \pm\infty$. The quadratic expression y' takes each of its values at two different points except its minimum value, which is achieved only at one point given by $y'' = 6x + 2a = 0$. Thus the tangent to the cubic at $x = -a/3$ is not parallel to any other tangent. This tangent has equation

$$y = -\frac{a^3}{27} + \frac{a^3}{9} - \frac{ab}{3} + c$$
$$+ \left(\frac{a^2}{3} - \frac{2a^2}{3} + b\right)\left(x + \frac{a}{3}\right)$$
$$= -\frac{a^3}{27} + c + \left(b - \frac{a^2}{3}\right)x.$$

10.

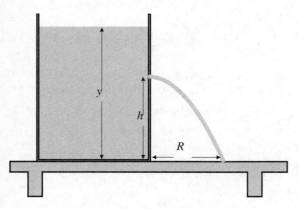

Fig. C-4.10

a) Let the origin be at the point on the table directly under the hole. If a water particle leaves the tank with horizontal velocity v, then its position $(X(t), Y(t))$, t seconds later, is given by

$$\frac{d^2X}{dt^2} = 0 \qquad \frac{d^2Y}{dt^2} = -g$$
$$\frac{dX}{dt} = v \qquad \frac{dY}{dt} = -gt$$
$$X = vt \qquad Y = -\frac{1}{2}gt^2 + h.$$

The range R of the particle (i.e., of the spurt) is the value of X when $Y = 0$, that is, at time $t = \sqrt{2h/g}$. Thus $R = v\sqrt{2h/g}$.

b) Since $v = k\sqrt{y - h}$, the range R is a function of y, the depth of water in the tank.

$$R = k\sqrt{\frac{2}{g}}\sqrt{h(y - h)}.$$

For a given depth y, R will be maximum if $h(y - h)$ is maximum. This occurs at the critical point $h = y/2$ of the quadratic $Q(h) = h(y - h)$.

c) By the result of part (c) of Problem 3 (with y replaced by $y - h$, the height of the surface of the water above the drain in the current problem), we have

$$y(t) - h = (y_0 - h)\left(1 - \frac{t}{T}\right)^2, \qquad \text{for } 0 \le t \le T.$$

As shown above, the range of the spurt at time t is

$$R(t) = k\sqrt{\frac{2}{g}}\sqrt{h\left(y(t) - h\right)}.$$

Since $R = R_0$ when $y = y_0$, we have

$$k = \frac{R_0}{\sqrt{\frac{2}{g}}\sqrt{h(y_0 - h)}}.$$

Therefore $R(t) = R_0\dfrac{\sqrt{h\left(y(t) - h\right)}}{\sqrt{h(y_0 - h)}} = R_0\left(1 - \dfrac{t}{T}\right).$

CHAPTER 5. INTEGRATION

Section 5.1   Sums and Sigma Notation
(page 278)

2. $\displaystyle\sum_{j=1}^{100} \frac{j}{j+1} = \frac{1}{2} + \frac{2}{3} + \frac{3}{4} + \cdots + \frac{100}{101}$

4. $\displaystyle\sum_{i=0}^{n-1} \frac{(-1)^i}{i+1} = 1 - \frac{1}{2} + \frac{1}{3} - \cdots + \frac{(-1)^{n-1}}{n}$

6. $\displaystyle\sum_{j=1}^{n} \frac{j^2}{n^3} = \frac{1}{n^3} + \frac{4}{n^3} + \frac{9}{n^3} + \cdots + \frac{n^2}{n^3}$

8. $2 + 2 + 2 + \cdots + 2$ (200 terms) equals $\displaystyle\sum_{i=1}^{200} 2$

10. $1 + 2x + 3x^2 + 4x^3 + \cdots + 100x^{99} = \displaystyle\sum_{i=1}^{100} i x^{i-1}$

12. $1 - x + x^2 - x^3 + \cdots + x^{2n} = \displaystyle\sum_{i=0}^{2n} (-1)^i x^i$

14. $\dfrac{1}{2} + \dfrac{2}{4} + \dfrac{3}{8} + \dfrac{4}{16} + \cdots + \dfrac{n}{2^n} = \displaystyle\sum_{i=1}^{n} \frac{i}{2^i}$

16. $\displaystyle\sum_{k=-5}^{m} \frac{1}{k^2+1} = \sum_{i=1}^{m+6} \frac{1}{((i-6)^2+1)}$

18. $\displaystyle\sum_{j=1}^{1,000} (2j+3) = \frac{2(1,000)(1,001)}{2} + 3,000 = 1,004,000$

20. $\displaystyle\sum_{i=1}^{n} (2^i - i^2) = 2^{n+1} - 2 - \frac{1}{6} n(n+1)(2n+1)$

22. $\displaystyle\sum_{i=0}^{n} e^{i/n} = \frac{e^{(n+1)/n} - 1}{e^{1/n} - 1}$

24. $1 + x + x^2 + \cdots + x^n = \begin{cases} \dfrac{1-x^{n+1}}{1-x} & \text{if } x \neq 1 \\ n+1 & \text{if } x = 1 \end{cases}$

26. Let $f(x) = 1 + x + x^2 + \cdots + x^{100} = \dfrac{x^{101} - 1}{x - 1}$ if $x \neq 1$. Then

$$f'(x) = 1 + 2x + 3x^2 + \cdots + 100x^{99}$$
$$= \frac{d}{dx} \frac{x^{101} - 1}{x - 1} = \frac{100x^{101} - 101x^{100} + 1}{(x-1)^2}.$$

28. Let $s = \dfrac{1}{2} + \dfrac{2}{4} + \dfrac{3}{8} + \cdots + \dfrac{n}{2^n}$. Then

$$\frac{s}{2} = \frac{1}{4} + \frac{2}{8} + \frac{3}{16} + \cdots + \frac{n}{2^{n+1}}.$$

Subtracting these two sums, we get

$$\frac{s}{2} = \frac{1}{2} + \frac{1}{4} + \frac{1}{8} + \cdots + \frac{1}{2^n} - \frac{n}{2^{n+1}}$$
$$= \frac{1}{2} \frac{1 - (1/2^n)}{1 - (1/2)} - \frac{n}{2^{n+1}}$$
$$= 1 - \frac{n+2}{2^{n+1}}.$$

Thus $s = 2 + (n+2)/2^n$.

30. $\displaystyle\sum_{n=1}^{10} (n^4 - (n-1)^4 = 10^4 - 0^4 = 10,000$

32. $\displaystyle\sum_{i=m}^{2m} \left(\frac{1}{i} - \frac{1}{i+1} \right) = \frac{1}{m} - \frac{1}{2m+1} = \frac{m+1}{m(2m+1)}$

34. The number of small shaded squares is $1 + 2 + \cdots + n$. Since each has area 1, the total area shaded is $\sum_{i=1}^{n} i$. But this area consists of a large right-angled triangle of area $n^2/2$ (below the diagonal), and n small triangles (above the diagonal) each of area 1/2. Equating these areas, we get

$$\sum_{i=1}^{n} i = \frac{n^2}{2} + n\frac{1}{2} = \frac{n(n+1)}{2}.$$

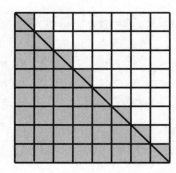

Fig. 5.1.34

36. The formula $\sum_{i=1}^{n} i = n(n+1)/2$ holds for $n = 1$, since it says $1 = 1$ in this case. Now assume that it holds for $n = $ some number $k \geq 1$; that is, $\sum_{i=1}^{k} i = k(k+1)/2$. Then for $n = k + 1$, we have

$$\sum_{i=1}^{k+1} i = \sum_{i=1}^{k} i + (k+1) = \frac{k(k+1)}{2} + (k+1) = \frac{(k+1)(k+2)}{2}.$$

Thus the formula also holds for $n = k + 1$. By induction, it holds for all positive integers n.

38. The formula $\sum_{i=1}^{n} r^{i-1} = (r^n - 1)/(r - 1)$ (for $r \neq 1$) holds for $n = 1$, since it says $1 = 1$ in this case. Now assume that it holds for $n =$ some number $k \geq 1$; that is, $\sum_{i=1}^{k} r^{i-1} = (r^k - 1)/(r - 1)$. Then for $n = k + 1$, we have

$$\sum_{i=1}^{k+1} r^{i-1} = \sum_{i=1}^{k} r^{i-1} + r^k = \frac{r^k - 1}{r - 1} + r^k = \frac{r^{k+1} - 1}{r - 1}.$$

Thus the formula also holds for $n = k + 1$. By induction, it holds for all positive integers n.

40. To show that

$$\sum_{j=1}^{n} j^3 = 1^3 + 2^3 + 3^3 + \cdots + n^3 = \frac{n^2(n + 1)^2}{4},$$

we write n copies of the identity

$$(k + 1)^4 - k^4 = 4k^3 + 6k^2 + 4k + 1,$$

one for each k from 1 to n:

$$2^4 - 1^4 = 4(1)^3 + 6(1)^2 + 4(1) + 1$$
$$3^4 - 2^4 = 4(2)^3 + 6(2)^2 + 4(2) + 1$$
$$4^4 - 3^4 = 4(3)^3 + 6(3)^2 + 4(3) + 1$$
$$\vdots$$
$$(n + 1)^4 - n^4 = 4(n)^3 + 6(n)^2 + 4(n) + 1.$$

Adding the left and right sides of these formulas we get

$$(n + 1)^4 - 1^4 = 4 \sum_{j=1}^{n} j^3 + 6 \sum_{j=1}^{n} j^2 + 4 \sum_{j=1}^{n} j + n$$

$$= 4 \sum_{j=1}^{n} j^3 + \frac{6n(n + 1)(2n + 1)}{6} + \frac{4n(n + 1)}{2} + n.$$

Hence,

$$4 \sum_{j=1}^{n} j^3 = (n + 1)^4 - 1 - n(n + 1)(2n + 1) - 2n(n + 1) - n$$

$$= n^2(n + 1)^2$$

so $\sum_{j=1}^{n} j^3 = \frac{n^2(n + 1)^2}{4}.$

42. To find $\sum_{j=1}^{n} j^4 = 1^4 + 2^4 + 3^4 + \cdots + n^4$, we write n copies of the identity

$$(k + 1)^5 - k^5 = 5k^4 + 10k^3 + 10k^2 + 5k + 1,$$

one for each k from 1 to n:

$$2^5 - 1^5 = 5(1)^4 + 10(1)^3 + 10(1)^2 + 5(1) + 1$$
$$3^5 - 2^5 = 5(2)^4 + 10(2)^3 + 10(2)^2 + 5(2) + 1$$
$$4^5 - 3^5 = 5(3)^4 + 10(3)^3 + 10(3)^2 + 5(3) + 1$$
$$\vdots$$
$$(n + 1)^5 - n^5 = 5(n)^4 + 10(n)^3 + 10(n)^2 + 5(n) + 1.$$

Adding the left and right sides of these formulas we get

$$(n + 1)^5 - 1^5 = 5 \sum_{j=1}^{n} j^4 + 10 \sum_{j=1}^{n} j^3 + 10 \sum_{j=1}^{n} j^2 + 5 \sum_{j=1}^{n} j + n.$$

Substituting the known formulas for all the sums except $\sum_{j=1}^{n} j^4$, and solving for this quantity, gives

$$\sum_{j=1}^{n} j^4 = \frac{n(n + 1)(2n + 1)(3n^2 + 3n - 1)}{30}.$$

Of course we got Maple to do the donkey work!

Section 5.2 Areas as Limits of Sums (page 284)

2. This is similar to #1; the rectangles now have width $3/n$ and the ith has height $2(3i/n)+1$, the value of $2x+1$ at $x = 3i/n$. The area is

$$A = \lim_{n \to \infty} \sum_{i=1}^{n} \frac{3}{n} \left(2\frac{3i}{n} + 1 \right)$$

$$= \lim_{n \to \infty} \frac{18}{n^2} \sum_{i=1}^{n} i + \frac{3}{n}n$$

$$= \lim_{n \to \infty} \frac{18}{n^2} \frac{n(n + 1)}{2} + 3 = 9 + 3 = 12 \text{sq. units.}$$

4. This is similar to #1; the rectangles have width $(2 - (-1))/n = 3/n$ and the ith has height the value of $3x + 4$ at $x = -1 + (3i/n)$. The area is

$$A = \lim_{n \to \infty} \sum_{i=1}^{n} \frac{3}{n} \left(-3 + 3\frac{3i}{n} + 4 \right)$$

$$= \lim_{n \to \infty} \frac{27}{n^2} \sum_{i=1}^{n} i + \frac{3}{n}n$$

$$= \lim_{n \to \infty} \frac{27}{n^2} \frac{n(n + 1)}{2} + 3 = \frac{27}{2} + 3 = \frac{33}{2} \text{sq. units.}$$

6. Divide $[0, a]$ into n equal subintervals of length $\Delta x = \dfrac{a}{n}$ by points $x_i = \dfrac{ia}{n}$, $(0 \leq i \leq n)$. Then

$$S_n = \sum_{i=1}^{n} \left(\frac{a}{n}\right) \left[\left(\frac{ia}{n}\right)^2 + 1 \right]$$

$$= \left(\frac{a}{n}\right)^3 \sum_{i=1}^{n} i^2 + \frac{a}{n} \sum_{i=1}^{n} (1)$$

(Use Theorem 1(a) and 1(c).)

$$= \left(\frac{a}{n}\right)^3 \frac{n(n+1)(2n+1)}{6} + \frac{a}{n}(n)$$

$$= \frac{a^3}{6} \frac{(n+1)(2n+1)}{n^2} + a.$$

$$\text{Area} = \lim_{n \to \infty} S_n = \frac{a^3}{3} + a \text{ sq. units.}$$

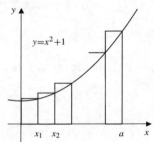

Fig. 5.2.6

8.

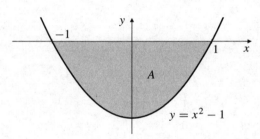

Fig. 5.2.8

The region in question lies between $x = -1$ and $x = 1$ and is symmetric about the y-axis. We can therefore double the area between $x = 0$ and $x = 1$. If we divide this interval into n equal subintervals of width $1/n$ and use the distance $0 - (x^2 - 1) = 1 - x^2$ between $y = 0$ and $y = x^2 - 1$ for the heights of rectangles, we find that the required area is

$$A = 2 \lim_{n \to \infty} \sum_{i=1}^{n} \frac{1}{n} \left(1 - \frac{i^2}{n^2} \right)$$

$$= 2 \lim_{n \to \infty} \sum_{i=1}^{n} \left(\frac{1}{n} - \frac{i^2}{n^3} \right)$$

$$= 2 \lim_{n \to \infty} \left(\frac{n}{n} - \frac{n(n+1)(2n+1)}{6n^3} \right) = 2 - \frac{4}{6} = \frac{4}{3} \text{ sq. units.}$$

10.

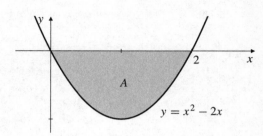

Fig. 5.2.10

The height of the region at position x is $0 - (x^2 - 2x) = 2x - x^2$. The "base" is an interval of length 2, so we approximate using n rectangles of width $2/n$. The shaded area is

$$A = \lim_{n \to \infty} \sum_{i=1}^{n} \frac{2}{n} \left(2\frac{2i}{n} - \frac{4i^2}{n^2} \right)$$

$$= \lim_{n \to \infty} \sum_{i=1}^{n} \left(\frac{8i}{n^2} - \frac{8i^2}{n^3} \right)$$

$$= \lim_{n \to \infty} \left(\frac{8}{n^2} \frac{n(n+1)}{2} - \frac{8}{n^3} \frac{n(n+1)(2n+1)}{6} \right)$$

$$= 4 - \frac{8}{3} = \frac{4}{3} \text{ sq. units.}$$

12. Divide $[0, b]$ into n equal subintervals of length $\Delta x = \dfrac{b}{n}$ by points $x_i = \dfrac{ib}{n}$, $(0 \leq i \leq n)$. Then

$$S_n = \sum_{i=1}^{n} \frac{b}{n} \left(e^{(ib/n)} \right) = \frac{b}{n} \sum_{i=1}^{n} \left(e^{(b/n)} \right)^i$$

$$= \frac{b}{n} e^{(b/n)} \sum_{i=1}^{n} \left(e^{(b/n)} \right)^{i-1} \quad \text{(Use Thm. 6.1.2(d).)}$$

$$= \frac{b}{n} e^{(b/n)} \frac{e^{(b/n)n} - 1}{e^{(b/n)} - 1}$$

$$= \frac{b}{n} e^{(b/n)} \frac{e^b - 1}{e^{(b/n)} - 1}.$$

Let $r = \dfrac{b}{n}$.

$$\text{Area} = \lim_{n \to \infty} S_n = (e^b - 1) \lim_{r \to 0+} e^r \lim_{r \to 0+} \frac{r}{e^r - 1} \quad \left[\frac{0}{0} \right]$$

$$= (e^b - 1)(1) \lim_{r \to 0+} \frac{1}{e^r} = e^b - 1 \text{ sq. units.}$$

14. $\text{Area} = \lim_{n \to \infty} \dfrac{b}{n}\left[\left(\dfrac{b}{n}\right)^3 + \left(\dfrac{2b}{n}\right)^3 + \cdots + \left(\dfrac{nb}{n}\right)^3\right]$

$= \lim_{n \to \infty} \dfrac{b^4}{n^4}(1^3 + 2^3 + 3^3 + \cdots + n^3)$

$= \lim_{n \to \infty} \dfrac{b^4}{n^4} \cdot \dfrac{n^2(n+1)^2}{4} = \dfrac{b^4}{4}\, \text{sq. units.}$

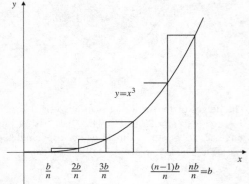

Fig. 5.2.14

16.

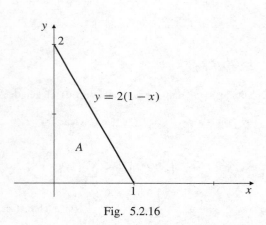

Fig. 5.2.16

$s_n = \displaystyle\sum_{i=1}^{n} \dfrac{2}{n}\left(1 - \dfrac{i}{n}\right)$ represents a sum of areas of n
rectangles each of width $1/n$ and having heights equal to
the height to the graph $y = 2(1-x)$ at the points
$x = i/n$. Thus $\lim_{n\to\infty} S_n$ is the area A of the triangle
in the figure above, and therefore has the value 1.

18.

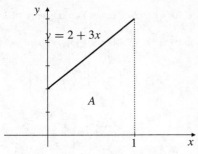

Fig. 5.2.18

$s_n = \displaystyle\sum_{i=1}^{n} \dfrac{2n + 3i}{n^2} = \sum_{i=1}^{n} \dfrac{1}{n}\left(2 + \dfrac{3i}{n}\right)$ represents a sum
of areas of n rectangles each of width $1/n$ and having
heights equal to the height to the graph $y = 2 + 3x$ at
the points $x = i/n$. Thus $\lim_{n\to\infty} S_n$ is the area of the
trapezoid in the figure above, and has the value
$1(2+5)/2 = 7/2$.

Section 5.3   The Definite Integral
(page 290)

2. $f(x) = x^2$ on $[0, 4]$, $n = 4$.

$L(f, P_4) = \left(\dfrac{4-0}{4}\right)[0 + (1)^2 + (2)^2 + (3)^2] = 14.$

$U(f, P_4) = \left(\dfrac{4-0}{4}\right)[(1)^2 + (2)^2 + (3)^2 + (4)^2] = 30.$

4. $f(x) = \ln x$ on $[1, 2]$, $n = 5$.

$L(f, P_5) = \left(\dfrac{2-1}{5}\right)\left[\ln 1 + \ln\dfrac{6}{5} + \ln\dfrac{7}{5} + \ln\dfrac{8}{5} + \ln\dfrac{9}{5}\right]$

$\approx 0.3153168.$

$U(f, P_5) = \left(\dfrac{2-1}{5}\right)\left[\ln\dfrac{6}{5} + \ln\dfrac{7}{5} + \ln\dfrac{8}{5} + \ln\dfrac{9}{5} + \ln 2\right]$

$\approx 0.4539462.$

6. $f(x) = \cos x$ on $[0, 2\pi]$, $n = 4$.

$L(f, P_4) = \left(\dfrac{2\pi}{4}\right)\left[\cos\dfrac{\pi}{2} + \cos\pi + \cos\pi + \cos\dfrac{3\pi}{2}\right] = -\pi$

$U(f, P_4) = \left(\dfrac{2\pi}{4}\right)\left[\cos 0 + \cos\dfrac{\pi}{2} + \cos\dfrac{3\pi}{2} + \cos 2\pi\right] = \pi.$

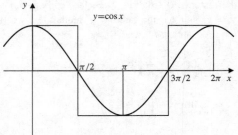

Fig. 5.3.6

8. $f(x) = 1 - x$ on $[0, 2]$. $P_n = \left\{0, \frac{2}{n}, \frac{4}{n}, \dots, \frac{2n-2}{n}, \frac{2n}{n}\right\}$.
We have

$$L(f, P_n) = \frac{2}{n}\left(\left(1 - \frac{2}{n}\right) + \left(1 - \frac{4}{n}\right) + \cdots + \left(1 - \frac{2n}{n}\right)\right)$$

$$= \frac{2}{n}n - \frac{4}{n^2}\sum_{i=1}^{n} i$$

$$= 2 - \frac{4}{n^2}\frac{n(n+1)}{2} = -\frac{2}{n} \to 0 \text{ as } n \to \infty,$$

$$U(f, P_n) = \frac{2}{n}\left(\left(1 - \frac{0}{n}\right) + \left(1 - \frac{2}{n}\right) + \cdots + \left(1 - \frac{2n-2}{n}\right)\right)$$

$$= \frac{2}{n}n - \frac{4}{n^2}\sum_{i=0}^{n-1} i$$

$$= 2 - \frac{4}{n^2}\frac{(n-1)n}{2} = \frac{2}{n} \to 0 \text{ as } n \to \infty.$$

Thus $\int_0^2 (1 - x)\, dx = 0$.

10. $f(x) = e^x$ on $[0, 3]$. $P_n = \left\{0, \frac{3}{n}, \frac{6}{n}, \dots, \frac{3n-3}{n}, \frac{3n}{n}\right\}$. We have (using the result of Exercise 51 (or 52) of Section 6.1)

$$L(f, P_n) = \frac{3}{n}\left(e^{0/n} + e^{3/n} + e^{6/n} + \cdots + e^{3(n-1)/n}\right)$$

$$= \frac{3}{n}\frac{e^{3n/n} - 1}{e^{3/n} - 1} = \frac{3(e^3 - 1)}{n(e^{3/n} - 1)},$$

$$U(f, P_n) = \frac{3}{n}\left(e^{3/n} + e^{6/n} + e^{9/n} + \cdots + e^{3n/n}\right) = e^{3/n}L(f, P_n).$$

By l'Hôpital's Rule,

$$\lim_{n\to\infty} n(e^{3/n} - 1) = \lim_{n\to\infty}\frac{e^{3/n} - 1}{1/n}$$

$$= \lim_{n\to\infty}\frac{e^{3/n}(-3/n^2)}{-1/n^2} = \lim_{n\to\infty}\frac{3e^{3/n}}{1} = 3.$$

Thus

$$\lim_{n\to\infty} L(f, P_n) = \lim_{n\to\infty} U(f, P_n) = e^3 - 1 = \int_0^3 e^x\, dx.$$

12. $\displaystyle\lim_{n\to\infty}\sum_{i=1}^{n}\frac{1}{n}\sqrt{\frac{i-1}{n}} = \int_0^1 \sqrt{x}\, dx$

14. $\displaystyle\lim_{n\to\infty}\sum_{i=1}^{n}\frac{2}{n}\ln\left(1 + \frac{2i}{n}\right) = \int_0^2 \ln(1+x)\, dx$

16. $\displaystyle\lim_{n\to\infty}\sum_{i=1}^{n}\frac{n}{n^2 + i^2} = \lim_{n\to\infty}\sum_{i=1}^{n}\frac{1}{n}\frac{1}{1 + (i/n)^2} = \int_0^1 \frac{dx}{1 + x^2}$

18. $P = \{x_0 < x_1 < \cdots < x_n\}$,
$P' = \{x_0 < x_1 < \cdots < x_{j-1} < x' < x_j < \cdots < x_n\}$.
Let m_i and M_i be, respectively, the minimum and maximum values of $f(x)$ on the interval $[x_{i-1}, x_i]$, for $1 \le i \le n$. Then

$$L(f, P) = \sum_{i=1}^{n} m_i(x_i - x_{i-1}),$$

$$U(f, P) = \sum_{i=1}^{n} M_i(x_i - x_{i-1}).$$

If m_j' and M_j' are the minimum and maximum values of $f(x)$ on $[x_{j-1}, x']$, and if m_j'' and M_j'' are the corresponding values for $[x', x_j]$, then

$$m_j' \ge m_j, \quad m_j'' \ge m_j, \quad M_j' \le M_j, \quad M_j'' \le M_j.$$

Therefore we have

$$m_j(x_j - x_{j-1}) \le m_j'(x' - x_{j-1}) + m_j''(x_j - x'),$$
$$M_j(x_j - x_{j-1}) \ge M_j'(x' - x_{j-1}) + M_j''(x_j - x').$$

Hence $L(f, P) \le L(f, P')$ and $U(f, P) \ge U(f, P')$.

If P'' is any refinement of P we can add the new points in P'' to those in P one at a time, and thus obtain

$$L(f, P) \le L(f, P''), \qquad U(f, P'') \le U(f, P).$$

Section 5.4 Properties of the Definite Integral (page 296)

2. $\displaystyle\int_0^2 3f(x)\, dx + \int_1^3 3f(x)\, dx - \int_0^3 2f(x)\, dx$

$$\qquad - \int_1^2 3f(x)\, dx$$

$$= \int_0^1 (3 - 2)f(x)\, dx + \int_1^2 (3 + 3 - 2 - 3)f(x)\, dx$$

$$\qquad + \int_2^3 (3 - 2)f(x)\, dx$$

$$= \int_0^3 f(x)\, dx$$

4. $\int_0^2 (3x + 1)\, dx = \text{shaded area} = \frac{1}{2}(1 + 7)(2) = 8$

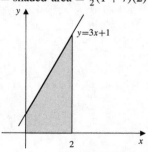

Fig. 5.4.4

6. $\int_{-1}^{2}(1-2x)\,dx = A_1 - A_2 = 0$

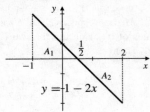

Fig. 5.4.6

8. $\int_{-\sqrt{2}}^{0}\sqrt{2-x^2}\,dx = $ quarter disk $= \frac{1}{4}\pi(\sqrt{2})^2 = \frac{\pi}{2}$

10. $\int_{-a}^{a}(a-|s|)\,ds = $ shaded area $= 2(\frac{1}{2}a^2) = a^2$

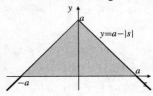

Fig. 5.4.10

12. Let $y = \sqrt{2x-x^2} \Rightarrow y^2 + (x-1)^2 = 1$.

$\int_{0}^{2}\sqrt{2x-x^2}\,dx = $ shaded area $= \frac{1}{2}\pi(1)^2 = \frac{\pi}{2}$.

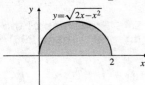

Fig. 5.4.12

14. $\int_{-3}^{3}(2+t)\sqrt{9-t^2}\,dt = 2\int_{-3}^{3}\sqrt{9-t^2}\,dt + \int_{-3}^{3}t\sqrt{9-t^2}\,dt$

$= 2\left(\frac{1}{2}\pi 3^2\right) + 0 = 9\pi$

16. $\int_{1}^{2}\sqrt{4-x^2}\,dx = $ area A_2 in figure above

$= $ area sector POQ $-$ area triangle POR

$= \frac{1}{6}(\pi 2^2) - \frac{1}{2}(1)\sqrt{3}$

$= \frac{2\pi}{3} - \frac{\sqrt{3}}{2}$

18. $\int_{2}^{3}(x^2-4)\,dx = \int_{0}^{3}x^2\,dx - \int_{0}^{2}x^2\,dx - 4(3-2)$

$= \frac{3^3}{3} - \frac{2^3}{3} - 4 = \frac{7}{3}$

20. $\int_{0}^{2}(v^2-v)\,dv = \frac{2^3}{3} - \frac{2^2}{2} = \frac{2}{3}$

22. $\int_{-6}^{6}x^2(2+\sin x)\,dx = \int_{-6}^{6}2x^2\,dx + \int_{-6}^{6}x^2\sin x\,dx$

$= 4\int_{0}^{6}x^2\,dx + 0 = \frac{4}{3}(6^3) = 288$

24. $\int_{2}^{4}\frac{1}{t}\,dt = \int_{1}^{4}\frac{1}{t}\,dt - \int_{1}^{2}\frac{1}{t}\,dt$

$= \ln 4 - \ln 2 = \ln(4/2) = \ln 2$

26. $\int_{1/4}^{3}\frac{1}{s}\,ds = \int_{1}^{3}\frac{1}{s}\,ds - \int_{1}^{1/4}\frac{1}{s}\,ds$

$= \ln 3 - \ln\frac{1}{4} = \ln 3 + \ln 4 = \ln 12$

28. Average $= \frac{1}{b-a}\int_{a}^{b}(x+2)\,dx$

$= \frac{1}{b-a}\left[\frac{1}{2}(b^2-a^2) + 2(b-a)\right]$

$= \frac{1}{2}(b+a) + 2 = \frac{4+a+b}{2}$

30. Average $= \frac{1}{3-0}\int_{0}^{3}x^2\,dx = \frac{1}{3}\frac{3^3}{3} = 3$

32. Average value $= \frac{1}{2-(1/2)}\int_{1/2}^{2}\frac{1}{s}\,ds$

$= \frac{2}{3}\left(\ln 2 - \ln\frac{1}{2}\right) = \frac{4}{3}\ln 2$

34. Let

$$f(x) = \begin{cases} 1+x & \text{if } x < 0 \\ 2 & \text{if } x \geq 0. \end{cases}$$

Then

$\int_{-3}^{2}f(x)\,dx = $ area(1) $+$ area(2) $-$ area(3)

$= (2\times 2) + \frac{1}{2}(1)(1) - \frac{1}{2}(2)(2) = 2\frac{1}{2}$.

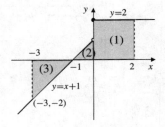

Fig. 5.4.34

36. $\int_{0}^{3}|2-x|\,dx = \int_{0}^{2}(2-x)\,dx + \int_{2}^{3}(x-2)\,dx$

$= \left(2x - \frac{x^2}{2}\right)\Big|_{0}^{2} + \left(\frac{x^2}{2} - 2x\right)\Big|_{2}^{3}$

$= 4 - 2 - 0 + \frac{9}{2} - 6 - 2 + 4 = \frac{5}{2}$

38. $\int_0^{3.5} [x]\, dx = $ shaded area $= 1 + 2 + 1.5 = 4.5.$

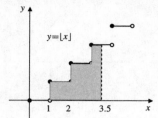

Fig. 5.4.38

40.

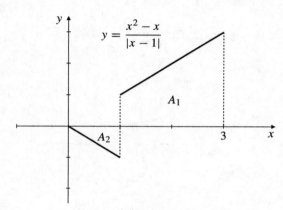

$y = \dfrac{x^2 - x}{|x - 1|}$

A_1

A_2

Fig. 5.4.40

$$\int_0^3 \frac{x^2 - x}{|x - 1|}\, dx$$
$$= \text{area } A_1 \ - \text{area } A_2$$
$$= \frac{1 + 3}{2}(2) - \frac{1}{2}(1)(1) = \frac{7}{2}$$

42. $\displaystyle\int_a^b \Big(f(x) - \bar{f}\Big)\, dx = \int_a^b f(x)\, dx - \int_a^b \bar{f}\, dx$

$$= (b - a)\bar{f} - \bar{f}\int_a^b dx$$
$$= (b - a)\bar{f} - (b - a)\bar{f} = 0$$

Section 5.5 The Fundamental Theorem of Calculus (page 301)

2. $\displaystyle\int_0^4 \sqrt{x}\, dx = \frac{2}{3}x^{3/2}\Big|_0^4 = \frac{16}{3}$

4. $\displaystyle\int_{-2}^{-1}\left(\frac{1}{x^2} - \frac{1}{x^3}\right) dx = \left(-\frac{1}{x} + \frac{1}{2x^2}\right)\Big|_{-2}^{-1}$
$$= 1 + \frac{1}{2} - \left(\frac{1}{2} + \frac{1}{8}\right) = \frac{7}{8}$$

6. $\displaystyle\int_1^2 \left(\frac{2}{x^3} - \frac{x^3}{2}\right) dx = \left(-\frac{1}{x^2} - \frac{x^4}{8}\right)\Big|_1^2 = -9/8$

8. $\displaystyle\int_4^9 \left(\sqrt{x} - \frac{1}{\sqrt{x}}\right) dx = \frac{2}{3}x^{3/2} - 2\sqrt{x}\,\Big|_4^9$
$$= \left[\frac{2}{3}(9)^{3/2} - 2\sqrt{9}\right] - \left[\frac{2}{3}(4)^{3/2} - 2\sqrt{4}\right] = \frac{32}{3}$$

10. $\displaystyle\int_0^{\pi/3} \sec^2\theta\, d\theta = \tan\theta\,\Big|_0^{\pi/3} = \tan\frac{\pi}{3} = \sqrt{3}$

12. $\displaystyle\int_0^{2\pi} (1 + \sin u)\, du = (u - \cos u)\Big|_0^{2\pi} = 2\pi$

14. $\displaystyle\int_{-2}^2 (e^x - e^{-x})\, dx = 0$ (odd function, symmetric interval)

16. $\displaystyle\int_{-1}^1 2^x\, dx = \frac{2^x}{\ln 2}\Big|_{-1}^1 = \frac{2}{\ln 2} - \frac{1}{2\ln 2} = \frac{3}{2\ln 2}$

18. $\displaystyle\int_0^{1/2} \frac{dx}{\sqrt{1 - x^2}} = \sin^{-1}x\,\Big|_0^{1/2} = \frac{\pi}{6}$

20. $\displaystyle\int_{-2}^0 \frac{dx}{4 + x^2} = \frac{1}{2}\tan^{-1}\frac{x}{2}\Big|_{-2}^0 = 0 - \frac{1}{2}\tan^{-1}(-1) = \frac{\pi}{8}$

22. Area $= \displaystyle\int_e^{e^2} \frac{1}{x}\, dx = \ln x\,\Big|_e^{e^2}$
$$= \ln e^2 - \ln e = 2 - 1 = 1 \text{ sq. units.}$$

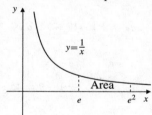

$y = \frac{1}{x}$

Area

Fig. 5.5.22

24. Since $y = 5 - 2x - 3x^2 = (5 + 3x)(1 - x)$, therefore $y = 0$ at $x = -\frac{5}{3}$ and 1, and $y > 0$ if $-\frac{5}{3} < x < 1$. Thus, the area is

$$\int_{-1}^1 (5 - 2x - 3x^2)\, dx = 2\int_0^1 (5 - 3x^2)\, dx$$
$$= 2(5x - x^3)\Big|_0^1$$
$$= 2(5 - 1) = 8 \text{ sq. units.}$$

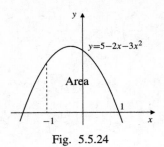

Fig. 5.5.24

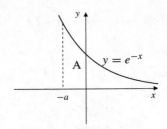

Fig. 5.5.30

26. Since $y = \sqrt{x}$ and $y = \dfrac{x}{2}$ intersect where $\sqrt{x} = \dfrac{x}{2}$, that is, at $x = 0$ and $x = 4$, thus,

$$\text{Area} = \int_0^4 \sqrt{x}\, dx - \int_0^4 \frac{x}{2}\, dx$$

$$= \frac{2}{3}x^{3/2}\Big|_0^4 - \frac{x^2}{4}\Big|_0^4$$

$$= \frac{16}{3} - \frac{16}{4} = \frac{4}{3} \text{ sq. units.}$$

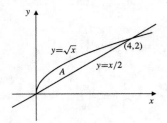

Fig. 5.5.26

28. The two graphs intersect at $(\pm 3, 3)$, thus

$$\text{Area} = 2\int_0^3 (12 - x^2)\, dx - 2\int_0^3 x\, dx$$

$$= 2\left(12x - \frac{1}{3}x^3\right)\Big|_0^3 - 2\left(\frac{1}{2}x^2\right)\Big|_0^3$$

$$= 2(36 - 9) - 9 = 45 \text{ sq. units.}$$

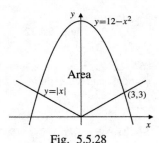

Fig. 5.5.28

30. $\text{Area} = \int_{-a}^0 e^{-x}\, dx = -e^{-x}\Big|_{-a}^0 = e^a - 1$ sq. units.

32. $\text{Area} = \displaystyle\int_1^{27} x^{-1/3}\, dx = \frac{3}{2}x^{2/3}\Big|_1^{27}$

$$= \frac{3}{2}(27)^{2/3} - \frac{3}{2} = 12 \text{ sq. units.}$$

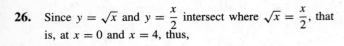

Fig. 5.5.32

34. $\displaystyle\int_1^3 \frac{\text{sgn}\,(x-2)}{x^2}\, dx = -\int_1^2 \frac{dx}{x^2} + \int_2^3 \frac{dx}{x^2}$

$$= \frac{1}{x}\Big|_1^2 - \frac{1}{x}\Big|_2^3 = -\frac{1}{3}$$

36. $\text{Average value} = \dfrac{1}{2-(-2)}\displaystyle\int_{-2}^2 e^{3x}\, dx$

$$= \frac{1}{4}\left(\frac{1}{3}e^{3x}\right)\Big|_{-2}^2$$

$$= \frac{1}{12}(e^6 - e^{-6}).$$

38. Since

$$g(t) = \begin{cases} 0, & \text{if } 0 \le t \le 1, \\ 1, & \text{if } 1 < t \le 3, \end{cases}$$

the average value of $g(t)$ over $[0,3]$ is

$$\frac{1}{3}\left[\int_0^1 (0)\, dt + \int_1^3 1\, dt\right] = \frac{1}{3}\left[0 + t\Big|_1^3\right]$$

$$= \frac{1}{3}(3 - 1) = \frac{2}{3}.$$

40. $\dfrac{d}{dt}\displaystyle\int_t^3 \frac{\sin x}{x}\, dx = \frac{d}{dt}\left[-\int_3^t \frac{\sin x}{x}\, dx\right] = -\frac{\sin t}{t}$

42. $\dfrac{d}{dx} x^2 \displaystyle\int_0^{x^2} \dfrac{\sin u}{u}\, du$

$= 2x \displaystyle\int_0^{x^2} \dfrac{\sin u}{u}\, du + x^2 \dfrac{d}{dx} \displaystyle\int_0^{x^2} \dfrac{\sin u}{u}\, du$

$= 2x \displaystyle\int_0^{x^2} \dfrac{\sin u}{u}\, du + x^2 \left[\dfrac{2x \sin x^2}{x^2} \right]$

$= 2x \displaystyle\int_0^{x^2} \dfrac{\sin u}{u}\, du + 2x \sin(x^2)$

44. $\dfrac{d}{d\theta} \displaystyle\int_{\sin\theta}^{\cos\theta} \dfrac{1}{1-x^2}\, dx$

$= \dfrac{d}{d\theta} \left[\displaystyle\int_a^{\cos\theta} \dfrac{1}{1-x^2}\, dx - \displaystyle\int_a^{\sin\theta} \dfrac{1}{1-x^2}\, dx \right]$

$= \dfrac{-\sin\theta}{1-\cos^2\theta} - \dfrac{\cos\theta}{1-\sin^2\theta}$

$= \dfrac{-1}{\sin\theta} - \dfrac{1}{\cos\theta} = -\csc\theta - \sec\theta$

46. $H(x) = 3x \displaystyle\int_4^{x^2} e^{-\sqrt{t}}\, dt$

$H'(x) = 3 \displaystyle\int_4^{x^2} e^{-\sqrt{t}}\, dt + 3x(2x e^{-|x|})$

$H'(2) = 3 \displaystyle\int_4^4 e^{-\sqrt{t}}\, dt + 3(2)(4e^{-2})$

$\qquad = 3(0) + 24e^{-2} = \dfrac{24}{e^2}$

48. $f(x) = 1 - \displaystyle\int_0^x f(t)\, dt$

$f'(x) = -f(x) \implies f(x) = Ce^{-x}$

$1 = f(0) = C$

$f(x) = e^{-x}.$

50. If $F(x) = \displaystyle\int_{17}^x \dfrac{\sin t}{1+t^2}\, dt$, then $F'(x) = \dfrac{\sin x}{1+x^2}$ and $F(17) = 0$.

52. $\displaystyle\lim_{n\to\infty} \dfrac{1}{n}\left[\left(1+\dfrac{1}{n}\right)^5 + \left(1+\dfrac{2}{n}\right)^5 + \cdots + \left(1+\dfrac{n}{n}\right)^5 \right]$

$= $ area below $y = x^5$, above $y = 0$,

$\qquad$ between $x = 1$ and $x = 2$

$= \displaystyle\int_1^2 x^5\, dx = \dfrac{1}{6} x^6 \Big|_1^2 = \dfrac{1}{6}(2^6 - 1) = \dfrac{21}{2}$

54. $\displaystyle\lim_{n\to\infty} \left(\dfrac{n}{n^2+1} + \dfrac{n}{n^2+4} + \dfrac{n}{n^2+9} + \cdots + \dfrac{n}{2n^2} \right)$

$= \displaystyle\lim_{n\to\infty} \dfrac{1}{n} \left(\dfrac{n^2}{n^2+1} + \dfrac{n^2}{n^2+4} + \dfrac{n^2}{n^2+9} + \cdots + \dfrac{n^2}{2n^2} \right)$

$= \displaystyle\lim_{n\to\infty} \dfrac{1}{n} \left(\dfrac{1}{1+\left(\dfrac{1}{n}\right)^2} + \dfrac{1}{1+\left(\dfrac{2}{n}\right)^2} + \cdots + \dfrac{1}{1+\left(\dfrac{n}{n}\right)^2} \right)$

$= $ area below $y = \dfrac{1}{1+x^2}$, above $y = 0$,

$\qquad$ between $x = 0$ and $x = 1$

$= \displaystyle\int_0^1 \dfrac{1}{1+x^2}\, dx = \tan^{-1} x \Big|_0^1 = \dfrac{\pi}{4}$

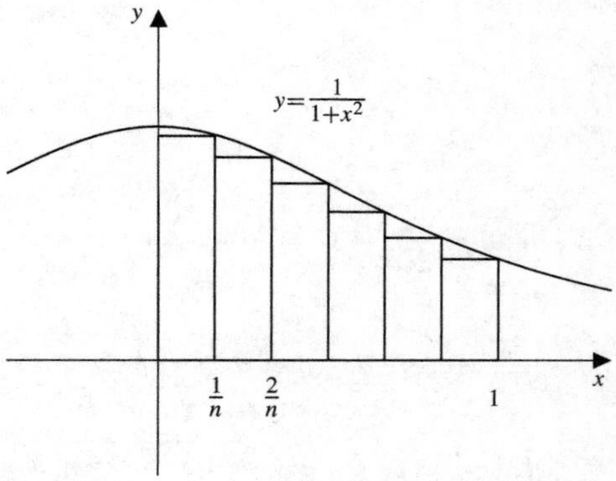

Fig. 5.5.54

Section 5.6 The Method of Substitution (page 308)

2. $\displaystyle\int \cos(ax+b)\, dx$ $\quad$ Let $u = ax + b$

$\qquad\qquad\qquad\qquad du = a\, dx$

$= \dfrac{1}{a} \displaystyle\int \cos u\, du = \dfrac{1}{a} \sin u + C$

$= \dfrac{1}{a} \sin(ax+b) + C.$

4. $\displaystyle\int e^{2x} \sin(e^{2x})\, dx$ $\quad$ Let $u = e^{2x}$

$\qquad\qquad\qquad\qquad du = 2e^{2x}\, dx$

$= \dfrac{1}{2} \displaystyle\int \sin u\, du = -\dfrac{1}{2} \cos u + C$

$= -\dfrac{1}{2} \cos(e^{2x}) + C.$

6. $\displaystyle\int \dfrac{\sin\sqrt{x}}{\sqrt{x}}\, dx$ $\quad$ Let $u = \sqrt{x}$

$\qquad\qquad\qquad\qquad du = \dfrac{dx}{2\sqrt{x}}$

$= 2 \displaystyle\int \sin u\, du = -2\cos u + C$

$= -2\cos\sqrt{x} + C.$

8. $\displaystyle\int x^2 2^{x^3+1}\,dx$　Let $u = x^3 + 1$

$du = 3x^2\,dx$

$\displaystyle= \frac{1}{3}\int 2^u\,du = \frac{1}{3}\frac{2^u}{\ln 2} + C$

$\displaystyle= \frac{2^{x^3+1}}{3\ln 2} + C.$

10. $\displaystyle\int \frac{\sec^2 x}{\sqrt{1-\tan^2 x}}\,dx$　Let $u = \tan x$

$du = \sec^2 x\,dx$

$\displaystyle= \int \frac{du}{\sqrt{1-u^2}}$

$= \sin^{-1} u + C$

$= \sin^{-1}(\tan x) + C.$

12. $\displaystyle\int \frac{\ln t}{t}\,dt$　Let $u = \ln t$

$\displaystyle du = \frac{dt}{t}$

$\displaystyle= \int u\,du = \frac{1}{2}u^2 + C = \frac{1}{2}(\ln t)^2 + C.$

14. $\displaystyle\int \frac{x+1}{\sqrt{x^2+2x+3}}\,dx$　Let $u = x^2 + 2x + 3$

$du = 2(x+1)\,dx$

$\displaystyle= \frac{1}{2}\int \frac{1}{\sqrt{u}}\,du = \sqrt{u} + C = \sqrt{x^2+2x+3} + C$

16. $\displaystyle\int \frac{x^2}{2+x^6}\,dx$　Let $u = x^3$

$du = 3x^2\,dx$

$\displaystyle= \frac{1}{3}\int \frac{du}{2+u^2} = \frac{1}{3\sqrt{2}}\tan^{-1}\left(\frac{u}{\sqrt{2}}\right) + C$

$\displaystyle= \frac{1}{3\sqrt{2}}\tan^{-1}\left(\frac{x^3}{\sqrt{2}}\right) + C.$

18. $\displaystyle\int \frac{dx}{e^x + e^{-x}} = \int \frac{e^x\,dx}{e^{2x}+1}$　Let $u = e^x$

$du = e^x\,dx$

$\displaystyle= \int \frac{du}{u^2+1} = \tan^{-1} u + C$

$= \tan^{-1} e^x + C.$

20. $\displaystyle\int \frac{x+1}{\sqrt{1-x^2}}\,dx$

$\displaystyle= \int \frac{x\,dx}{\sqrt{1-x^2}} + \int \frac{dx}{\sqrt{1-x^2}}$　Let $u = 1 - x^2$

$du = -2x\,dx$

in the first integral only

$\displaystyle= -\frac{1}{2}\int \frac{du}{\sqrt{u}} + \sin^{-1} x = -\sqrt{u} + \sin^{-1} x + C$

$= -\sqrt{1-x^2} + \sin^{-1} x + C.$

22. $\displaystyle\int \frac{dx}{\sqrt{4+2x-x^2}} = \int \frac{dx}{\sqrt{5-(1-x)^2}}$　Let $u = 1 - x$

$du = -dx$

$\displaystyle= -\int \frac{du}{\sqrt{5-u^2}} = -\sin^{-1}\left(\frac{u}{\sqrt{5}}\right) + C$

$\displaystyle= -\sin^{-1}\left(\frac{1-x}{\sqrt{5}}\right) + C = \sin^{-1}\left(\frac{x-1}{\sqrt{5}}\right) + C.$

24. $\displaystyle\int \sin^4 t \cos^5 t\,dt$

$\displaystyle= \int \sin^4 t(1-\sin^2 t)^2 \cos t\,dt$　Let $u = \sin t$

$du = \cos t\,dt$

$\displaystyle= \int (u^4 - 2u^6 + u^8)\,du = \frac{u^5}{5} - \frac{2u^7}{7} + \frac{u^9}{9} + C$

$\displaystyle= \frac{1}{5}\sin^5 t - \frac{2}{7}\sin^7 t + \frac{1}{9}\sin^9 t + C.$

26. $\displaystyle\int \sin^2 x \cos^2 x\,dx = \int \left(\frac{\sin 2x}{2}\right)^2 dx$

$\displaystyle= \frac{1}{4}\int \frac{1-\cos 4x}{2}\,dx = \frac{x}{8} - \frac{\sin 4x}{32} + C.$

28. $\displaystyle\int \cos^4 x\,dx = \int \frac{[1+\cos(2x)]^2}{4}\,dx$

$\displaystyle= \frac{1}{4}\int [1 + 2\cos(2x) + \cos^2(2x)]\,dx$

$\displaystyle= \frac{x}{4} + \frac{\sin(2x)}{4} + \frac{1}{8}\int 1 + \cos(4x)\,dx$

$\displaystyle= \frac{x}{4} + \frac{\sin(2x)}{4} + \frac{x}{8} + \frac{\sin(4x)}{32} + C$

$\displaystyle= \frac{3x}{8} + \frac{\sin(2x)}{4} + \frac{\sin(4x)}{32} + C.$

30. $\displaystyle\int \sec^6 x \tan^2 x\,dx$

$\displaystyle= \int \sec^2 x \tan^2 x(1+\tan^2 x)^2\,dx$　Let $u = \tan x$

$du = \sec^2 x\,dx$

$\displaystyle= \int (u^2 + 2u^4 + u^6)\,du = \frac{1}{3}u^3 + \frac{2}{5}u^5 + \frac{1}{7}u^7 + C$

$\displaystyle= \frac{1}{3}\tan^3 x + \frac{2}{5}\tan^5 x + \frac{1}{7}\tan^7 x + C.$

32. $\displaystyle\int \sin^{-2/3} x \cos^3 x\,dx$　Let $u = \sin x$

$du = \cos x\,dx$

$\displaystyle= \int \frac{1-u^2}{u^{2/3}}\,du = 3u^{1/3} - \frac{3}{7}u^{7/3} + C$

$\displaystyle= 3\sin^{1/3} x - \frac{3}{7}\sin^{7/3} x + C.$

34. $\displaystyle\int \frac{\sin^3(\ln x)\cos^3(\ln x)}{x}\,dx$ Let $u = \sin(\ln x)$

$$du = \frac{\cos(\ln x)}{x}\,dx$$

$$= \int u^3(1-u^2)\,du = \frac{1}{4}u^4 - \frac{1}{6}u^6 + C$$

$$= \frac{1}{4}\sin^4(\ln x) - \frac{1}{6}\sin^6(\ln x) + C.$$

36. $\displaystyle\int \frac{\sin^3 x}{\cos^4 x}\,dx = \int \tan^3 x \sec x\,dx$

$$= \int (\sec^2 x - 1)\sec x \tan x\,dx \quad \text{Let } u = \sec x$$

$$du = \sec x \tan x\,dx$$

$$= \int (u^2 - 1)\,du = \tfrac{1}{3}u^3 - u + C$$

$$= \tfrac{1}{3}\sec^3 x - \sec x + C.$$

38. $\displaystyle\int \frac{\cos^4 x}{\sin^8 x}\,dx = \int \cot^4 x \csc^4 x\,dx$

$$= \int \cot^4 x(1 + \cot^2 x)\csc^2 x\,dx \quad \text{Let } u = \cot x$$

$$du = -\csc^2 x\,dx$$

$$= -\int u^4(1 + u^2)\,du = -\frac{u^5}{5} - \frac{u^7}{7} + C$$

$$= -\frac{1}{5}\cot^5 x - \frac{1}{7}\cot^7 x + C.$$

40. $\displaystyle\int_1^{\sqrt{e}} \frac{\sin(\pi \ln x)}{x}\,dx$ Let $u = \pi \ln x$

$$du = \frac{\pi}{x}\,dx$$

$$= \frac{1}{\pi}\int_0^{\pi/2} \sin u\,du = -\frac{1}{\pi}\cos u \Big|_0^{\pi/2}$$

$$= -\frac{1}{\pi}(0 - 1) = \frac{1}{\pi}.$$

42. $\displaystyle\int_{\pi/4}^{\pi} \sin^5 x\,dx$

$$= \int_{\pi/4}^{\pi} (1 - \cos^2 x)^2 \sin x\,dx \quad \text{Let } u = \cos x$$

$$du = -\sin x\,dx$$

$$= -\int_{1/\sqrt{2}}^{-1} (1 - 2u^2 + u^4)\,du = u - \frac{2}{3}u^3 + \frac{1}{5}u^5 \Big|_{-1}^{1/\sqrt{2}}$$

$$= \frac{1}{\sqrt{2}} - \frac{1}{3\sqrt{2}} + \frac{1}{20\sqrt{2}} - \left(-1 + \frac{2}{3} - \frac{1}{5}\right) = \frac{43}{60\sqrt{2}} + \frac{8}{15}.$$

44. $\displaystyle\int_{\pi^2/16}^{\pi^2/9} \frac{2^{\sin\sqrt{x}}\cos\sqrt{x}}{\sqrt{x}}\,dx$ Let $u = \sin\sqrt{x}$

$$du = \frac{\cos\sqrt{x}}{2\sqrt{x}}\,dx$$

$$= 2\int_{1/\sqrt{2}}^{\sqrt{3}/2} 2^u\,du = \frac{2(2^u)}{\ln 2} \Big|_{1/\sqrt{2}}^{\sqrt{3}/2}$$

$$= \frac{2}{\ln 2}(2^{\sqrt{3}/2} - 2^{1/\sqrt{2}}).$$

46. $\displaystyle\text{Area} = \int_0^2 \frac{x}{x^2 + 16}\,dx$ Let $u = x^2 + 16$

$$du = 2x\,dx$$

$$= \frac{1}{2}\int_{16}^{20} \frac{du}{u} = \frac{1}{2}\ln u \Big|_{16}^{20}$$

$$= \frac{1}{2}(\ln 20 - \ln 16) = \frac{1}{2}\ln\left(\frac{5}{4}\right) \text{ sq. units.}$$

48. The area bounded by the ellipse $(x^2/a^2) + (y^2/b^2) = 1$ is

$$4\int_0^a b\sqrt{1 - \frac{x^2}{a^2}}\,dx \quad \text{Let } x = au$$

$$dx = a\,du$$

$$= 4ab\int_0^1 \sqrt{1 - u^2}\,du.$$

The integral is the area of a quarter circle of radius 1. Hence

$$\text{Area} = 4ab\left(\frac{\pi(1)^2}{4}\right) = \pi ab \text{ sq. units.}$$

50. We have

$$\int \cos ax \cos bx\,dx$$

$$= \frac{1}{2}\int [\cos(ax - bx) + \cos(ax + bx)]\,dx$$

$$= \frac{1}{2}\int \cos[(a - b)x]\,dx + \frac{1}{2}\int \cos[(a + b)x]\,dx$$

Let $u = (a - b)x$, $du = (a - b)\,dx$ in the first integral; let $v = (a + b)x$, $dv = (a + b)\,dx$ in the second integral.

$$= \frac{1}{2(a - b)}\int \cos u\,du + \frac{1}{2(a + b)}\int \cos v\,dv$$

$$= \frac{1}{2}\left[\frac{\sin[(a - b)x]}{(a - b)} + \frac{\sin[(a + b)x]}{(a + b)}\right] + C.$$

$$\int \sin ax \sin bx\,dx$$

$$= \frac{1}{2}\int [\cos(ax - bx) - \cos(ax + bx)]\,dx$$

$$= \frac{1}{2}\left[\frac{\sin[(a - b)x]}{(a - b)} - \frac{\sin[(a + b)x]}{(a + b)}\right] + C.$$

$$\int \sin ax \cos bx\,dx$$

$$= \frac{1}{2}\int [\sin(ax + bx) + \sin(ax - bx)]\,dx$$

$$= \frac{1}{2}\left[\int \sin[(a + b)x]\,dx + \int \sin[(a - b)x]\,dx\right]$$

$$= -\frac{1}{2}\left[\frac{\cos[(a + b)x]}{(a + b)} + \frac{\cos[(a - b)x]}{(a - b)}\right] + C.$$

52. If $1 \le m \le k$, we have

$$\int_{-\pi}^{\pi} f(x) \cos mx \, dx = \frac{a_0}{2} \int_{-\pi}^{\pi} \cos mx \, dx$$

$$+ \sum_{n=1}^{k} a_n \int_{-\pi}^{\pi} \cos nx \, \cos mx \, dx$$

$$+ \sum_{n=1}^{k} b_n \int_{-\pi}^{\pi} \sin nx \, \cos mx \, dx.$$

By the previous exercise, all the integrals on the right side are zero except the one in the first sum having $n = m$. Thus the whole right side reduces to

$$a_m \int_{-\pi}^{\pi} \cos^2 (mx) \, dx = a_m \int_{-\pi}^{\pi} \frac{1 + \cos(2mx)}{2} \, dx$$

$$= \frac{a_m}{2} (2\pi + 0) = \pi a_m.$$

Thus

$$a_m = \frac{1}{\pi} \int_{-\pi}^{\pi} f(x) \cos mx \, dx.$$

A similar argument shows that

$$b_m = \frac{1}{\pi} \int_{-\pi}^{\pi} f(x) \sin mx \, dx.$$

For $m = 0$ we have

$$\int_{-\pi}^{\pi} f(x) \cos mx \, dx = \int_{-\pi}^{\pi} f(x) \, dx$$

$$= \frac{a_0}{2} \int_{-\pi}^{\pi} dx$$

$$+ \sum_{n=1}^{k} (a_n \cos(nx) + b_n \sin(nx)) \, dx$$

$$= \frac{a_0}{2} (2\pi) + 0 + 0 = a_0 \pi,$$

so the formula for a_m holds for $m = 0$ also.

Section 5.7   Areas of Plane Regions (page 313)

2. Area of $R = \int_0^1 (\sqrt{x} - x^2) \, dx$

$$= \left(\frac{2}{3} x^{3/2} - \frac{1}{3} x^3 \right) \Big|_0^1 = \frac{2}{3} - \frac{1}{3} = \frac{1}{3} \text{ sq. units.}$$

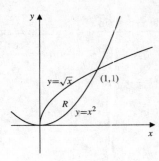

Fig. 5.7.2

4. For intersections:
$x^2 - 2x = 6x - x^2 \Rightarrow 2x^2 - 8x = 0$
i.e., $x = 0$ or 4.

Area of $R = \int_0^4 \left[6x - x^2 - (x^2 - 2x) \right] dx$

$$= \int_0^4 (8x - 2x^2) \, dx$$

$$= \left(4x^2 - \frac{2}{3} x^3 \right) \Big|_0^4 = \frac{64}{3} \text{ sq. units.}$$

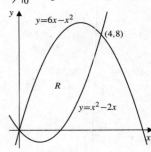

Fig. 5.7.4

6. For intersections:
$7 + y = 2y^2 - y + 3 \Rightarrow 2y^2 - 2y - 4 = 0$
$2(y - 2)(y + 1) = 0 \Rightarrow$ i.e., $y = -1$ or 2.

Area of $R = \int_{-1}^{2} [(7 + y) - (2y^2 - y + 3)] \, dy$

$$= 2 \int_{-1}^{2} (2 + y - y^2) \, dy$$

$$= 2 \left(2y + \frac{1}{2} y^2 - \frac{1}{3} y^3 \right) \Big|_{-1}^{2} = 9 \text{ sq. units.}$$

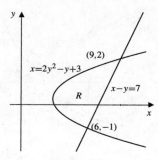

Fig. 5.7.6

8. 4 Shaded area $= \int_0^1 (x^2 - x^3)\, dx$

$= \left(\dfrac{1}{3} x^3 - \dfrac{1}{4} x^4 \right) \Big|_0^1 = \dfrac{1}{12}$ sq. units.

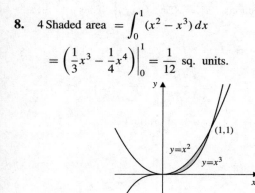

Fig. 5.7.8

10. For intersections:

$y^2 = 2y^2 - y - 2 \Rightarrow y^2 - y - 2 = 0$

$(y - 2)(y + 1) = 0 \Rightarrow$ i.e., $y = -1$ or 2.

Area of $R = \int_{-1}^2 [y^2 - (2y^2 - y - 2)]\, dy$

$= \int_{-1}^2 [2 + y - y^2]\, dy = \left(2y + \dfrac{1}{2} y^2 - \dfrac{1}{3} y^3 \right) \Big|_{-1}^2$

$= \dfrac{9}{2}$ sq. units.

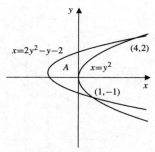

Fig. 5.7.10

12. Area of shaded region $= 2 \int_0^1 [(1 - x^2) - (x^2 - 1)^2]\, dx$

$= 2 \int_0^1 (x^2 - x^4)\, dx = 2 \left(\dfrac{1}{3} x^3 - \dfrac{1}{5} x^5 \right) \Big|_0^1 = \dfrac{4}{15}$ sq. units.

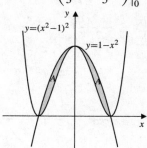

Fig. 5.7.12

14. For intersections:

$\dfrac{4x}{3 + x^2} = 1 \Rightarrow x^2 - 4x + 3 = 0$

i.e., $x = 1$ or 3.

Shaded area $= \int_1^3 \left[\dfrac{4x}{3 + x^2} - 1 \right] dx$

$= [2 \ln(3 + x^2) - x] \Big|_1^3 = 2 \ln 3 - 2$ sq. units.

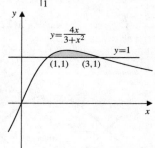

Fig. 5.7.14

16. Area $A = \int_{-\pi}^{\pi} (\sin y - (y^2 - \pi^2))\, dy$

$= \left(-\cos y + \pi^2 y - \dfrac{y^3}{3} \right) \Big|_{-\pi}^{\pi} = \dfrac{4\pi^3}{3}$ sq. units.

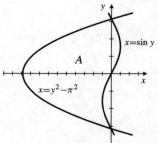

Fig. 5.7.16

18. Area $= \displaystyle\int_{-\pi/2}^{\pi/2} (1 - \sin^2 x)\, dx$

$\qquad = 2 \displaystyle\int_0^{\pi/2} \dfrac{1 + \cos(2x)}{2}\, dx$

$\qquad = \left(x + \dfrac{\sin(2x)}{2} \right)\Big|_0^{\pi/2} = \dfrac{\pi}{2}$ sq. units.

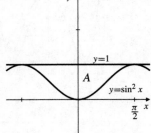

Fig. 5.7.18

20. Area $A = 2 \displaystyle\int_0^{\pi/4} (\cos^2 x - \sin^2 x)\, dx$

$\qquad = 2 \displaystyle\int_0^{\pi/4} \cos(2x)\, dx = \sin(2x)\Big|_0^{\pi/4} = 1$ sq. units.

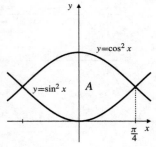

Fig. 5.7.20

22. For intersections: $x^{1/3} = \tan(\pi x/4)$. Thus $x = \pm 1$.

$\quad$ Area $A = 2 \displaystyle\int_0^1 \left(x^{1/3} - \tan\dfrac{\pi x}{4} \right) dx$

$\qquad = 2 \left(\dfrac{3}{4} x^{4/3} - \dfrac{4}{\pi} \ln\left| \sec \dfrac{\pi x}{4} \right| \right)\Big|_0^1$

$\qquad = \dfrac{3}{2} - \dfrac{8}{\pi} \ln\sqrt{2} = \dfrac{3}{2} - \dfrac{4}{\pi} \ln 2$ sq. units.

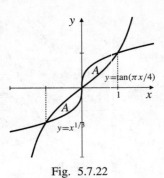

Fig. 5.7.22

24. For intersections: $|x| = \sqrt{2}\cos(\pi x/4)$. Thus $x = \pm 1$.

$\quad$ Area $A = 2 \displaystyle\int_0^1 \left(\sqrt{2}\cos\dfrac{\pi x}{4} - x \right) dx$

$\qquad = \left(\dfrac{8\sqrt{2}}{\pi} \sin\dfrac{\pi x}{4} - x^2 \right)\Big|_0^1$

$\qquad = \dfrac{8}{\pi} - 1$ sq. units.

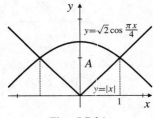

Fig. 5.7.24

26. For intersections: $e^x = x + 2$. There are two roots, both of which must be found numerically. We used a TI-85 solve routine to get $x_1 \approx -1.841406$ and $x_2 \approx 1.146193$. Thus

$\quad$ Area $A = \displaystyle\int_{x_1}^{x_2} \left(x + 2 - e^x \right) dx$

$\qquad = \left(\dfrac{x^2}{2} + 2x - e^x \right)\Big|_{x_1}^{x_2}$

$\qquad \approx 1.949091$ sq. units.

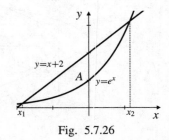

Fig. 5.7.26

28. Loop area $= 2 \int_{-2}^{0} x^2 \sqrt{2+x}\, dx$ Let $u^2 = 2 + x$
$$2u\, du = dx$$
$$= 2 \int_{0}^{\sqrt{2}} (u^2 - 2)^2 u(2u)\, du = 4 \int_{0}^{\sqrt{2}} (u^6 - 4u^4 + 4u^2)\, du$$
$$= 4\left(\frac{1}{7}u^7 - \frac{4}{5}u^5 + \frac{4}{3}u^3\right)\Big|_{0}^{\sqrt{2}} = \frac{256\sqrt{2}}{105} \text{ sq. units.}$$

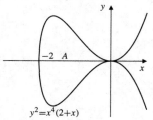

Fig. 5.7.28

30. The tangent line to $y = x^3$ at $(1, 1)$ is $y - 1 = 3(x - 1)$, or $y = 3x - 2$. The intersections of $y = x^3$ and this tangent line occur where $x^3 - 3x + 2 = 0$. Of course $x = 1$ is a (double) root of this cubic equation, which therefore factors to $(x - 1)^2(x + 2) = 0$. The other intersection is at $x = -2$. Thus

$$\text{Area of } R = \int_{-2}^{1} (x^3 - 3x + 2)\, dx$$
$$= \left(\frac{x^4}{4} - \frac{3x^2}{2} + 2x\right)\Big|_{-2}^{1}$$
$$= -\frac{15}{4} - \frac{3}{2} + 6 + 2 + 4 = \frac{27}{4} \text{ sq. units.}$$

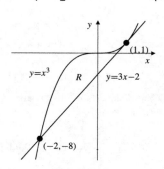

Fig. 5.7.30

Review Exercises 5   (page 314)

2. The number of balls is

$$40 \times 30 + 39 \times 29 + \cdots + 12 \times 2 + 11 \times 1$$
$$= \sum_{i=1}^{30} i(i + 10) = \frac{(30)(31)(61)}{6} + 10\frac{(30)(31)}{2} = 14,105.$$

4. $R_n = \sum_{i=1}^{n} (1/n)\sqrt{1 + (i/n)}$ is a Riemann sum for $f(x) = \sqrt{1 + x}$ on the interval $[0, 1]$. Thus

$$\lim_{n \to \infty} R_n = \int_{0}^{1} \sqrt{1 + x}\, dx$$
$$= \frac{2}{3}(1 + x)^{3/2}\Big|_{0}^{1} = \frac{4\sqrt{2} - 2}{3}.$$

6. $\int_{0}^{\sqrt{5}} \sqrt{5 - x^2}\, dx = 1/4$ of the area of a circle of radius $\sqrt{5}$
$$= \frac{1}{4}\pi(\sqrt{5})^2 = \frac{5\pi}{4}$$

8. $\int_{0}^{\pi} \cos x\, dx = \text{area } A_1 - \text{area } A_2 = 0$

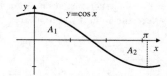

Fig. R-5.8

10. $\bar{h} = \frac{1}{3} \int_{0}^{3} |x - 2|\, dx = \frac{1}{3}\frac{5}{2} = \frac{5}{6}$ (via #9)

12. $f(x) = \int_{-13}^{\sin x} \sqrt{1 + t^2}\, dt, \quad f'(x) = \sqrt{1 + \sin^2 x}(\cos x)$

14. $g(\theta) = \int_{e^{\sin\theta}}^{e^{\cos\theta}} \ln x\, dx$
$$g'(\theta) = (\ln(e^{\cos\theta}))e^{\cos\theta}(-\sin\theta) - (\ln(e^{\sin\theta}))e^{\sin\theta}\cos\theta$$
$$= -\sin\theta\cos\theta(e^{\cos\theta} + e^{\sin\theta})$$

16. $I = \int_{0}^{\pi} x f(\sin x)\, dx$ Let $x = \pi - u$
$$dx = -du$$
$$= -\int_{\pi}^{0} (\pi - u) f(\sin(\pi - u))\, du \quad \text{(but } \sin(\pi - u) = \sin u\text{)}$$
$$= \pi \int_{0}^{\pi} f(\sin u)\, du - \int_{0}^{\pi} u f(\sin u)\, du$$
$$= \pi \int_{0}^{\pi} f(\sin x)\, dx - I.$$

Now, solving for I, we get

$$\int_{0}^{\pi} x f(\sin x)\, dx = I = \frac{\pi}{2} \int_{0}^{\pi} f(\sin x)\, dx.$$

18. The area bounded by $y = (x - 1)^2$, $y = 0$, and $x = 0$ is

$$\int_{0}^{1} (x - 1)^2\, dx = \frac{(x - 1)^3}{3}\Big|_{0}^{1} = \frac{1}{3} \text{ sq. units..}$$

20. $y = 4x - x^2$ and $y = 3$ meet where $x^2 - 4x + 3 = 0$, that is, at $x = 1$ and $x = 3$. Since $4x - x^2 \geq 3$ on $[1, 3]$, the required area is

$$\int_1^3 (4x - x^2 - 3)\, dx = \left(2x^2 - \frac{x^3}{3} - 3x \right)\Big|_1^3 = \frac{4}{3} \text{ sq. units.}$$

22. $y = 5 - x^2$ and $y = 4/x^2$ meet where $5 - x^2 = 4/x^2$, that is, where

$$x^4 - 5x^2 + 4 = 0$$
$$(x^2 - 1)(x^2 - 4) = 0.$$

There are four intersections: $x = \pm 1$ and $x = \pm 2$. By symmetry (see the figure) the total area bounded by the curves is

$$2 \int_1^2 \left(5 - x^2 - \frac{4}{x^2} \right) dx = 2 \left(5x - \frac{x^3}{3} + \frac{4}{x} \right)\Big|_1^2 = \frac{4}{3} \text{ sq. units.}$$

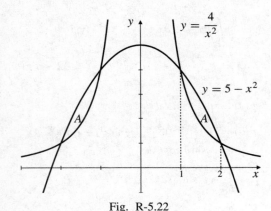

Fig. R-5.22

24. $\displaystyle\int_1^e \frac{\ln x}{x}\, dx$ Let $u = \ln x$
$$du = dx/x$$
$$= \int_0^1 u\, du = \frac{u^2}{2}\Big|_0^1 = \frac{1}{2}$$

26. $\displaystyle\int \sin^3(\pi x)\, dx$
$$= \int \sin(\pi x)\left(1 - \cos^2(\pi x) \right) dx \quad \text{Let } u = \cos(\pi x)$$
$$du = -\pi \sin(\pi x)\, dx$$
$$= -\frac{1}{\pi} \int (1 - u^2)\, du$$
$$= \frac{1}{\pi} \left(\frac{u^3}{3} - u \right) + C = \frac{1}{3\pi} \cos^3(\pi x) - \frac{1}{\pi} \cos(\pi x) + C$$

28. $\displaystyle\int_1^{\sqrt[4]{e}} \frac{\tan^2(\pi \ln x)}{x}\, dx$ Let $u = \pi \ln x$
$$du = (\pi/x)\, dx$$
$$= \frac{1}{\pi} \int_0^{\pi/4} \tan^2 u\, du = \frac{1}{\pi} \int_0^{\pi/4} (\sec^2 u - 1)\, du$$
$$= \frac{1}{\pi} (\tan u - u)\Big|_0^{\pi/4} = \frac{1}{\pi} - \frac{1}{4}$$

30. $\displaystyle\int \cos^2 \frac{t}{5} \sin^2 \frac{t}{5}\, dt = \frac{1}{4} \int \sin^2 \frac{2t}{5}\, dt$
$$= \frac{1}{8} \int \left(1 - \cos \frac{4t}{5} \right) dt$$
$$= \frac{1}{8} \left(t - \frac{5}{4} \sin \frac{4t}{5} \right) + C$$

32. $f(x) = 4x - x^2 \geq 0$ if $0 \leq x \leq 4$, and $f(x) < 0$ otherwise. If $a < b$, then $\int_a^b f(x)\, dx$ will be maximum if $[a, b] = [0, 4]$; extending the interval to the left of 0 or to the right of 4 will introduce negative contributions to the integral. The maximum value is

$$\int_0^4 (4x - x^2)\, dx = \left(2x^2 - \frac{x^3}{3} \right)\Big|_0^4 = \frac{32}{3}.$$

34.   If $y(t)$ is the distance the object falls in t seconds from its release time, then

$$y''(t) = g, \quad y(0) = 0, \quad \text{and } y'(0) = 0.$$

Antidifferentiating twice and using the initial conditions leads to

$$y(t) = \frac{1}{2} g t^2.$$

The average height during the time interval $[0, T]$ is

$$\frac{1}{T} \int_0^T \frac{1}{2} g t^2\, dt = \frac{g}{2T} \frac{T^3}{3} = \frac{g T^2}{6} = y\left(\frac{T}{\sqrt{3}} \right).$$

Challenging Problems 5 (page 315)

2. a) $\cos\left((j + \frac{1}{2})t\right) - \cos\left((j - \frac{1}{2})t\right)$

$= \cos(jt)\cos(\frac{1}{2}t) - \sin(jt)\sin(\frac{1}{2}t)$

$\quad - \cos(jt)\cos(\frac{1}{2}t) - \sin(jt)\sin(\frac{1}{2}t)$

$= -2\sin(jt)\sin(\frac{1}{2}t)$.

Therefore, we obtain a telescoping sum:

$\sum_{j=1}^{n} \sin(jt)$

$= -\dfrac{1}{2\sin(\frac{1}{2}t)} \sum_{j=1}^{n} \left[\cos\left((j + \frac{1}{2})t\right) - \cos\left((j - \frac{1}{2})t\right)\right]$

$= -\dfrac{1}{2\sin(\frac{1}{2}t)} \left[\cos\left((n + \frac{1}{2})t\right) - \cos(\frac{1}{2}t)\right]$

$= \dfrac{1}{2\sin(\frac{1}{2}t)} \left[\cos(\frac{1}{2}t) - \cos\left((n + \frac{1}{2})t\right)\right]$.

b) Let $P_n = \{0, \frac{\pi}{2n}, \frac{2\pi}{2n}, \frac{3\pi}{2n}, \ldots \frac{n\pi}{2n}\}$ be the partition of $[0, \pi/2]$ into n subintervals of equal length $\Delta x = \pi/2n$. Using $t = \pi/2n$ in the formula obtained in part (a), we get

$\displaystyle\int_0^{\pi/2} \sin x \, dx$

$= \lim_{n\to\infty} \sum_{j=1}^{n} \sin\left(\dfrac{j\pi}{2n}\right) \dfrac{\pi}{2n}$

$= \lim_{n\to\infty} \dfrac{\pi}{2n} \dfrac{1}{2\sin(\pi/(4n))} \left(\cos\dfrac{\pi}{4n} - \cos\dfrac{(2n+1)\pi}{4n}\right)$

$= \lim_{n\to\infty} \dfrac{\pi/(4n)}{\sin(\pi/(4n))} \lim_{n\to\infty} \left(\cos\dfrac{\pi}{4n} - \cos\dfrac{(2n+1)\pi}{4n}\right)$

$= 1 \times \left(\cos 0 - \cos\dfrac{\pi}{2}\right) = 1$.

4. $f(x) = 1/x^2$, $1 = x_0 < x_1 < x_2 < \cdots < x_n = 2$. If $c_i = \sqrt{x_{i-1}x_i}$, then

$$x_{i-1}^2 < x_{i-1}x_i = c_i^2 < x_i^2,$$

so $x_{i-1} < c_i < x_i$. We have

$\displaystyle\sum_{i=1}^{n} f(c_i)\,\Delta x_i = \sum_{i=1}^{n} \dfrac{1}{x_{i-1}x_i}(x_i - x_{i-1})$

$= \displaystyle\sum_{i=1}^{n} \left(\dfrac{1}{x_{i-1}} - \dfrac{1}{x_i}\right)$ (telescoping)

$= \dfrac{1}{x_0} - \dfrac{1}{x_n} = 1 - \dfrac{1}{2} = \dfrac{1}{2}$.

Thus $\displaystyle\int_1^2 \dfrac{dx}{x^2} = \lim_{n\to\infty} \sum_{i=1}^{n} f(c_i)\,\Delta x_i = \dfrac{1}{2}$.

6. Let $f(x) = ax^3 + bx^2 + cx + d$. We used Maple to calculate the following:

The tangent to $y = f(x)$ at $P = (p, f(p))$ has equation

$y = g(x) = ap^3 + bp^2 + cp + d + (3ap^2 + 2bp + c)(x - p)$.

This line intersects $y = f(x)$ at $x = p$ (double root) and at $x = q$, where

$$q = -\dfrac{2ap + b}{a}.$$

Similarly, the tangent to $y = f(x)$ at $x = q$ has equation

$y = h(x) = aq^3 + bq^2 + cq + d + (3aq^2 + 2bq + c)(x - q)$,

and intersects $y = f(x)$ at $x = q$ (double root) and $x = r$, where

$$r = -\dfrac{2aq + b}{a} = \dfrac{4ap + b}{a}.$$

The area between $y = f(x)$ and the tangent line at P is the absolute value of

$\displaystyle\int_p^q (f(x) - g(x)\, dx$

$= -\dfrac{1}{12}\left(\dfrac{81a^4p^4 + 108a^3bp^3 + 54a^2b^2p^2 + 12ab^3p + b^4}{a^3}\right)$.

The area between $y = f(x)$ and the tangent line at $Q = (q, f(q))$ is the absolute value of

$\displaystyle\int_q^r (f(x) - h(x)\, dx$

$= -\dfrac{4}{3}\left(\dfrac{81a^4p^4 + 108a^3bp^3 + 54a^2b^2p^2 + 12ab^3p + b^4}{a^3}\right)$,

which is 16 times the area between $y = f(x)$ and the tangent at P.

8. Let $f(x) = ax^4 + bx^3 + cx^2 + dx + e$. The tangent to $y = f(x)$ at $P = (p, f(p))$ has equation

$y = g(x) = ap^4 + bp^3 + cp^2 + dp + e + (4ap^3 + 3bp^2 + 2cp + d)(x - p)$,

and intersects $y = f(x)$ at $x = p$ (double root) and at the two points

$$x = \dfrac{-2ap - b \pm \sqrt{b^2 - 4ac - 4abp - 8a^2p^2}}{2a}.$$

If these latter two points coincide, then the tangent is a "double tangent." This happens if

$$8a^2p^2 + 4abp + 4ac - b^2 = 0,$$

which has two solutions, which we take to be p and q:

$$p = \frac{-b + \sqrt{3b^2 - 8ac}}{4a}$$

$$q = \frac{-b - \sqrt{3b^2 - 8ac}}{4a} = -p - \frac{b}{2a}.$$

(Both roots exist and are distinct provided $3b^2 > 8ac$.)
The point T corresponds to $x = t = (p+q)/2 = -b/4a$.
The tangent to $y = f(x)$ at $x = t$ has equation

$$y = h(x) = -\frac{3b^4}{256a^3} + \frac{b^2 c}{16a^2} - \frac{bd}{4a} + e + \left(\frac{b^3}{8a^2} - \frac{bc}{2a} + d\right)\left(x + \frac{b}{4a}\right)$$

and it intersects $y = f(x)$ at the points U and V with x-coordinates

$$u = \frac{-b - \sqrt{2}\sqrt{3b^2 - 8ac}}{4a},$$

$$v = \frac{-b + \sqrt{2}\sqrt{3b^2 - 8ac}}{4a}.$$

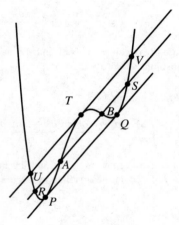

Fig. C-5.8

a) The areas between the curve $y = f(x)$ and the lines PQ and UV are, respectively, the absolute values of

$$A_1 = \int_p^q (f(x) - g(x))\, dx \quad \text{and} \quad A_2 = \int_u^v (h(x) - f(x))\, dx.$$

Maple calculates these two integrals and simplifies the ratio A_1/A_2 to be $1/\sqrt{2}$.

b) The two inflection points A and B of f have x-coordinates shown by Maple to be

$$\alpha = \frac{-3b - \sqrt{3(3b^2 - 8ac)}}{12a} \quad \text{and}$$

$$\beta = \frac{-3b + \sqrt{3(3b^2 - 8ac)}}{12a}.$$

It then determines the four points of intersection of the line $y = k(x)$ through these inflection points and the curve. The other two points have x-coordinates

$$r = \frac{-3b - \sqrt{15(3b^2 - 8ac)}}{12a} \quad \text{and}$$

$$s = \frac{-3b + \sqrt{15(3b^2 - 8ac)}}{12a}.$$

The region bounded by RS and the curve $y = f(x)$ is divided into three parts by A and B. The areas of these three regions are the absolute values of

$$A_1 = \int_r^\alpha (k(x) - f(x))\, dx$$

$$A_2 = \int_\alpha^\beta (f(x) - k(x))\, dx$$

$$A_3 = \int_\beta^s (k(x) - f(x))\, dx.$$

The expressions calculated by Maple for $k(x)$ and for these three areas are very complicated, but Maple simplifies the rations A_3/A_1 and A_2/A_1 to 1 and 2 respectively, as was to be shown.

CHAPTER 6.   TECHNIQUES OF INTEGRATION

Section 6.1   Integration by Parts (page 321)

2. $\displaystyle\int (x+3)e^{2x}\,dx$

$$U = x + 3 \quad dV = e^{2x}\,dx$$
$$dU = dx \quad V = \tfrac{1}{2}e^{2x}$$
$$= \frac{1}{2}(x+3)e^{2x} - \frac{1}{2}\int e^{2x}\,dx$$
$$= \frac{1}{2}(x+3)e^{2x} - \frac{1}{4}e^{2x} + C.$$

4. $\displaystyle\int (x^2 - 2x)e^{kx}\,dx$

$$U = x^2 - 2x \qquad dV = e^{kx}$$
$$dU = (2x - 2)\,dx \qquad V = \frac{1}{k}e^{kx}$$
$$= \frac{1}{k}(x^2 - 2x)e^{kx} - \frac{1}{k}\int (2x - 2)e^{kx}\,dx$$
$$U = x - 1 \quad dV = e^{kx}\,dx$$
$$dU = dx \qquad V = \frac{1}{k}e^{kx}$$
$$= \frac{1}{k}(x^2 - 2x)e^{kx} - \frac{2}{k}\left[\frac{1}{k}(x-1)e^{kx} - \frac{1}{k}\int e^{kx}\,dx\right]$$
$$= \frac{1}{k}(x^2 - 2x)e^{kx} - \frac{2}{k^2}(x-1)e^{kx} + \frac{2}{k^3}e^{kx} + C.$$

6. $\displaystyle\int x(\ln x)^3\,dx = I_3$ where

$$I_n = \int x(\ln x)^n\,dx$$
$$U = (\ln x)^n \qquad dV = x\,dx$$
$$dU = \frac{n}{x}(\ln x)^{n-1}\,dx \qquad V = \frac{1}{2}x^2$$
$$= \frac{1}{2}x^2(\ln x)^n - \frac{n}{2}\int x(\ln x)^{n-1}\,dx$$
$$= \frac{1}{2}x^2(\ln x)^n - \frac{n}{2}I_{n-1}$$
$$I_3 = \frac{1}{2}x^2(\ln x)^3 - \frac{3}{2}I_2$$
$$= \frac{1}{2}x^2(\ln x)^3 - \frac{3}{2}\left[\frac{1}{2}x^2(\ln x)^2 - \frac{2}{2}I_1\right]$$
$$= \frac{1}{2}x^2(\ln x)^3 - \frac{3}{4}x^2(\ln x)^2 + \frac{3}{2}\left[\frac{1}{2}x^2(\ln x) - \frac{1}{2}I_0\right]$$
$$= \frac{1}{2}x^2(\ln x)^3 - \frac{3}{4}x^2(\ln x)^2 + \frac{3}{4}x^2(\ln x) - \frac{3}{4}\int x\,dx$$
$$= \frac{x^2}{2}\left[(\ln x)^3 - \frac{3}{2}(\ln x)^2 + \frac{3}{2}(\ln x) - \frac{3}{4}\right] + C.$$

8. $\displaystyle\int x^2 \tan^{-1}x\,dx$

$$U = \tan^{-1}x \quad dV = x^2\,dx$$
$$dU = \frac{dx}{1 + x^2} \quad V = \frac{x^3}{3}$$
$$= \frac{x^3}{3}\tan^{-1}x - \frac{1}{3}\int \frac{x^3}{1 + x^2}\,dx$$
$$= \frac{x^3}{3}\tan^{-1}x - \frac{1}{3}\int\left(x - \frac{x}{1+x^2}\right)dx$$
$$= \frac{x^3}{3}\tan^{-1}x - \frac{x^2}{6} + \frac{1}{6}\ln(1 + x^2) + C.$$

10. $\displaystyle\int x^5 e^{-x^2}\,dx = I_2$ where

$$I_n = \int x^{(2n+1)} e^{-x^2}\,dx$$

$$U = x^{2n} \qquad dV = xe^{-x^2}\,dx$$
$$dU = 2nx^{(2n-1)}\,dx \qquad V = -\tfrac{1}{2}e^{-x^2}$$
$$= -\frac{1}{2}x^{2n}e^{-x^2} + n\int x^{(2n-1)}e^{-x^2}\,dx$$
$$= -\frac{1}{2}x^{2n}e^{-x^2} + nI_{n-1}$$
$$I_2 = -\frac{1}{2}x^4 e^{-x^2} + 2\left[-\frac{1}{2}x^2 e^{-x^2} + \int xe^{-x^2}\,dx\right]$$
$$= -\frac{1}{2}e^{-x^2}(x^4 + 2x^2 + 2) + C.$$

12. $\displaystyle I = \int \tan^2 x \sec x\,dx$

$$U = \tan x \qquad dV = \sec x \tan x\,dx$$
$$dU = \sec^2 x\,dx \qquad V = \sec x$$
$$= \sec x \tan x - \int \sec^3 x\,dx$$
$$= \sec x \tan x - \int (1 + \tan^2 x)\sec x\,dx$$
$$= \sec x \tan x - \ln|\sec x + \tan x| - I$$

Thus, $I = \tfrac{1}{2}\sec x \tan x - \tfrac{1}{2}\ln|\sec x + \tan x| + C.$

14. $\displaystyle I = \int xe^{\sqrt{x}}\,dx$ Let $x = w^2$
$$dx = 2w\,dw$$
$$= 2\int w^3 e^w\,dw = 2I_3 \text{ where}$$
$$I_n = \int w^n e^w\,dw$$
$$U = w^n \qquad dV = e^w\,dw$$
$$dU = nw^{n-1}\,dw \qquad V = e^w$$
$$= w^n e^w - nI_{n-1}.$$
$$I = 2I_3 = 2w^3 e^w - 6[w^2 e^w - 2(we^w - I_0)]$$
$$= e^{\sqrt{x}}(2x\sqrt{x} - 6x + 12\sqrt{x} - 12) + C.$$

16. $\displaystyle\int_0^1 \sqrt{x}\,\sin(\pi\sqrt{x})\,dx$ Let $x = w^2$
$$dx = 2w\,dw$$

$= 2\displaystyle\int_0^1 w^2\sin(\pi w)\,dw$

$\quad U = w^2 \qquad dV = \sin(\pi w)\,dw$
$\quad dU = 2w\,dw \qquad V = -\dfrac{\cos(\pi w)}{\pi}$

$= -\dfrac{2}{\pi}w^2\cos(\pi w)\Big|_0^1 + \dfrac{4}{\pi}\displaystyle\int_0^1 w\cos(\pi w)\,dw$

$\quad U = w \qquad dV = \cos(\pi w)\,dw$
$\quad dU = dw \qquad V = \dfrac{\sin(\pi w)}{\pi}$

$= \dfrac{2}{\pi} + \dfrac{4}{\pi}\left[\dfrac{w}{\pi}\sin(\pi w)\right]\Big|_0^1 - \dfrac{4}{\pi^2}\displaystyle\int_0^1\sin(\pi w)\,dw$

$= \dfrac{2}{\pi} + \dfrac{4}{\pi^3}\cos(\pi w)\Big|_0^1 = \dfrac{2}{\pi} + \dfrac{4}{\pi^3}(-2) = \dfrac{2}{\pi} - \dfrac{8}{\pi^3}.$

18. $\displaystyle\int x\sin^2 x\,dx = \dfrac{1}{2}\int(x - x\cos 2x)\,dx$

$= \dfrac{x^2}{4} - \dfrac{1}{2}\displaystyle\int x\cos 2x\,dx$

$\quad U = x \qquad dV = \cos 2x\,dx$
$\quad dU = dx \qquad V = \dfrac{1}{2}\sin 2x$

$= \dfrac{x^2}{4} - \dfrac{1}{2}\left[\dfrac{1}{2}x\sin 2x - \dfrac{1}{2}\displaystyle\int\sin 2x\,dx\right]$

$= \dfrac{x^2}{4} - \dfrac{x}{4}\sin 2x - \dfrac{1}{8}\cos 2x + C.$

20. $I = \displaystyle\int_1^e \sin(\ln x)\,dx$

$\quad U = \sin(\ln x) \qquad dV = dx$
$\quad dU = \dfrac{\cos(\ln x)}{x}\,dx \qquad V = x$

$= x\sin(\ln x)\Big|_1^e - \displaystyle\int_1^e\cos(\ln x)\,dx$

$\quad U = \cos(\ln x) \qquad dV = dx$
$\quad dU = -\dfrac{\sin(\ln x)}{x}\,dx \qquad V = x$

$= e\sin(1) - \left[x\cos(\ln x)\Big|_1^e + I\right]$

Thus, $I = \dfrac{1}{2}[e\sin(1) - e\cos(1) + 1].$

22. $\displaystyle\int_0^4 \sqrt{x}\,e^{\sqrt{x}}\,dx$ Let $x = w^2$
$$dx = 2w\,dw$$

$= 2\displaystyle\int_0^2 w^2 e^w\,dw = 2I_2$

See solution #16 for the formula
$I_n = \displaystyle\int w^n e^w\,dw = w^n e^w - nI_{n-1}.$

$= 2\left(w^2 e^w\Big|_0^2 - 2I_1\right) = 8e^2 - 4\left(we^w\Big|_0^2 - I_0\right)$

$= 8e^2 - 8e^2 + 4\displaystyle\int_0^2 e^w\,dw = 4(e^2 - 1).$

24. $\displaystyle\int x\sec^{-1} x\,dx$

$\quad U = \sec^{-1} x \qquad dV = x\,dx$
$\quad dU = \dfrac{dx}{|x|\sqrt{x^2-1}} \qquad V = \dfrac{1}{2}x^2$

$= \dfrac{1}{2}x^2\sec^{-1} x - \dfrac{1}{2}\displaystyle\int\dfrac{|x|}{\sqrt{x^2-1}}\,dx$

$= \dfrac{1}{2}x^2\sec^{-1} x - \dfrac{1}{2}\operatorname{sgn}(x)\sqrt{x^2-1} + C.$

26. $\displaystyle\int(\sin^{-1} x)^2\,dx$ Let $x = \sin\theta$
$$dx = \cos\theta\,d\theta$$

$= \displaystyle\int\theta^2\cos\theta\,d\theta$

$\quad U = \theta^2 \qquad dV = \cos\theta\,d\theta$
$\quad dU = 2\theta\,d\theta \qquad V = \sin\theta$

$= \theta^2\sin\theta - 2\displaystyle\int\theta\sin\theta\,d\theta$

$\quad U = \theta \qquad dV = \sin\theta\,d\theta$
$\quad dU = d\theta \qquad V = -\cos\theta$

$= \theta^2\sin\theta - 2\left(-\theta\cos\theta + \displaystyle\int\cos\theta\,d\theta\right)$

$= \theta^2\sin\theta + 2\theta\cos\theta - 2\sin\theta + C$

$= x(\sin^{-1} x)^2 + 2\sqrt{1-x^2}(\sin^{-1} x) - 2x + C.$

28. By the procedure used in Example 4 of Section 7.1,
$$\int e^x\cos x\,dx = \tfrac{1}{2}e^x(\sin x + \cos x) + C;$$
$$\int e^x\sin x\,dx = \tfrac{1}{2}e^x(\sin x - \cos x) + C.$$

Now

$\displaystyle\int xe^x\cos x\,dx$

$\quad U = x \qquad dV = e^x\cos x\,dx$
$\quad dU = dx \qquad V = \tfrac{1}{2}e^x(\sin x + \cos x)$

$= \tfrac{1}{2}xe^x(\sin + \cos x) - \tfrac{1}{2}\displaystyle\int e^x(\sin x + \cos x)\,dx$

$= \tfrac{1}{2}xe^x(\sin + \cos x)$
$\qquad - \tfrac{1}{4}e^x(\sin x - \cos x + \sin x + \cos x) + C$

$= \tfrac{1}{2}xe^x(\sin x + \cos x) - \tfrac{1}{2}e^x\sin x + C.$

30. The tangent line to $y = \ln x$ at $x = 1$ is $y = x - 1$, Hence,

$$\text{Shaded area} = \frac{1}{2}(1)(1) + (1)(e-2) - \int_1^e \ln x \, dx$$

$$= e - \frac{3}{2} - (x \ln x - x)\Big|_1^e$$

$$= e - \frac{3}{2} - e + e + 0 - 1 = e - \frac{5}{2} \text{ sq. units.}$$

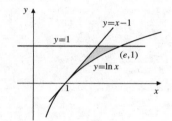

Fig. 6.1.30

32. $I_n = \int_0^{\pi/2} x^n \sin x \, dx$

$$U = x^n \qquad\qquad dV = \sin x \, dx$$
$$dU = nx^{n-1} dx \qquad V = -\cos x$$

$$= -x^n \cos x \Big|_0^{\pi/2} + n \int_0^{\pi/2} x^{n-1} \cos x \, dx$$

$$U = x^{n-1} \qquad\qquad dV = \cos x \, dx$$
$$dU = (n-1)x^{n-2} dx \qquad V = \sin x$$

$$= n\left[x^{n-1} \sin x \Big|_0^{\pi/2} - (n-1) \int_0^{\pi/2} x^{n-2} \sin x \, dx \right]$$

$$= n\left(\frac{\pi}{2}\right)^{n-1} - n(n-1)I_{n-2}, \qquad (n \geq 2).$$

$$I_0 = \int_0^{\pi/2} \sin x \, dx = -\cos x \Big|_0^{\pi/2} = 1.$$

$$I_6 = 6\left(\frac{\pi}{2}\right)^5 - 6(5)\left\{4\left(\frac{\pi}{2}\right)^3 - 4(3)\left[2\left(\frac{\pi}{2}\right) - 2(1)I_0\right]\right\}$$

$$= \frac{3}{16}\pi^5 - 15\pi^3 + 360\pi - 720.$$

34. We have

$$I_n = \int \sec^n x \, dx \qquad (n \geq 3)$$

$$U = \sec^{n-2} x \qquad\qquad dV = \sec^2 x \, dx$$
$$dU = (n-2)\sec^{n-2} x \tan x \, dx \qquad V = \tan x$$

$$= \sec^{n-2} x \tan x - (n-2) \int \sec^{n-2} x \tan^2 x \, dx$$

$$= \sec^{n-2} x \tan x - (n-2) \int \sec^{n-2} x (\sec^2 x - 1) \, dx$$

$$= \sec^{n-2} x \tan x - (n-2)I_n + (n-2)I_{n-2} + C$$

$$I_n = \frac{1}{n-1}(\sec^{n-2} x \tan x) + \frac{n-2}{n-1}I_{n-2} + C.$$

$$I_1 = \int \sec x \, dx = \ln|\sec x + \tan x| + C;$$

$$I_2 = \int \sec^2 x \, dx = \tan x + C.$$

$$I_6 = \frac{1}{5}(\sec^4 x \tan x) + \frac{4}{5}\left(\frac{1}{3}\sec^2 x \tan x + \frac{2}{3}I_2\right) + C$$

$$= \frac{1}{5}\sec^4 x \tan x + \frac{4}{15}\sec^2 x \tan x + \frac{8}{15}\tan x + C.$$

$$I_7 = \frac{1}{6}(\sec^5 x \tan x) + \frac{5}{6}\left[\frac{1}{4}\sec^3 x \tan x + \right.$$

$$\left. \frac{3}{4}\left(\frac{1}{2}\sec x \tan x + \frac{1}{2}I_1\right)\right] + C$$

$$= \frac{1}{6}\sec^5 x \tan x + \frac{5}{24}\sec^3 x \tan x + \frac{15}{48}\sec x \tan x + $$

$$\frac{15}{48}\ln|\sec x + \tan x| + C.$$

36. Given that $f(a) = f(b) = 0$.

$$\int_a^b (x-a)(b-x)f''(x) \, dx$$

$$U = (x-a)(b-x) \qquad dV = f''(x) \, dx$$
$$dU = (b+a-2x) \, dx \qquad V = f'(x)$$

$$= (x-a)(b-x)f'(x)\Big|_a^b - \int_a^b (b+a-2x)f'(x) \, dx$$

$$U = b+a-2x \qquad dV = f'(x) \, dx$$
$$dU = -2dx \qquad\qquad V = f(x)$$

$$= 0 - \left[(b+a-2x)f(x)\Big|_a^b + 2\int_a^b f(x) \, dx\right]$$

$$= -2\int_a^b f(x) \, dx.$$

38. $I_n = \int_0^{\pi/2} \cos^n x \, dx.$

a) For $0 \leq x \leq \pi/2$ we have $0 \leq \cos x \leq 1$, and so $0 \leq \cos^{2n+2} x \leq \cos^{2n+1} x \leq \cos^{2n} x$. Therefore $0 \leq I_{2n+2} \leq I_{2n+1} \leq I_{2n}$.

b) Since $I_n = \dfrac{n-1}{n}I_{n-2}$, we have $I_{2n+2} = \dfrac{2n+1}{2n+2}I_{2n}$. Combining this with part (a), we get

$$\frac{2n+1}{2n+2} = \frac{I_{2n+2}}{I_{2n}} \leq \frac{I_{2n+1}}{I_{2n}} \leq 1.$$

The left side approaches 1 as $n \to \infty$, so, by the Squeeze Theorem,

$$\lim_{n\to\infty} \frac{I_{2n+1}}{I_{2n}} = 1.$$

c) By Example 6 we have, since $2n + 1$ is odd and $2n$ is even,

$$I_{2n+1} = \frac{2n}{2n+1} \cdot \frac{2n-2}{2n-1} \cdots \frac{4}{5} \cdot \frac{2}{3}$$

$$I_{2n} = \frac{2n-1}{2n} \cdot \frac{2n-3}{2n-2} \cdots \frac{3}{4} \cdot \frac{1}{2} \cdot \frac{\pi}{2}.$$

Multiplying the expression for I_{2n+1} by $\pi/2$ and dividing by the expression for I_{2n}, we obtain, by part (b),

$$\lim_{n\to\infty} \frac{\dfrac{2n}{2n+1} \cdot \dfrac{2n-2}{2n-1} \cdots \dfrac{4}{5} \cdot \dfrac{2}{3} \cdot \dfrac{\pi}{2}}{\dfrac{2n-1}{2n} \cdot \dfrac{2n-3}{2n-2} \cdots \dfrac{3}{4} \cdot \dfrac{1}{2} \cdot \dfrac{\pi}{2}} = \frac{\pi}{2} \times 1 = \frac{\pi}{2},$$

or, rearranging the factors on the left,

$$\lim_{n\to\infty} \frac{2}{1} \cdot \frac{2}{3} \cdot \frac{4}{3} \cdot \frac{4}{5} \cdots \frac{2n}{2n-1} \cdot \frac{2n}{2n+1} = \frac{\pi}{2}.$$

Section 6.2   Inverse Substitutions
(page 328)

2. $\displaystyle\int \frac{x^2\,dx}{\sqrt{1-4x^2}}$ Let $2x = \sin u$

$\qquad\qquad\qquad\qquad 2\,dx = \cos u\,du$

$\displaystyle = \frac{1}{8}\int \frac{\sin^2 u\cos u\,du}{\cos u}$

$\displaystyle = \frac{1}{16}\int (1-\cos 2u)\,du = \frac{u}{16} - \frac{\sin 2u}{32} + C$

$\displaystyle = \frac{1}{16}\sin^{-1} 2x - \frac{1}{16}\sin u\cos u + C$

$\displaystyle = \frac{1}{16}\sin^{-1} 2x - \frac{1}{8}x\sqrt{1-4x^2} + C.$

4. $\displaystyle\int \frac{dx}{x\sqrt{1-4x^2}}$ Let $x = \frac{1}{2}\sin\theta$

$\qquad\qquad\qquad\qquad dx = \frac{1}{2}\cos\theta\,d\theta$

$\displaystyle = \int \frac{\cos\theta\,d\theta}{\sin\theta\sqrt{1-\sin^2\theta}} = \int \csc\theta\,d\theta$

$\displaystyle = \ln|\csc\theta - \cot\theta| + C = \ln\left|\frac{1}{2x} - \frac{\sqrt{1-4x^2}}{2x}\right| + C$

$\displaystyle = \ln\left|\frac{1 - \sqrt{1-4x^2}}{x}\right| + C_1.$

6. $\displaystyle\int \frac{dx}{x\sqrt{9-x^2}}$ Let $x = 3\sin\theta$

$\qquad\qquad\qquad\qquad dx = 3\cos\theta\,d\theta$

$\displaystyle = \int \frac{3\cos\theta\,d\theta}{3\sin\theta\,3\cos\theta} = \frac{1}{3}\int \csc\theta\,d\theta$

$\displaystyle = \frac{1}{3}\ln|\csc\theta - \cot\theta| + C = \frac{1}{3}\ln\left|\frac{3}{x} - \frac{\sqrt{9-x^2}}{x}\right| + C$

$\displaystyle = \frac{1}{3}\ln\left|\frac{3 - \sqrt{9-x^2}}{x}\right| + C.$

8. $\displaystyle\int \frac{dx}{\sqrt{9+x^2}}$ Let $x = 3\tan\theta$

$\qquad\qquad\qquad\qquad dx = 3\sec^2\theta\,d\theta$

$\displaystyle = \int \frac{3\sec^2\theta\,d\theta}{3\sec\theta} = \int \sec\theta\,d\theta$

$\displaystyle = \ln|\sec\theta + \tan\theta| + C = \ln(x + \sqrt{9+x^2}) + C_1.$

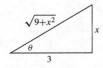

Fig. 6.2.8

10. $\displaystyle\int \frac{\sqrt{9+x^2}}{x^4}\,dx$ Let $x = 3\tan\theta$

$\qquad\qquad\qquad\qquad\quad dx = 3\sec^2\theta\,d\theta$

$\displaystyle = \int \frac{(3\sec\theta)(3\sec^2\theta)\,d\theta}{81\tan^4\theta}$

$\displaystyle = \frac{1}{9}\int \frac{\sec^3\theta}{\tan^4\theta}\,d\theta = \frac{1}{9}\int \frac{\cos\theta}{\sin^4\theta}\,d\theta$ Let $u = \sin\theta$

$\qquad\qquad\qquad\qquad\qquad\qquad\qquad\qquad du = \cos\theta\,d\theta$

$\displaystyle = \frac{1}{9}\int \frac{du}{u^4} = -\frac{1}{27u^3} + C = -\frac{1}{27\sin^3\theta} + C$

$\displaystyle = -\frac{(9+x^2)^{3/2}}{27x^3} + C.$

12. $\displaystyle\int \frac{dx}{(a^2+x^2)^{3/2}}$ Let $x = a\tan\theta$

$\qquad\qquad\qquad\qquad\qquad dx = a\sec^2\theta\,d\theta$

$\displaystyle = \int \frac{a\sec^2\theta\,d\theta}{(a^2 + a^2\tan^2\theta)^{3/2}} = \int \frac{a\sec^2\theta\,d\theta}{a^3\sec^3\theta}$

$\displaystyle = \frac{1}{a^2}\int \cos\theta\,d\theta = \frac{1}{a^2}\sin\theta + C = \frac{x}{a^2\sqrt{a^2+x^2}} + C.$

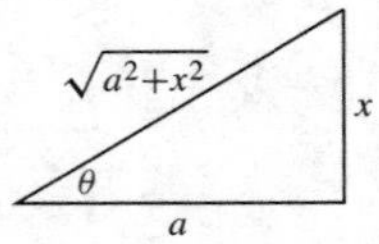

Fig. 6.2.12

14. $\displaystyle\int \frac{dx}{(1 + 2x^2)^{5/2}}$ Let $x = \dfrac{1}{\sqrt{2}}\tan\theta$

$$dx = \frac{1}{\sqrt{2}}\sec^2\theta\,d\theta$$

$$= \frac{1}{\sqrt{2}}\int \frac{\sec^2\theta\,d\theta}{(1 + \tan^2\theta)^{5/2}} = \frac{1}{\sqrt{2}}\int \cos^3\theta\,d\theta$$

$$= \frac{1}{\sqrt{2}}\int (1 - \sin^2\theta)\cos\theta\,d\theta \quad \text{Let } u = \sin\theta$$
$$du = \cos\theta\,d\theta$$

$$= \frac{1}{\sqrt{2}}\int (1 - u^2)\,du = \frac{1}{\sqrt{2}}\left(u - \frac{1}{3}u^3\right) + C$$

$$= \frac{1}{\sqrt{2}}\sin\theta - \frac{1}{3\sqrt{2}}\sin^3\theta + C$$

$$= \frac{\sqrt{2}x}{\sqrt{2}\sqrt{1 + 2x^2}} - \frac{1}{3\sqrt{2}}\left(\frac{\sqrt{2}x}{\sqrt{1 + 2x^2}}\right)^3 + C$$

$$= \frac{4x^3 + 3x}{3(1 + 2x^2)^{3/2}} + C.$$

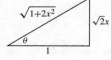

Fig. 6.2.14

16. $\displaystyle\int \frac{dx}{x^2\sqrt{x^2 - a^2}}$ Let $x = a\sec\theta$ $(a > 0)$

$$dx = a\sec\theta\,\tan\theta\,d\theta$$

$$= \int \frac{a\sec\theta\,\tan\theta\,d\theta}{a^2\sec^2\theta\,a\tan\theta}$$

$$= \frac{1}{a^2}\int \cos\theta\,d\theta = \frac{1}{a^2}\sin\theta + C$$

$$= \frac{1}{a^2}\frac{\sqrt{x^2 - a^2}}{x} + C.$$

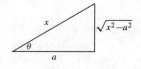

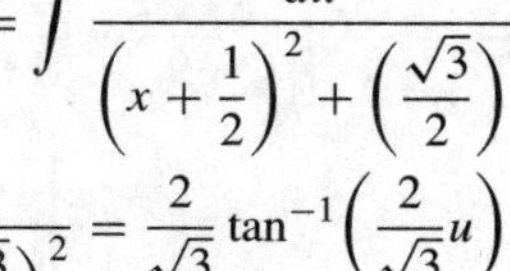

Fig. 6.2.16

18. $\displaystyle\int \frac{dx}{x^2 + x + 1} = \int \frac{dx}{\left(x + \dfrac{1}{2}\right)^2 + \left(\dfrac{\sqrt{3}}{2}\right)^2}$ Let $u = x + \dfrac{1}{2}$
$$du = dx$$

$$= \int \frac{du}{u^2 + \left(\dfrac{\sqrt{3}}{2}\right)^2} = \frac{2}{\sqrt{3}}\tan^{-1}\left(\frac{2}{\sqrt{3}}u\right) + C$$

$$= \frac{2}{\sqrt{3}}\tan^{-1}\left(\frac{2x + 1}{\sqrt{3}}\right) + C.$$

20. $\displaystyle\int \frac{x\,dx}{x^2 - 2x + 3} = \int \frac{(x - 1) + 1}{(x - 1)^2 + 2}\,dx$ Let $u = x - 1$
$$du = dx$$

$$= \int \frac{u\,du}{u^2 + 2} + \int \frac{du}{u^2 + 2}$$

$$= \frac{1}{2}\ln(u^2 + 2) + \frac{1}{\sqrt{2}}\tan^{-1}\left(\frac{u}{\sqrt{2}}\right) + C$$

$$= \frac{1}{2}\ln(x^2 - 2x + 3) + \frac{1}{\sqrt{2}}\tan^{-1}\left(\frac{x - 1}{\sqrt{2}}\right) + C.$$

22. $\displaystyle\int \frac{dx}{(4x - x^2)^{3/2}}$

$$= \int \frac{dx}{[4 - (2 - x)^2]^{3/2}} \quad \text{Let } 2 - x = 2\sin u$$
$$-dx = 2\cos u\,du$$

$$= -\int \frac{2\cos u\,du}{8\cos^3 u} = -\frac{1}{4}\int \sec^2 u\,du$$

$$= -\frac{1}{4}\tan u + C = \frac{1}{4}\frac{x - 2}{\sqrt{4x - x^2}} + C.$$

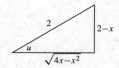

Fig. 6.2.22

24. $\displaystyle\int \frac{dx}{(x^2 + 2x + 2)^2} = \int \frac{dx}{[(x + 1)^2 + 1]^2}$ Let $x + 1 = \tan u$
$$dx = \sec^2 u\,du$$

$$= \int \frac{\sec^2 u\,du}{\sec^4 u} = \int \cos^2 u\,du$$

$$= \frac{1}{2}\int (1 + \cos 2u)\,du = \frac{u}{2} + \frac{\sin 2u}{4} + C$$

$$= \frac{1}{2}\tan^{-1}(x + 1) + \frac{1}{2}\sin u\,\cos u + C$$

$$= \frac{1}{2}\tan^{-1}(x + 1) + \frac{1}{2}\frac{x + 1}{x^2 + 2x + 2} + C.$$

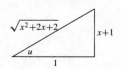

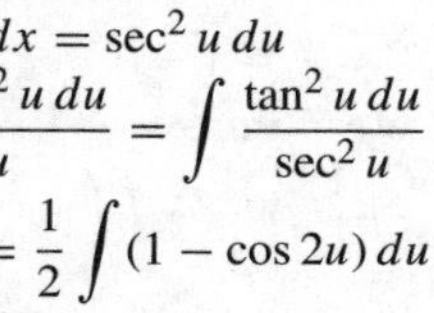

Fig. 6.2.24

26. $\displaystyle\int \frac{x^2\,dx}{(1 + x^2)^2}$ Let $x = \tan u$
$$dx = \sec^2 u\,du$$

$$= \int \frac{\tan^2 u\,\sec^2 u\,du}{\sec^4 u} = \int \frac{\tan^2 u\,du}{\sec^2 u}$$

$$= \int \sin^2 u\,du = \frac{1}{2}\int (1 - \cos 2u)\,du$$

$$= \frac{u}{2} - \frac{\sin u\,\cos u}{2} + C$$

$$= \frac{1}{2}\tan^{-1}x - \frac{1}{2}\frac{x}{1 + x^2} + C.$$

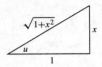

Fig. 6.2.26

28. $I = \displaystyle\int \sqrt{9 + x^2}\, dx$ Let $x = 3 \tan \theta$
$$dx = 3 \sec^2 \theta \, d\theta$$

$$= \int 3 \sec \theta \, 3 \sec^2 \theta \, d\theta$$

$$= 9 \int \sec^3 \theta \, d\theta$$

$$U = \sec \theta \qquad\qquad dV = \sec^2 \theta \, d\theta$$
$$dU = \sec \theta \tan \theta \, d\theta \qquad V = \tan \theta$$

$$= 9 \sec \theta \tan \theta - 9 \int \sec \theta \tan^2 \theta \, d\theta$$

$$= 9 \sec \theta \tan \theta - 9 \int \sec \theta (\sec^2 \theta - 1) \, d\theta$$

$$= 9 \sec \theta \tan \theta + 9 \int \sec \theta \, d\theta - 9 \int \sec^3 \theta \, d\theta$$

$$= 9 \sec \theta \tan \theta + 9 \ln |\sec \theta + \tan \theta| - I$$

$$I = \frac{9}{2}\left[\left(\frac{\sqrt{9 + x^2}}{3} \right) \left(\frac{x}{3} \right) \right] + \frac{9}{2} \ln \left| \frac{\sqrt{9 + x^2}}{3} + \frac{x}{3} \right| + C$$

$$= \frac{1}{2} x \sqrt{9 + x^2} + \frac{9}{2} \ln \left(\sqrt{9 + x^2} + x \right) + C_1.$$

$$\text{(where } C_1 = C - \frac{9}{2} \ln 3)$$

30. $\displaystyle\int \frac{dx}{1 + x^{1/3}}$ Let $x = u^3$
$$dx = 3 u^2 \, du$$

$$= 3 \int \frac{u^2 \, du}{1 + u}$$ Let $v = 1 + u$
$$dv = du$$

$$= 3 \int \frac{v^2 - 2v + 1}{v} \, dv = 3 \int \left(v - 2 + \frac{1}{v} \right) dv$$

$$= 3 \left(\frac{v^2}{2} - 2v + \ln |v| \right) + C$$

$$= \frac{3}{2}(1 + x^{1/3})^2 - 6(1 + x^{1/3}) + 3 \ln |1 + x^{1/3}| + C.$$

32. $\displaystyle\int \frac{x \sqrt{2 - x^2}}{\sqrt{x^2 + 1}} \, dx$ Let $u^2 = x^2 + 1$
$$2u \, du = 2x \, dx$$

$$= \int \frac{u \sqrt{3 - u^2} \, du}{u}$$

$$= \int \sqrt{3 - u^2} \, du$$ Let $u = \sqrt{3} \sin v$
$$du = \sqrt{3} \cos v \, dv$$

$$= \int (\sqrt{3} \cos v) \sqrt{3} \cos v \, dv = 3 \int \cos^2 v \, dv$$

$$= \frac{3}{2}(v + \sin v \, \cos v) + C$$

$$= \frac{3}{2} \sin^{-1} \left(\frac{u}{\sqrt{3}} \right) + \frac{3}{2} \frac{u \sqrt{3 - u^2}}{3} + C$$

$$= \frac{3}{2} \sin^{-1} \left(\sqrt{\frac{x^2 + 1}{3}} \right) + \frac{1}{2} \sqrt{(x^2 + 1)(2 - x^2)} + C.$$

34. $\displaystyle\int_0^{\pi/2} \frac{\cos x}{\sqrt{1 + \sin^2 x}} \, dx$ Let $u = \sin x$
$$du = \cos x \, dx$$

$$= \int_0^1 \frac{du}{\sqrt{1 + u^2}}$$ Let $u = \tan w$
$$du = \sec^2 w \, dw$$

$$= \int_0^{\pi/4} \frac{\sec^2 w \, dw}{\sec w} = \int_0^{\pi/4} \sec w \, dw$$

$$= \ln |\sec w + \tan w| \Big|_0^{\pi/4}$$

$$= \ln |\sqrt{2} + 1| - \ln |1 + 0| = \ln(\sqrt{2} + 1).$$

36. $\displaystyle\int_1^2 \frac{dx}{x^2 \sqrt{9 - x^2}}$ Let $x = 3 \sin u$
$$dx = 3 \cos u \, du$$

$$= \int_{x=1}^{x=2} \frac{3 \cos u \, du}{9 \sin^2 u (3 \cos u)} = \frac{1}{9} \int_{x=1}^{x=2} \csc^2 u \, du$$

$$= \frac{1}{9} (-\cot u) \Big|_{x=1}^{x=2} = -\frac{1}{9} \left(\frac{\sqrt{9 - x^2}}{x} \right) \Big|_{x=1}^{x=2}$$

$$= -\frac{1}{9} \left(\frac{\sqrt{5}}{2} - \frac{\sqrt{8}}{1} \right) = \frac{2\sqrt{2}}{9} - \frac{\sqrt{5}}{18}.$$

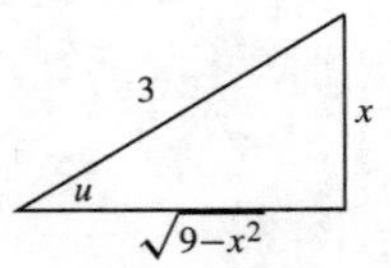

Fig. 6.2.36

38. $\displaystyle\int_0^{\pi/2} \frac{d\theta}{1+\cos\theta+\sin\theta}$ Let $x = \tan\dfrac{\theta}{2}$, $d\theta = \dfrac{2}{1+x^2}\,dx$,

$$\cos\theta = \frac{1-x^2}{1+x^2}, \quad \sin\theta = \frac{2x}{1+x^2}.$$

$$= \int_0^1 \frac{\left(\dfrac{2}{1+x^2}\right) dx}{1+\left(\dfrac{1-x^2}{1+x^2}\right)+\left(\dfrac{2x}{1+x^2}\right)}$$

$$= 2\int_0^1 \frac{dx}{2+2x} = \int_0^1 \frac{dx}{1+x}$$

$$= \ln|1+x|\Big|_0^1 = \ln 2.$$

40. Area $= \displaystyle\int_{1/2}^1 \frac{dx}{\sqrt{2x-x^2}} = \int_{1/2}^1 \frac{dx}{\sqrt{1-(x-1)^2}}$

Let $u = x-1$

$du = dx$

$$= \int_{-1/2}^0 \frac{du}{\sqrt{1-u^2}} = \sin^{-1} u\Big|_{-1/2}^0$$

$$= 0 - \left(-\frac{\pi}{6}\right) = \frac{\pi}{6} \text{ sq. units.}$$

42. Average value $= \dfrac{1}{4}\displaystyle\int_0^4 \frac{dx}{(x^2-4x+8)^{3/2}}$

$$= \frac{1}{4}\int_0^4 \frac{dx}{[(x-2)^2+4]^{3/2}}$$

Let $x-2 = 2\tan u$

$dx = 2\sec^2 u\,du$

$$= \frac{1}{4}\int_{-\pi/4}^{\pi/4} \frac{2\sec^2 u\,du}{8\sec^3 u}$$

$$= \frac{1}{16}\int_{-\pi/4}^{\pi/4} \cos u\,du = \frac{1}{16}\sin u\Big|_{-\pi/4}^{\pi/4}$$

$$= \frac{1}{16}\left(\frac{1}{\sqrt{2}}+\frac{1}{\sqrt{2}}\right) = \frac{\sqrt{2}}{16}.$$

44. The circles intersect at $x = \frac{1}{4}$, so the common area is

$A_1 + A_2$ where

$$A_1 = 2\int_{1/4}^1 \sqrt{1-x^2}\,dx \quad \text{Let } x = \sin u$$
$$dx = \cos u\,du$$

$$= 2\int_{x=1/4}^{x=1} \cos^2 u\,du$$

$$= (u+\sin u\cos u)\Big|_{x=1/4}^{x=1}$$

$$= (\sin^{-1} x + x\sqrt{1-x^2})\Big|_{x=1/4}^{x=1}$$

$$= \frac{\pi}{2} - \sin^{-1}\frac{1}{4} - \frac{\sqrt{15}}{16} \text{ sq. units.}$$

$$A_2 = 2\int_0^{1/4} \sqrt{4-(x-2)^2}\,dx \quad \text{Let } x-2 = 2\sin v$$
$$dx = 2\cos v\,dv$$

$$= 8\int_{x=0}^{x=1/4} \cos^2 v\,dv$$

$$= 4(v+\sin v\cos v)\Big|_{x=0}^{x=1/4}$$

$$= 4\left[\sin^{-1}\left(\frac{x-2}{2}\right)+\left(\frac{x-2}{2}\right)\frac{\sqrt{4x-x^2}}{2}\right]\Bigg|_{x=0}^{x=1/4}$$

$$= 4\left[\sin^{-1}\left(-\frac{7}{8}\right)-\frac{7\sqrt{15}}{64}+\frac{\pi}{2}\right]$$

$$= -4\sin^{-1}\left(\frac{7}{8}\right)-\frac{7\sqrt{15}}{16}+2\pi \text{ sq. units.}$$

Hence, the common area is

$$A_1 + A_2 = \frac{5\pi}{2} - \frac{\sqrt{15}}{2}$$
$$- \sin^{-1}\left(\frac{1}{4}\right) - 4\sin^{-1}\left(\frac{7}{8}\right) \text{ sq. units.}$$

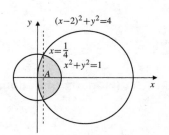

Fig. 6.2.44

46. Shaded area $= 2 \int_c^a b\sqrt{1 - \left(\dfrac{x}{a}\right)^2}\, dx$ Let $x = a \sin u$

$$dx = a \cos u\, du$$

$$= 2ab \int_{x=c}^{x=a} \cos^2 u\, du$$

$$= ab(u + \sin u \cos u)\Big|_{x=c}^{x-a}$$

$$= \left(ab \sin^{-1} \frac{x}{a} + \frac{b}{a} x\sqrt{a^2 - x^2}\right)\Big|_c^a$$

$$= ab\left(\frac{\pi}{2} - \sin^{-1} \frac{c}{a}\right) - \frac{cb}{a}\sqrt{a^2 - c^2} \text{ sq. units.}$$

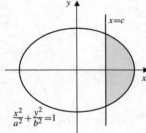

Fig. 6.2.46

48. $\displaystyle\int \frac{dx}{\sqrt{x^2 - a^2}}$ Let $x = a \cosh u$

$$dx = a \sinh u\, du$$

$$= \int \frac{a \sinh u\, du}{a \sinh u} = u + C$$

$$= \cosh^{-1} \frac{x}{a} + C = \ln(x + \sqrt{x^2 - a^2}) + C, \quad (x \geq a).$$

$$\int \frac{dx}{x^2\sqrt{x^2 - a^2}} = \int \frac{a \sinh u\, du}{a^2 \cosh^2 u\, a \sinh u}$$

$$= \frac{1}{a^2} \int \operatorname{sech}^2 u\, du = \frac{1}{a^2} \tanh u + C$$

$$= \frac{1}{a^2} \tanh \left(\cosh^{-1} \frac{x}{a}\right) + C$$

$$= \frac{1}{a^2} \cdot \frac{\dfrac{x}{a} + \sqrt{\dfrac{x^2}{a^2} - 1} - \dfrac{1}{\dfrac{x}{a} - \sqrt{\dfrac{x^2}{a^2} - 1}}}{\dfrac{x}{a} + \sqrt{\dfrac{x^2}{a^2} - 1} + \dfrac{1}{\dfrac{x}{a} - \sqrt{\dfrac{x^2}{a^2} - 1}}} + C$$

$$= \frac{\sqrt{x^2 - a^2}}{a^2 x} + C_1.$$

Section 6.3   Integrals of Rational Functions (page 336)

2. $\displaystyle\int \frac{dx}{5 - 4x} = -\frac{1}{4} \ln |5 - 4x| + C.$

4. $\displaystyle\int \frac{x^2}{x - 4}\, dx = \int \left(x + 4 + \frac{16}{x - 4}\right) dx$

$$= \frac{x^2}{2} + 4x + 16 \ln |x - 4| + C.$$

6. $\displaystyle\frac{1}{5 - x^2} = \frac{A}{\sqrt{5} - x} + \frac{B}{\sqrt{5} + x}$

$$= \frac{(A + B)\sqrt{5} + (A - B)x}{5 - x^2}$$

$$\Rightarrow \begin{cases} A + B = \dfrac{1}{\sqrt{5}} \\ A - B = 0 \end{cases} \Rightarrow A = B = \frac{1}{2\sqrt{5}}.$$

$$\int \frac{1}{5 - x^2}\, dx = \frac{1}{2\sqrt{5}} \int \left(\frac{1}{\sqrt{5} - x} + \frac{1}{\sqrt{5} + x}\right) dx$$

$$= \frac{1}{2\sqrt{5}}\left(-\ln |\sqrt{5} - x| + \ln |\sqrt{5} + x|\right) + C$$

$$= \frac{1}{2\sqrt{5}} \ln \left|\frac{\sqrt{5} + x}{\sqrt{5} - x}\right| + C.$$

8. $\displaystyle\frac{1}{b^2 - a^2 x^2} = \frac{A}{b - ax} + \frac{B}{b + ax}$

$$= \frac{(A + B)b + (A - B)ax}{b^2 - a^2 x^2}$$

$$\Rightarrow A = B = \frac{1}{2b}$$

$$\int \frac{dx}{b^2 - a^2 x^2} = \frac{1}{2b} \int \left(\frac{1}{b - ax} + \frac{1}{b + ax}\right) dx$$

$$= \frac{1}{2b}\left(\frac{-\ln |b - ax|}{a} + \frac{\ln |b + ax|}{a}\right) + C$$

$$= \frac{1}{2ab} \ln \left|\frac{b + ax}{b - ax}\right| + C.$$

10. $\displaystyle\frac{x}{3x^2 + 8x - 3} = \frac{A}{3x - 1} + \frac{B}{x + 3}$

$$= \frac{(A + 3B)x + (3A - B)}{3x^2 + 8x - 3}$$

$$\Rightarrow \begin{cases} A + 3B = 1 \\ 3A - B = 0 \end{cases} \Rightarrow A = \frac{1}{10}, \ B = \frac{3}{10}.$$

$$\int \frac{x\, dx}{3x^2 + 8x - 3} = \frac{1}{10} \int \left(\frac{1}{3x - 1} + \frac{3}{x + 3}\right) dx$$

$$= \frac{1}{30} \ln |3x - 1| + \frac{3}{10} \ln |x + 3| + C.$$

12. $\displaystyle\frac{1}{x^3 + 9x} = \frac{A}{x} + \frac{Bx + C}{x^2 + 9}$

$$= \frac{Ax^2 + 9A + Bx^2 + Cx}{x^3 + 9x}$$

$$\Rightarrow \begin{cases} A + B = 0 \\ C = 0 \\ 9A = 1 \end{cases} \Rightarrow A = \frac{1}{9}, \ B = -\frac{1}{9}, \ C = 0.$$

$$\int \frac{dx}{x^3 + 9x} = \frac{1}{9} \int \left(\frac{1}{x} - \frac{x}{x^2 + 9}\right) dx$$

$$= \frac{1}{9} \ln |x| - \frac{1}{18} \ln(x^2 + 9) + K.$$

14. $\int \dfrac{x}{2+6x+9x^2}\,dx = \int \dfrac{x}{(3x+1)^2+1}\,dx$ Let $u=3x+1$

$\qquad\qquad\qquad\qquad\qquad\qquad\qquad\qquad du = 3\,dx$

$\dfrac{1}{9}\int \dfrac{u-1}{u^2+1}\,du = \dfrac{1}{9}\int \dfrac{u}{u^2+1}\,du - \dfrac{1}{9}\int \dfrac{1}{u^2+1}\,du$

$\qquad = \dfrac{1}{18}\ln(u^2+1) - \dfrac{1}{9}\tan^{-1} u + C$

$\qquad = \dfrac{1}{18}\ln(2+6x+9x^2) - \dfrac{1}{9}\tan^{-1}(3x+1) + C.$

16. First divide to obtain

$$\dfrac{x^3+1}{x^2+7x+12} = x-7+\dfrac{37x+85}{(x+4)(x+3)}$$

$$\dfrac{37x+85}{(x+4)(x+3)} = \dfrac{A}{x+4}+\dfrac{B}{x+3}$$

$$= \dfrac{(A+B)x+3A+4B}{x^2+7x+12}$$

$$\Rightarrow \begin{cases} A+B=37 \\ 3A+4B=85 \end{cases} \Rightarrow A=63,\ B=-26.$$

Now we have

$$\int \dfrac{x^3+1}{12+7x+x^2}\,dx = \int\left(x-7+\dfrac{63}{x+4}-\dfrac{26}{x+3}\right)dx$$

$$= \dfrac{x^2}{2}-7x+63\ln|x+4|-26\ln|x+3|+C.$$

18. The partial fraction decomposition is

$$\dfrac{1}{x^4-a^4} = \dfrac{A}{x-a}+\dfrac{B}{x+a}+\dfrac{Cx+D}{x^2+a^2}$$

$$= \dfrac{A(x^3+ax^2+a^2x+a^3)+B(x^3-ax^2+a^2x-a^3)}{x^4-a^4}$$

$$+ \dfrac{C(x^3-a^2x)+D(x^2-a^2)}{x^4-a^4}$$

$$\Rightarrow \begin{cases} A+B+C=0 \\ aA-aB+D=0 \\ a^2A+a^2B-a^2C=0 \\ a^3A-a^3B-a^2D=1 \end{cases}$$

$$\Rightarrow A=\dfrac{1}{4a^3},\ B=-\dfrac{1}{4a^3},\ C=0,\ D=-\dfrac{1}{2a^2}.$$

$$\int \dfrac{dx}{x^4-a^4} = \dfrac{1}{4a^3}\int\left(\dfrac{1}{x-a}-\dfrac{1}{x+a}-\dfrac{2a}{x^2+a^2}\right)dx$$

$$= \dfrac{1}{4a^3}\ln\left|\dfrac{x-a}{x+a}\right|-\dfrac{1}{2a^3}\tan^{-1}\left(\dfrac{x}{a}\right)+K.$$

20. Here the expansion is

$$\dfrac{1}{x^3+2x^2+2x} = \dfrac{A}{x}+\dfrac{Bx+C}{x^2+2x+2}$$

$$= \dfrac{A(x^2+2x+2)+Bx^2+Cx}{x^3+2x^2+2}$$

$$\Rightarrow \begin{cases} A+B=0 \\ 2A+C=0 \\ 2A=1 \end{cases} \Rightarrow A=-B=\dfrac{1}{2},\ C=-1,$$

so we have

$$\int \dfrac{dx}{x^3+2x^2+2x} = \dfrac{1}{2}\int \dfrac{dx}{x}-\dfrac{1}{2}\int \dfrac{x+2}{x^2+2x+2}\,dx$$

$\qquad\qquad$ Let $u=x+1$

$\qquad\qquad\qquad du=dx$

$$= \dfrac{1}{2}\ln|x|-\dfrac{1}{2}\int \dfrac{u+1}{u^2+1}\,du$$

$$= \dfrac{1}{2}\ln|x|-\dfrac{1}{4}\ln(u^2+1)-\dfrac{1}{2}\tan^{-1} u + K$$

$$= \dfrac{1}{2}\ln|x|-\dfrac{1}{4}\ln(x^2+2x+2)-\dfrac{1}{2}\tan^{-1}(x+1)+K.$$

22. Here the expansion is

$$\dfrac{x^2+1}{x^3+8} = \dfrac{A}{x+2}+\dfrac{Bx+C}{x^2-2x+4}$$

$$= \dfrac{A(x^2-2x+4)+B(x^2+2x)+C(x+2)}{x^3+8}$$

$$\Rightarrow \begin{cases} A+B=1 \\ -2A+2B+C=0 \\ 4A+2C=1 \end{cases} \Rightarrow A=\dfrac{5}{12}\ B=\dfrac{7}{12},\ C=-\dfrac{1}{3},$$

so we have

$$\int \dfrac{x^2+1}{x^3+8}\,dx = \dfrac{5}{12}\int \dfrac{dx}{x+2}+\dfrac{1}{12}\int \dfrac{7x-4}{(x-1)^2+3}\,dx$$

$\qquad\qquad$ Let $u=x-1$

$\qquad\qquad\qquad du=dx$

$$= \dfrac{5}{12}\ln|x+2|+\dfrac{1}{12}\int \dfrac{7u+3}{u^2+3}\,du$$

$$= \dfrac{5}{12}\ln|x+2|+\dfrac{7}{24}\ln(x^2-2x+4)$$

$$+ \dfrac{1}{4\sqrt{3}}\tan^{-1}\dfrac{x-1}{\sqrt{3}}+K.$$

24. The expansion is

$$\dfrac{x^2}{(x^2-1)(x^2-4)} = \dfrac{A}{x-1}+\dfrac{B}{x+1}+\dfrac{C}{x-2}+\dfrac{D}{x+2}$$

$$A=\lim_{x\to 1}\dfrac{x^2}{(x+1)(x^2-4)}=\dfrac{1}{2(-3)}=-\dfrac{1}{6}$$

$$B=\lim_{x\to -1}\dfrac{x^2}{(x-1)(x^2-4)}=\dfrac{1}{-2(-3)}=\dfrac{1}{6}$$

$$C=\lim_{x\to 2}\dfrac{x^2}{(x^2-1)(x+2)}=\dfrac{4}{3(4)}=\dfrac{1}{3}$$

$$D=\lim_{x\to -2}\dfrac{x^2}{(x^2-1)(x-2)}=\dfrac{4}{3(-4)}=-\dfrac{1}{3}.$$

Therefore

$$\int \frac{x^2}{(x^2 - 1)(x^2 - 4)} \, dx = -\frac{1}{6} \ln |x - 1| + \frac{1}{6} \ln |x + 1| + \frac{1}{3} \ln |x - 2| - \frac{1}{3} \ln |x + 2| + K.$$

26. We have

$$\int \frac{x \, dx}{(x^2 - x + 1)^2} = \int \frac{x \, dx}{\left[(x - \frac{1}{2})^2 + \frac{3}{4} \right]^2} \quad \begin{matrix} \text{Let } u = x - \frac{1}{2} \\ du = dx \end{matrix}$$

$$= \int \frac{u \, du}{(u^2 + \frac{3}{4})^2} + \frac{1}{2} \int \frac{du}{(u^2 + \frac{3}{4})^2}$$

Let $u = \frac{\sqrt{3}}{2} \tan v$,

$du = \frac{\sqrt{3}}{2} \sec^2 v \, dv$ in the second integral.

$$= -\frac{1}{2} \left(\frac{1}{u^2 + \frac{3}{4}} \right) + \frac{1}{2} \int \frac{\frac{\sqrt{3}}{2} \sec^2 v \, dv}{\frac{9}{16} \sec^4 v}$$

$$= \frac{-1}{2(x^2 - x + 1)} + \frac{4}{3\sqrt{3}} \int \cos^2 v \, dv$$

$$= \frac{-1}{2(x^2 - x + 1)} + \frac{2}{3\sqrt{3}} (v + \sin v \cos v) + C$$

$$= \frac{-1}{2(x^2 - x + 1)} + \frac{2}{3\sqrt{3}} \tan^{-1} \frac{2x - 1}{\sqrt{3}} + \frac{2}{3\sqrt{3}} \frac{2(x - \frac{1}{2})\sqrt{3}}{(2\sqrt{x^2 - x + 1})^2} + C$$

$$= \frac{2}{3\sqrt{3}} \tan^{-1} \frac{2x - 1}{\sqrt{3}} + \frac{x - 2}{3(x^2 - x + 1)} + C.$$

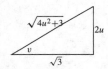

Fig. 6.3.26

28. We have

$$\int \frac{dt}{(t - 1)(t^2 - 1)^2}$$

$$= \int \frac{dt}{(t - 1)^3 (t + 1)^2} \quad \begin{matrix} \text{Let } u = t - 1 \\ du = dt \end{matrix}$$

$$= \int \frac{du}{u^3 (u + 2)^2}$$

$$\frac{1}{u^3 (u + 2)^2} = \frac{A}{u} + \frac{B}{u^2} + \frac{C}{u^3} + \frac{D}{u + 2} + \frac{E}{(u + 2)^2}$$

$$= \frac{A(u^4 + 4u^3 + 4u^2) + B(u^3 + 4u^2 + 4u)}{u^3 (u + 2)^2}$$

$$\frac{C(u^2 + 4u + 4) + D(u^4 + 2u^3) + Eu^3}{u^3 (u + 2)^2}$$

$$\Rightarrow \begin{cases} A + D = 0 \\ 4A + B + 2D + E = 0 \\ 4A + 4B + C = 0 \\ 4B + 4C = 0 \\ 4C = 1 \end{cases}$$

$$\Rightarrow A = \frac{3}{16}, \ B = -\frac{1}{4}, \ C = \frac{1}{4}, \ D = -\frac{3}{16}, \ E = -\frac{1}{8}.$$

$$\int \frac{du}{u^3 (u + 2)^2}$$

$$= \frac{3}{16} \int \frac{du}{u} - \frac{1}{4} \int \frac{du}{u^2} + \frac{1}{4} \int \frac{du}{u^3}$$

$$- \frac{3}{16} \int \frac{du}{u + 2} - \frac{1}{8} \int \frac{du}{(u + 2)^2}$$

$$= \frac{3}{16} \ln |t - 1| + \frac{1}{4(t - 1)} - \frac{1}{8(t - 1)^2} - \frac{3}{16} \ln |t + 1| + \frac{1}{8(t + 1)} + K.$$

30. $\displaystyle \int \frac{dx}{e^{2x} - 4e^x + 4} = \int \frac{dx}{(e^x - 2)^2} \quad \begin{matrix} \text{Let } u = e^x \\ du = e^x \, dx \end{matrix}$

$$= \int \frac{du}{u(u - 2)^2}$$

$$\frac{1}{u(u - 2)^2} = \frac{A}{u} + \frac{B}{u - 2} + \frac{C}{(u - 2)^2}$$

$$= \frac{A(u^2 - 4u + 4) + B(u^2 - 2u) + + Cu}{u(u - 2)^2}$$

$$\Rightarrow \begin{cases} A + B = 0 \\ -4A - 2B + C = 0 \Rightarrow A = \frac{1}{4}, \ B = -\frac{1}{4}, \ C = \frac{1}{2}. \\ 4A = 1 \end{cases}$$

$$\int \frac{du}{u(u - 2)^2} = \frac{1}{4} \int \frac{du}{u} - \frac{1}{4} \int \frac{du}{u - 2} + \frac{1}{2} \int \frac{du}{(u - 2)^2}$$

$$= \frac{1}{4} \ln |u| - \frac{1}{4} \ln |u - 2| - \frac{1}{2} \frac{1}{(u - 2)} + K$$

$$= \frac{x}{4} - \frac{1}{4} \ln |e^x - 2| - \frac{1}{2(e^x - 2)} + K.$$

32. We have

$$I = \int \frac{dx}{x(1 - x^2)^{3/2}} \quad \begin{matrix} \text{Let } u^2 = 1 - x^2 \\ 2u \, du = -2x \, dx \end{matrix}$$

$$= -\int \frac{u \, du}{(1 - u^2) u^3} = -\int \frac{du}{(1 - u^2) u^2}$$

$$\frac{1}{u^2 (1 - u^2)} = \frac{A}{u} + \frac{B}{u^2} + \frac{C}{1 - u} + \frac{D}{1 + u}$$

$$= \frac{A(u - u^3) + B(1 - u^2) + C(u^2 + u^3) + D(u^2 - u^3)}{u^2 (1 - u^2)}$$

$$\Rightarrow \begin{cases} -A + C - D = 0 \\ -B + C + D = 0 \\ A = 0 \\ B = 1 \end{cases}$$

$$\Rightarrow A = 0, \ B = 1, \ C = \frac{1}{2}, \ D = \frac{1}{2}.$$

$$I = -\int \frac{du}{(1-u^2)u^2} = -\int \frac{du}{u^2} - \frac{1}{2}\int \frac{du}{1-u} - \frac{1}{2}\int \frac{du}{1+u}$$

$$= \frac{1}{u} + \frac{1}{2}\ln|1-u| - \frac{1}{2}\ln|1+u| + K$$

$$= \frac{1}{\sqrt{1-x^2}} + \frac{1}{2}\ln\left|\frac{1-\sqrt{1-x^2}}{1+\sqrt{1-x^2}}\right| + K$$

$$= \frac{1}{\sqrt{1-x^2}} + \ln\left(1-\sqrt{1-x^2}\right) - \ln|x| + K.$$

34. $\displaystyle\int \frac{d\theta}{\cos\theta(1+\sin\theta)}$ Let $u = \sin\theta$
$du = \cos\theta\, d\theta$

$$= \int \frac{du}{(1-u^2)(1+u)} = \int \frac{du}{(1-u)(1+u)^2}$$

$$\frac{1}{(1-u)(1+u)^2} = \frac{A}{1-u} + \frac{B}{1+u} + \frac{C}{(1+u)^2}$$

$$= \frac{A(1+2u+u^2) + B(1-u^2) + C(1-u)}{(1-u)(1+u)^2}$$

$$\Rightarrow \begin{cases} A - B = 0 \\ 2A - C = 0 \\ A + B + C = 1 \end{cases} \Rightarrow A = \frac{1}{4},\ B = \frac{1}{4},\ C = \frac{1}{2}.$$

$$\int \frac{du}{(1-u)(1+u)^2}$$

$$= \frac{1}{4}\int \frac{du}{1-u} + \frac{1}{4}\int \frac{du}{1+u} + \frac{1}{2}\int \frac{du}{(1+u)^2}$$

$$= \frac{1}{4}\ln\left|\frac{1+\sin\theta}{1-\sin\theta}\right| - \frac{1}{2(1+\sin\theta)} + C.$$

Section 6.4 Integration Using Computer Algebra or Tables (page 340)

2. According to Maple

$$\int \frac{1+x+x^2}{(x^4-1)(x^4-16)^2}\, dx$$

$$= \frac{\ln(x-1)}{300} - \frac{\ln(x+1)}{900} - \frac{7}{15,360(x-2)}$$

$$\quad - \frac{613}{460,800}\ln(x-2) - \frac{1}{5,120(x+2)} + \frac{79}{153,600}\ln(x+2)$$

$$\quad - \frac{\ln(x^2+1)}{900} + \frac{47}{115,200}\ln(x^2+4)$$

$$\quad - \frac{23}{25,600}\tan^{-1}(x/2) - \frac{6x+8}{15,360(x^2+4)}$$

One suspects it has forgotten to use absolute values in some of the logarithms.

4. Maple, Mathematica, and Derive readily gave

$$\int_0^1 \frac{1}{(x^2+1)^3}\, dx = \frac{3\pi}{32} + \frac{1}{4}.$$

6. Use the last integral in the list involving $\sqrt{x^2 \pm a^2}$.

$$\int \sqrt{(x^2+4)^3}\, dx = \frac{x}{4}(x^2+10)\sqrt{x^2+4} + 6\ln|x+\sqrt{x^2+4}| + C$$

8. Use the 8th integral in the miscellaneous algebraic set.

$$\int \frac{dt}{t\sqrt{3t-5}} = \frac{2}{\sqrt{5}}\tan^{-1}\sqrt{\frac{3t-5}{5}} + C$$

10. We make a change of variable and then use the first two integrals in the exponential/logarithmic set.

$$\int x^7 e^{x^2}\, dx \quad \text{Let } u = x^2$$
$$\qquad\qquad\qquad du = 2x\, dx$$

$$= \frac{1}{2}\int u^3 e^u\, du$$

$$= \frac{1}{2}\left(u^3 e^u - 3\int u^2 e^u\, du\right)$$

$$= \frac{u^3 e^u}{2} - \frac{3}{2}\left(u^2 e^u - 2\int u e^u\, du\right)$$

$$= \left(\frac{u^3}{2} - \frac{3u^2}{2} + 3(u-1)\right)e^u + C$$

$$= \left(\frac{x^6}{2} - \frac{3x^4}{2} + 3x^2 - 3\right)e^{x^2} + C$$

12. Use integrals 17 and 16 in the miscellaneous algebraic set.

$$\int \frac{\sqrt{2x-x^2}}{x^2}\, dx$$

$$= -\frac{(2x-x^2)^{3/2}}{x^2} - \frac{1}{1}\int \frac{\sqrt{2x-x^2}}{x}\, dx$$

$$= -\frac{(2x-x^2)^{3/2}}{x^2} - \sqrt{2x-x^2} - \sin^{-1}(x-1) + C$$

14. Use the last integral in the miscellaneous algebraic set. Then complete the square, change variables, and use the second last integral in the elementary list.

$$\int \frac{dx}{(\sqrt{4x-x^2})^4}$$

$$= \frac{x-2}{8}(\sqrt{4x-x^2})^{-2} + \frac{1}{8}\int \frac{dx}{4x-x^2}$$

$$= \frac{x-2}{8(4x-x^2)} + \frac{1}{8}\int \frac{dx}{4-(x-2)^2} \quad \text{Let } u = x-2$$
$$\qquad\qquad\qquad\qquad\qquad\qquad du = dx$$

$$= \frac{x-2}{8(4x-x^2)} + \frac{1}{8}\int \frac{du}{4-u^2}$$

$$= \frac{x-2}{8(4x-x^2)} + \frac{1}{32}\ln\left|\frac{u+2}{u-2}\right| + C$$

$$= \frac{x-2}{8(4x-x^2)} + \frac{1}{32}\ln\left|\frac{x}{x-4}\right| + C$$

Section 6.5 Improper Integrals (page 347)

2. $\displaystyle\int_3^\infty \frac{1}{(2x-1)^{2/3}}\,dx$ Let $u = 2x - 1$

$$du = 2\,dx$$

$$= \frac{1}{2}\int_5^\infty \frac{du}{u^{2/3}} = \frac{1}{2}\lim_{R\to\infty}\int_5^R u^{-2/3}\,du$$

$$= \frac{1}{2}\lim_{R\to\infty} 3u^{1/3}\Big|_5^R = \infty \quad \text{(diverges)}$$

4. $\displaystyle\int_{-\infty}^{-1} \frac{dx}{x^2+1} = \lim_{R\to -\infty}\int_R^{-1}\frac{dx}{x^2+1}$

$$= \lim_{R\to -\infty}\left[\tan^{-1}(-1) - \tan^{-1}(R)\right]$$

$$= -\frac{\pi}{4} - \left(-\frac{\pi}{2}\right) = \frac{\pi}{4}.$$

This integral converges.

6. $\displaystyle\int_0^a \frac{dx}{a^2 - x^2} = \lim_{C\to a-}\int_0^C \frac{dx}{a^2 - x^2}$

$$= \lim_{C\to a-}\frac{1}{2a}\ln\left|\frac{a+x}{a-x}\right|\,\Big|_0^C$$

$$= \lim_{C\to a-}\frac{1}{2a}\ln\frac{a+C}{a-C} = \infty.$$

The integral diverges to infinity.

8. $\displaystyle\int_0^1 \frac{dx}{x\sqrt{1-x}}$ Let $u^2 = 1 - x$

$$2u\,du = -dx$$

$$= \int_0^1 \frac{2u\,du}{(1-u^2)u} = 2\lim_{c\to 1-}\int_0^c \frac{du}{1-u^2}$$

$$= 2\lim_{c\to 1-}\frac{1}{2}\ln\left|\frac{u+1}{u-1}\right|\,\Big|_0^c = \infty \quad \text{(diverges)}$$

10. $\displaystyle\int_0^\infty xe^{-x}\,dx$

$$= \lim_{R\to\infty}\int_0^R xe^{-x}\,dx$$

$$U = x \qquad dV = e^{-x}\,dx$$

$$dU = dx \qquad V = -e^{-x}$$

$$= \lim_{R\to\infty}\left(-xe^{-x}\Big|_0^R + \int_0^R e^{-x}\,dx\right)$$

$$= \lim_{R\to\infty}\left(-\frac{R}{e^R} - \frac{1}{e^R} + 1\right) = 1.$$

The integral converges.

12. $\displaystyle\int_0^\infty \frac{x}{1+2x^2}\,dx = \lim_{R\to\infty}\int_0^R \frac{x}{1+2x^2}\,dx$

$$= \lim_{R\to\infty}\frac{1}{4}\ln(1+2x^2)\,\Big|_0^R$$

$$= \lim_{R\to\infty}\left[\frac{1}{4}\ln(1+2R^2) - \frac{1}{4}\ln 1\right] = \infty.$$

This integral diverges to infinity.

14. $\displaystyle\int_0^{\pi/2}\sec x\,dx = \lim_{C\to(\pi/2)-}\ln|\sec x + \tan x|\,\Big|_0^C$

$$= \lim_{C\to(\pi/2)-}\ln|\sec C + \tan C| = \infty.$$

This integral diverges to infinity.

16. $\displaystyle\int_e^\infty \frac{dx}{x(\ln x)}$ Let $u = \ln x$

$$du = \frac{dx}{x}$$

$$= \lim_{R\to\infty}\int_1^{\ln R}\frac{du}{u} = \lim_{R\to\infty}\ln|u|\,\Big|_1^{\ln R}$$

$$= \lim_{R\to\infty}\ln(\ln R) - \ln 1 = \infty.$$

This integral diverges to infinity.

18. $\displaystyle\int_e^\infty \frac{dx}{x(\ln x)^2}$ Let $u = \ln x$

$$du = \frac{dx}{x}$$

$$= \lim_{R\to\infty}\int_1^{\ln R}\frac{du}{u^2} = \lim_{R\to\infty}\left(-\frac{1}{\ln R} + 1\right) = 1.$$

The integral converges.

20. $\displaystyle I = \int_{-\infty}^\infty \frac{x\,dx}{1+x^4} = \int_{-\infty}^0 + \int_0^\infty = I_1 + I_2$

$$I_2 = \int_0^\infty \frac{x\,dx}{1+x^4} \quad \text{Let } u = x^2$$

$$du = 2x\,dx$$

$$= \frac{1}{2}\int_0^\infty \frac{du}{1+u^2} = \frac{1}{2}\lim_{R\to\infty}\tan^{-1}u\,\Big|_0^R = \frac{\pi}{4}$$

Similarly, $I_1 = -\dfrac{\pi}{4}$. Therefore, $I = 0$.

22. $\displaystyle I = \int_{-\infty}^\infty e^{-|x|}\,dx = \int_{-\infty}^0 e^x\,dx + \int_0^\infty e^{-x}\,dx = I_1 + I_2$

$$I_2 = \int_0^\infty e^{-x}\,dx = 1$$

Similarly, $I_1 = 1$. Therefore, $I = 2$.

24. Area of shaded region $= \displaystyle\int_0^\infty (e^{-x} - e^{-2x})\,dx$

$$= \lim_{R\to\infty}\left(-e^{-x} + \frac{1}{2}e^{-2x}\right)\,\Big|_0^R$$

$$= \lim_{R\to\infty}\left(-e^{-R} + \frac{1}{2}e^{-2R} + 1 - \frac{1}{2}\right) = \frac{1}{2} \text{ sq. units.}$$

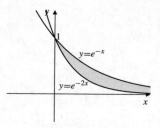

Fig. 6.5.24

26. The required area is

$$\text{Area} = \int_0^\infty x^{-2} e^{-1/x} \, dx$$

$$= \int_0^1 x^{-2} e^{-1/x} \, dx + \int_1^\infty x^{-2} e^{-1/x} \, dx$$

$$= I_1 + I_2.$$

Then let $u = -\dfrac{1}{x}$ and $du = x^{-2} \, dx$ in both I_1 and I_2:

$$I_1 = \lim_{C \to 0+} \int_C^1 x^{-2} e^{-1/x} \, dx = \lim_{C \to 0+} \int_{-1/C}^{-1} e^u \, du$$

$$= \lim_{C \to 0+} (e^{-1} - e^{-1/C}) = \frac{1}{e}.$$

$$I_2 = \lim_{R \to \infty} \int_1^R x^{-2} e^{-1/x} \, dx = \lim_{R \to \infty} \int_{-1}^{-1/R} e^u \, du$$

$$= \lim_{R \to \infty} (e^{-1/R} - e^{-1}) = 1 - \frac{1}{e}.$$

Hence, the total area is $I_1 + I_2 = 1$ square unit.

28. $\displaystyle\int_{-1}^1 \frac{x \operatorname{sgn} x}{x + 2} \, dx = \int_{-1}^0 \frac{-x}{x + 2} \, dx + \int_0^1 \frac{x}{x + 2} \, dx$

$$= \int_{-1}^0 \left(-1 + \frac{2}{x + 2} \right) dx + \int_0^1 \left(1 - \frac{2}{x + 2} \right) dx$$

$$= (-x + 2 \ln |x + 2|) \Big|_{-1}^0 + (x - 2 \ln |x + 2|) \Big|_0^1 = \ln \frac{16}{9}.$$

30. Since $\dfrac{x^2}{x^5 + 1} \leq \dfrac{1}{x^3}$ for all $x \geq 0$, therefore

$$I = \int_0^\infty \frac{x^2}{x^5 + 1} \, dx$$

$$= \int_0^1 \frac{x^2}{x^5 + 1} \, dx + \int_1^\infty \frac{x^2}{x^5 + 1} \, dx$$

$$\leq \int_0^1 \frac{x^2}{x^5 + 1} \, dx + \int_1^\infty \frac{dx}{x^3}$$

$$= I_1 + I_2.$$

Since I_1 is a proper integral (finite) and I_2 is a convergent improper integral, (see Theorem 2), therefore I converges.

32. Since $\dfrac{x \sqrt{x}}{x^2 - 1} \geq \dfrac{1}{\sqrt{x}}$ for all $x > 1$, therefore

$$I = \int_2^\infty \frac{x \sqrt{x}}{x^2 - 1} \, dx \geq \int_2^\infty \frac{dx}{\sqrt{x}} = I_1 = \infty.$$

Since I_1 is a divergent improper integral, I diverges.

34. On $[0,1]$, $\dfrac{1}{\sqrt{x} + x^2} \leq \dfrac{1}{\sqrt{x}}$. On $[1, \infty)$, $\dfrac{1}{\sqrt{x} + x^2} \leq \dfrac{1}{x^2}$. Thus,

$$\int_0^1 \frac{dx}{\sqrt{x} + x^2} \leq \int_0^1 \frac{dx}{\sqrt{x}}$$

$$\int_1^\infty \frac{dx}{\sqrt{x} + x^2} \leq \int_1^\infty \frac{dx}{x^2}.$$

Since both of these integrals are convergent, therefore so is their sum $\displaystyle\int_0^\infty \frac{dx}{\sqrt{x} + x^2}$.

36. Since $\sin x \leq x$ for all $x \geq 0$, thus $\dfrac{\sin x}{x} \leq 1$. Then

$$I = \int_0^\pi \frac{\sin x}{x} \, dx = \lim_{\epsilon \to 0+} \int_\epsilon^\pi \frac{\sin x}{x} \, dx \leq \int_0^\pi (1) \, dx = \pi.$$

Hence, I converges.

38. Since $0 \leq 1 - \cos \sqrt{x} = 2 \sin^2 \left(\dfrac{\sqrt{x}}{2} \right) \leq 2 \left(\dfrac{\sqrt{x}}{2} \right)^2 = \dfrac{x}{2}$,

for $x \geq 0$, therefore $\displaystyle\int_0^{\pi^2} \frac{dx}{1 - \cos \sqrt{x}} \geq 2 \int_0^{\pi^2} \frac{dx}{x}$, which diverges to infinity.

40. Since $\ln x$ grows more slowly than any positive power of x, therefore we have $\ln x \leq k x^{1/4}$ for some constant k and every $x \geq 2$. Thus, $\dfrac{1}{\sqrt{x} \ln x} \geq \dfrac{1}{k x^{3/4}}$ for $x \geq 2$ and $\displaystyle\int_2^\infty \frac{dx}{\sqrt{x} \ln x}$ diverges to infinity by comparison with $\dfrac{1}{k} \displaystyle\int_2^\infty \frac{dx}{x^{3/4}}$.

42. We are given that $\displaystyle\int_0^\infty e^{-x^2} \, dx = \frac{1}{2} \sqrt{\pi}$.

a) First we calculate

$$\int_0^\infty x^2 e^{-x^2} \, dx = \lim_{R \to \infty} \int_0^R x^2 e^{-x^2} \, dx$$

$$U = x \qquad dV = x e^{-x^2} \, dx$$

$$dU = dx \qquad V = -\frac{1}{2} e^{-x^2}$$

$$= \lim_{R \to \infty} \left[-\frac{1}{2} x e^{-x^2} \Big|_0^R + \frac{1}{2} \int_0^R e^{-x^2} \, dx \right]$$

$$= -\frac{1}{2} \lim_{R \to \infty} R e^{-R^2} + \frac{1}{2} \int_0^\infty e^{-x^2} \, dx$$

$$= 0 + \frac{1}{4} \sqrt{\pi} = \frac{1}{4} \sqrt{\pi}.$$

b) Similarly,

$$\int_0^\infty x^4 e^{-x^2}\, dx = \lim_{R\to\infty} \int_0^R x^4 e^{-x^2}\, dx$$

$$U = x^3 \qquad\qquad dV = xe^{-x^2}\, dx$$
$$dU = 3x^2\, dx \qquad V = -\tfrac{1}{2}e^{-x^2}$$

$$= \lim_{R\to\infty}\left[-\frac{1}{2}x^3 e^{-x^2}\Big|_0^R + \frac{3}{2}\int_0^R x^2 e^{-x^2}\, dx \right]$$

$$= -\frac{1}{2}\lim_{R\to\infty} R^3 e^{-R^2} + \frac{3}{2}\int_0^\infty x^2 e^{-x^2}\, dx$$

$$= 0 + \frac{3}{2}\left(\frac{1}{4}\sqrt{\pi}\right) = \frac{3}{8}\sqrt{\pi}.$$

44. $\Gamma(x) = \displaystyle\int_0^\infty t^{x-1} e^{-t}\, dt.$

a) Since $\lim_{t\to\infty} t^{x-1} e^{-t/2} = 0$, there exists $T > 0$ such that $t^{x-1} e^{-t/2} \le 1$ if $t \ge T$. Thus

$$0 \le \int_T^\infty t^{x-1} e^{-t}\, dt \le \int_T^\infty e^{-t}\, dt = 2e^{-T/2}$$

and $\displaystyle\int_T^\infty t^{x-1} e^{-t}\, dt$ converges by the comparison theorem.

If $x > 0$, then

$$0 \le \int_0^T t^{x-1} e^{-t}\, dt < \int_0^T t^{x-1}\, dt$$

converges by Theorem 2(b). Thus the integral defining $\Gamma(x)$ converges.

b) $\Gamma(x+1) = \displaystyle\int_0^\infty t^x e^{-t}\, dt$

$$= \lim_{\substack{c\to 0+ \\ R\to\infty}} \int_c^R t^x e^{-t}\, dt$$

$$U = t^x \qquad\qquad dV = e^{-t}\, dt$$
$$dU = xt^{x-1}\, dx \qquad V = -e^{-t}$$

$$= \lim_{\substack{c\to 0+ \\ R\to\infty}} \left(-t^x e^{-t}\Big|_c^R + x\int_c^R t^{x-1} e^{-t}\, dt \right)$$

$$= 0 + x\int_0^\infty t^{x-1} e^{-t}\, dt = x\Gamma(x).$$

c) $\Gamma(1) = \displaystyle\int_0^\infty e^{-t}\, dt = 1 = 0!.$
By (b), $\Gamma(2) = 1\Gamma(1) = 1 \times 1 = 1 = 1!.$
In general, if $\Gamma(k+1) = k!$ for some positive integer k, then
$\Gamma(k+2) = (k+1)\Gamma(k+1) = (k+1)k! = (k+1)!.$
Hence $\Gamma(n+1) = n!$ for all integers $n \ge 0$, by induction.

d) $\Gamma\left(\dfrac{1}{2}\right) = \displaystyle\int_0^\infty t^{-1/2} e^{-t}\, dt$ Let $t = x^2$
$$dt = 2x\, dx$$

$$= \int_0^\infty \frac{1}{x} e^{-x^2}\, 2x\, dx = 2\int_0^\infty e^{-x^2}\, dx = \sqrt{\pi}$$

$$\Gamma\left(\frac{3}{2}\right) = \frac{1}{2}\Gamma\left(\frac{1}{2}\right) = \frac{1}{2}\sqrt{\pi}.$$

Section 6.6 The Trapezoid and Midpoint Rules (page 354)

2. The exact value of I is

$$I = \int_0^1 e^{-x}\, dx = -e^{-x}\Big|_0^1$$

$$= 1 - \frac{1}{e} \approx 0.6321206.$$

The approximations are

$$T_4 = \tfrac{1}{4}\left(\tfrac{1}{2}e^0 + e^{-1/4} + e^{-1/2} + e^{-3/4} + \tfrac{1}{2}e^{-1}\right)$$
$$\approx 0.6354094$$
$$M_4 = \tfrac{1}{4}\left(e^{-1/8} + e^{-3/8} + e^{-5/8} + e^{-7/8}\right)$$
$$\approx 0.6304774$$
$$T_8 = \tfrac{1}{2}(T_4 + M_4) \approx 0.6329434$$
$$M_8 = \tfrac{1}{8}\left(e^{-1/16} + e^{-3/16} + e^{-5/16} + e^{-7/16} + \right.$$
$$\left. e^{-9/16} + e^{-11/16} + e^{-13/16} + e^{-15/16}\right)$$
$$\approx 0.6317092$$
$$T_{16} = \tfrac{1}{2}(T_8 + M_8) \approx 0.6323263.$$

The exact errors are

$$I - T_4 = -0.0032888; \qquad I - M_4 = 0.0016432;$$
$$I - T_8 = -0.0008228; \qquad I - M_8 = 0.0004114;$$
$$I - T_{16} = -0.0002057.$$

If $f(x) = e^{-x}$, then $f^{(2)}(x) = e^{-x}$. On $[0,1]$, $|f^{(2)}(x)| \le 1$. Therefore, the error bounds are:

Trapezoid : $|I - T_n| \le \dfrac{1}{12}\left(\dfrac{1}{n}\right)^2$

$$|I - T_4| \le \frac{1}{12}\left(\frac{1}{16}\right) \approx 0.0052083;$$
$$|I - T_8| \le \frac{1}{12}\left(\frac{1}{64}\right) \approx 0.001302;$$
$$|I - T_{16}| \le \frac{1}{12}\left(\frac{1}{256}\right) \approx 0.0003255.$$

Midpoint : $|I - M_n| \le \dfrac{1}{24}\left(\dfrac{1}{n}\right)^2$

$$|I - M_4| \le \frac{1}{24}\left(\frac{1}{16}\right) \approx 0.0026041;$$
$$|I - M_8| \le \frac{1}{24}\left(\frac{1}{64}\right) \approx 0.000651.$$

Note that the actual errors satisfy these bounds.

4. The exact value of I is

$$I = \int_0^1 \frac{dx}{1+x^2} = \tan^{-1} x \Big|_0^1 = \frac{\pi}{4} \approx 0.7853982.$$

The approximations are

$$T_4 = \frac{1}{4}\left[\frac{1}{2}(1) + \frac{16}{17} + \frac{4}{5} + \frac{16}{25} + \frac{1}{2}\left(\frac{1}{2}\right)\right]$$
$$\approx 0.7827941$$

$$M_4 = \frac{1}{4}\left[\frac{64}{65} + \frac{64}{73} + \frac{64}{89} + \frac{64}{113}\right]$$
$$\approx 0.7867001$$

$$T_8 = \frac{1}{2}(T_4 + M_4) \approx 0.7847471$$

$$M_8 = \frac{1}{8}\left[\frac{256}{257} + \frac{256}{265} + \frac{256}{281} + \frac{256}{305} + \right.$$
$$\left. \frac{256}{337} + \frac{256}{377} + \frac{256}{425} + \frac{256}{481}\right]$$
$$\approx 0.7857237$$

$$T_{16} = \frac{1}{2}(T_8 + M_8) \approx 0.7852354.$$

The exact errors are

$$I - T_4 = 0.0026041; \quad I - M_4 = -0.0013019;$$
$$I - T_8 = 0.0006511; \quad I - M_8 = -0.0003255;$$
$$I - T_{16} = 0.0001628.$$

Since $f(x) = \dfrac{1}{1+x^2}$, then $f'(x) = \dfrac{-2x}{(1+x^2)^2}$ and $f''(x) = \dfrac{6x^2 - 2}{(1+x^2)^3}$. On $[0,1]$, $|f''(x)| \le 4$. Therefore, the error bounds are

$$\text{Trapezoid}: |I - T_n| \le \frac{4}{12}\left(\frac{1}{n}\right)^2$$

$$|I - T_4| \le \frac{4}{12}\left(\frac{1}{16}\right) \approx 0.0208333;$$

$$|I - T_8| \le \frac{4}{12}\left(\frac{1}{64}\right) \approx 0.0052083;$$

$$|I - T_{16}| \le \frac{4}{12}\left(\frac{1}{256}\right) \approx 0.001302.$$

$$\text{Midpoint}: |I - M_n| \le \frac{4}{24}\left(\frac{1}{n}\right)^2$$

$$|I - M_4| \le \frac{4}{24}\left(\frac{1}{16}\right) \approx 0.0104167;$$

$$|I - M_8| \le \frac{4}{24}\left(\frac{1}{64}\right) \approx 0.0026042.$$

The exact errors are much smaller than these bounds. In part, this is due to very crude estimates made for $|f''(x)|$.

6. $M_4 = 2(3.8 + 6.7 + 8 + 5.2) = 47.4$

8. $M_4 = 100 \times 2(4 + 5.5 + 5.5 + 4) = 3,800 \text{ km}^2$

10. The approximations for $I = \int_0^1 e^{-x^2}\, dx$ are

$$M_8 = \frac{1}{8}\left(e^{-1/256} + e^{-9/256} + e^{-25/256} + e^{-49/256} + \right.$$
$$\left. e^{-81/256} + e^{-121/256} + e^{-169/256} + e^{-225/256}\right)$$
$$\approx 0.7473$$

$$T_{16} = \frac{1}{16}\left[\frac{1}{2}(1) + e^{-1/256} + e^{-1/64} + e^{-9/256} + e^{-1/16} + \right.$$
$$e^{-25/256} + e^{-9/64} + e^{-49/256} + e^{-1/4} + e^{-81/256} +$$
$$e^{-25/64} + e^{-121/256} + e^{-9/16} + e^{-169/256} + e^{-49/64} +$$
$$\left. e^{-225/256} + \frac{1}{2}e^{-1}\right]$$
$$\approx 0.74658.$$

Since $f(x) = e^{-x^2}$, we have $f'(x) = -2xe^{-x^2}$, $f''(x) = 2(2x^2 - 1)e^{-x^2}$, and $f'''(x) = 4x(3 - 2x^2)e^{-x^2}$. Since $f'''(x) \ne 0$ on $(0,1)$, therefore the maximum value of $|f''(x)|$ on $[0,1]$ must occur at an endpoint of that interval. We have $f''(0) = -2$ and $f''(1) = 2/e$, so $|f''(x)| \le 2$ on $[0,1]$. The error bounds are

$$|I - M_n| \le \frac{2}{24}\left(\frac{1}{n}\right)^2 \Rightarrow |I - M_8| \le \frac{2}{24}\left(\frac{1}{64}\right)$$
$$\approx 0.00130.$$

$$|I - T_n| \le \frac{2}{12}\left(\frac{1}{n}\right)^2 \Rightarrow |I - T_{16}| \le \frac{2}{12}\left(\frac{1}{256}\right)$$
$$\approx 0.000651.$$

According to the error bounds,

$$\int_0^1 e^{-x^2}\, dx = 0.747,$$

accurate to two decimal places, with error no greater than 1 in the third decimal place.

12. The exact value of I is

$$I = \int_0^1 x^2\, dx = \frac{x^3}{3}\Big|_0^1 = \frac{1}{3}.$$

The approximation is

$$T_1 = (1)\left[\frac{1}{2}(0)^2 + \frac{1}{2}(1)^2\right] = \frac{1}{2}.$$

The actual error is $I - T_1 = -\frac{1}{6}$. However, since $f(x) = x^2$, then $f''(x) = 2$ on $[0,1]$, so the error estimate here gives

$$|I - T_1| \le \frac{2}{12}(1)^2 = \frac{1}{6}.$$

Since this is the actual size of the error in this case, the constant "12" in the error estimate cannot be improved (i.e., cannot be made larger).

14. Let $y = f(x)$. We are given that m_1 is the midpoint of $[x_0, x_1]$ where $x_1 - x_0 = h$. By tangent line approximate in the subinterval $[x_0, x_1]$,

$$f(x) \approx f(m_1) + f'(m_1)(x - m_1).$$

The error in this approximation is

$$E(x) = f(x) - f(m_1) - f'(m_1)(x - m_1).$$

If $f''(t)$ exists for all t in $[x_0, x_1]$ and $|f''(t)| \le K$ for some constant K, then by Theorem 4 of Section 3.5,

$$|E(x)| \le \frac{K}{2}(x - m_1)^2.$$

Hence,

$$|f(x) - f(m_1) - f'(m_1)(x - m_1)| \le \frac{K}{2}(x - m_1)^2.$$

We integrate both sides of this inequlity. Noting that $x_1 - m_1 = m_1 - x_0 = \frac{1}{2}h$, we obtain for the left side

$$\left| \int_{x_0}^{x_1} f(x)\,dx - \int_{x_0}^{x_1} f(m_1)\,dx \right.$$
$$\left. - \int_{x_0}^{x_1} f'(m_1)(x - m_1)\,dx \right|$$
$$= \left| \int_{x_0}^{x_1} f(x)\,dx - f(m_1)h - f'(m_1)\frac{(x - m_1)^2}{2}\bigg|_{x_0}^{x_1} \right|$$
$$= \left| \int_{x_0}^{x_1} f(x)\,dx - f(m_1)h \right|.$$

Integrating the right-hand side, we get

$$\int_{x_0}^{x_1} \frac{K}{2}(x - m_1)^2\,dx = \frac{K}{2}\frac{(x - m_1)^3}{3}\bigg|_{x_0}^{x_1}$$
$$= \frac{K}{6}\left(\frac{h^3}{8} + \frac{h^3}{8}\right) = \frac{K}{24}h^3.$$

Hence,

$$\left| \int_{x_0}^{x_1} f(x)\,dx - f(m_1)h \right|$$
$$= \left| \int_{x_0}^{x_1} [f(x) - f(m_1) - f'(m_1)(x - m_1)]\,dx \right|$$
$$\le \frac{K}{24}h^3.$$

A similar estimate holds on each subinterval $[x_{j-1}, x_j]$ for $1 \le j \le n$. Therefore,

$$\left| \int_a^b f(x)\,dx - M_n \right| = \left| \sum_{j=1}^n \left(\int_{x_{j-1}}^{x_j} f(x)\,dx - f(m_j)h \right) \right|$$
$$\le \sum_{j=1}^n \left| \int_{x_{j-1}}^{x_j} f(x)\,dx - f(m_j)h \right|$$
$$\le \sum_{j=1}^n \frac{K}{24}h^3 = \frac{K}{24}nh^3 = \frac{K(b-a)}{24}h^2.$$

because $nh = b - a$.

Section 6.7 Simpson's Rule (page 359)

2. The exact value of I is

$$I = \int_0^1 e^{-x}\,dx = -e^{-x}\bigg|_0^1$$
$$= 1 - \frac{1}{e} \approx 0.6321206.$$

The approximations are

$$S_4 = \frac{1}{12}(e^0 + 4e^{-1/4} + 2e^{-1/2} + 4e^{-3/4} + e^{-1})$$
$$\approx 0.6321342$$
$$S_8 = \frac{1}{24}(e^0 + 4e^{-1/8} + 2e^{-1/4} + 4e^{-3/8} +$$
$$2e^{-1/2} + 4e^{-5/8} + 2e^{-3/4} + 4e^{-7/8} + e^{-1})$$
$$\approx 0.6321214.$$

The actual errors are

$$I - S_4 = -0.0000136; \quad I - S_8 = -0.0000008.$$

These errors are evidently much smaller than the corresponding errors for the corresponding Trapezoid Rule approximations.

4. The exact value of I is

$$I = \int_0^1 \frac{dx}{1 + x^2} = \tan^{-1} x\bigg|_0^1 = \frac{\pi}{4} \approx 0.7853982.$$

The approximations are

$$S_4 = \frac{1}{12}\left[1 + 4\left(\frac{16}{17}\right) + 2\left(\frac{4}{5}\right) + 4\left(\frac{16}{25}\right) + \frac{1}{2}\right]$$
$$\approx 0.7853922$$
$$S_8 = \frac{1}{24}\left[1 + 4\left(\frac{64}{65}\right) + 2\left(\frac{16}{17}\right) + 4\left(\frac{64}{73}\right) + \right.$$
$$\left. 2\left(\frac{4}{5}\right) + 4\left(\frac{64}{89}\right) + 2\left(\frac{16}{25}\right) + 4\left(\frac{64}{113}\right) + \frac{1}{2}\right]$$
$$\approx 0.7853981.$$

The actual errors are

$$I - S_4 = 0.0000060; \quad I - S_8 = 0.0000001,$$

accurate to 7 decimal places. These errors are evidently much smaller than the corresponding errors for the corresponding Trapezoid Rule approximation.

6. $S_8 = 100 \times \frac{1}{3}[0 + 4(4 + 5.5 + 5.5 + 4) + 2(5.5 + 5 + 4.5) + 0]$
$$\approx 3,533 \text{ km}^2$$

8. Let $I = \displaystyle\int_a^b f(x)\,dx$, and the interval $[a, b]$ be subdivided into $2n$ subintervals of equal length $h = (b-a)/2n$. Let $y_j = f(x_j)$ and $x_j = a + jh$ for $0 \le j \le 2n$, then

$$S_{2n} = \frac{1}{3}\left(\frac{b-a}{2n}\right)\left[y_0 + 4y_1 + 2y_2 + \cdots\right.$$
$$\left. + 2y_{2n-2} + 4y_{2n-1} + y_{2n}\right]$$
$$= \frac{1}{3}\left(\frac{b-a}{2n}\right)\left[y_0 + 4\sum_{j=1}^{2n-1} y_j - 2\sum_{j=1}^{n-1} y_{2j} + y_{2n}\right]$$

and

$$T_{2n} = \frac{1}{2}\left(\frac{b-a}{2n}\right)\left(y_0 + 2\sum_{j=1}^{2n-1} y_j + y_{2n}\right)$$
$$T_n = \frac{1}{2}\left(\frac{b-a}{n}\right)\left(y_0 + 2\sum_{j=1}^{n-1} y_{2j} + y_{2n}\right).$$

Since $T_{2n} = \frac{1}{2}(T_n + M_n) \Rightarrow M_n = 2T_{2n} - T_n$, then

$$\frac{T_n + 2M_n}{3} = \frac{T_n + 2(2T_{2n} - T_n)}{3} = \frac{4T_{2n} - T_n}{3}$$
$$\frac{2T_{2n} + M_n}{3} = \frac{2T_{2n} + 2T_{2n} - T_n}{3} = \frac{4T_{2n} - T_n}{3}.$$

Hence,

$$\frac{T_n + 2M_n}{3} = \frac{2T_{2n} + M_n}{3} = \frac{4T_{2n} - T_n}{3}.$$

Using the formulas of T_{2n} and T_n obtained above,

$$\frac{4T_{2n} - T_n}{3}$$
$$= \frac{1}{3}\left[\frac{4}{2}\left(\frac{b-a}{2n}\right)\left(y_0 + 2\sum_{j=1}^{2n-1} y_j + y_{2n}\right)\right.$$
$$\left. - \frac{1}{2}\left(\frac{b-a}{n}\right)\left(y_0 + 2\sum_{j=1}^{n-1} y_{2j} + y_{2n}\right)\right]$$
$$= \frac{1}{3}\left(\frac{b-a}{2n}\right)\left[y_0 + 4\sum_{j=1}^{2n-1} y_j - 2\sum_{j=1}^{n-1} y_{2j} + y_{2n}\right]$$
$$= S_{2n}.$$

Hence,

$$S_{2n} = \frac{4T_{2n} - T_n}{3} = \frac{T_n + 2M_n}{3} = \frac{2T_{2n} + M_n}{3}.$$

10. The approximations for $I = \displaystyle\int_0^1 e^{-x^2}\,dx$ are

$$S_8 = \frac{1}{3}\left(\frac{1}{8}\right)\left[1 + 4\left(e^{-1/64} + e^{-9/64} + e^{-25/64} + \right.\right.$$
$$\left.\left. e^{-49/64}\right) + 2\left(e^{-1/16} + e^{-1/4} + e^{-9/16}\right) + e^{-1}\right]$$
$$\approx 0.7468261$$
$$S_{16} = \frac{1}{3}\left(\frac{1}{16}\right)\left[1 + 4\left(e^{-1/256} + e^{-9/256} + e^{-25/256} + \right.\right.$$
$$e^{-49/256} + e^{-81/256} + e^{-121/256} + e^{-169/256} +$$
$$\left. e^{-225/256}\right) + 2\left(e^{-1/64} + e^{-1/16} + e^{-9/64} + e^{-1/4} + \right.$$
$$\left.\left. e^{-25/64} + e^{-9/16} + e^{-49/64}\right) + e^{-1}\right]$$
$$\approx 0.7468243.$$

If $f(x) = e^{-x^2}$, then $f^{(4)}(x) = 4e^{-x^2}(4x^4 - 12x^2 + 3)$. On $[0,1]$, $|f^{(4)}(x)| \le 12$, and the error bounds are

$$|I - S_n| \le \frac{12(1)}{180}\left(\frac{1}{n}\right)^4$$
$$|I - S_8| \le \frac{12}{180}\left(\frac{1}{8}\right)^4 \approx 0.0000163$$
$$|I - S_{16}| \le \frac{12}{180}\left(\frac{1}{16}\right)^4 \approx 0.0000010.$$

Comparing the two approximations,

$$I = \int_0^1 e^{-x^2}\,dx = 0.7468,$$

accurate to 4 decimal places.

12. The exact value of I is

$$I = \int_0^1 x^3 \, dx = \left. \frac{x^4}{4} \right|_0^1 = \frac{1}{4}.$$

The approximation is

$$S_2 = \frac{1}{3}\left(\frac{1}{2}\right)\left[0^3 + 4\left(\frac{1}{2}\right)^3 + 1^3\right] = \frac{1}{4}.$$

The actual error is zero. Hence, Simpson's Rule is exact for the cubic function $f(x) = x^3$. Since it is evidently exact for quadratic functions $f(x) = Bx^2 + Cx + D$, it must also be exact for arbitrary cubics $f(x) = Ax^3 + Bx^2 + Cx + D$.

Section 6.8 Other Aspects of Approximate Integration   (page 364)

2. $\displaystyle\int_0^1 \frac{e^x}{\sqrt{1-x}} \, dx$ Let $t^2 = 1 - x$

$$2t \, dt = -dx$$

$$= -\int_1^0 \frac{e^{1-t^2}}{t} 2t \, dt = 2\int_0^1 e^{1-t^2} \, dt.$$

4. $\displaystyle\int_1^\infty \frac{dx}{x^2 + \sqrt{x} + 1}$ Let $x = \frac{1}{t^2}$

$$dx = -\frac{2\,dt}{t^3}$$

$$= \int_1^0 \frac{1}{\left(\frac{1}{t^2}\right)^2 + \sqrt{\frac{1}{t^2}} + 1}\left(-\frac{2\,dt}{t^3}\right)$$

$$= 2\int_0^1 \frac{t\,dt}{t^4 + t^3 + 1}.$$

6. Let

$$\int_0^\infty \frac{dx}{x^4 + 1} = \int_0^1 \frac{dx}{x^4 + 1} + \int_1^\infty \frac{dx}{x^4 + 1} = I_1 + I_2.$$

Let $x = \frac{1}{t}$ and $dx = -\frac{dt}{t^2}$ in I_2, then

$$I_2 = \int_1^0 \frac{1}{\left(\frac{1}{t}\right)^4 + 1}\left(-\frac{dt}{t^2}\right) = \int_0^1 \frac{t^2}{1 + t^4} \, dt.$$

Hence,

$$\int_0^\infty \frac{dx}{x^4 + 1} = \int_0^1 \left(\frac{1}{x^4 + 1} + \frac{x^2}{1 + x^4}\right) dx$$

$$= \int_0^1 \frac{x^2 + 1}{x^4 + 1} \, dx.$$

8. Let

$$I = \int_1^\infty e^{-x^2} \, dx \quad \text{Let } x = \frac{1}{t}$$

$$dx = -\frac{dt}{t^2}$$

$$= \int_1^0 e^{-(1/t)^2}\left(-\frac{1}{t^2}\right) dt = \int_0^1 \frac{e^{-1/t^2}}{t^2} \, dt.$$

Observe that

$$\lim_{t\to 0+} \frac{e^{-1/t^2}}{t^2} = \lim_{t\to 0+} \frac{t^{-2}}{e^{1/t^2}} \quad \left[\frac{\infty}{\infty}\right]$$

$$= \lim_{t\to 0+} \frac{-2t^{-3}}{e^{1/t^2}(-2t^{-3})}$$

$$= \lim_{t\to 0+} \frac{1}{e^{1/t^2}} = 0.$$

Hence,

$$S_2 = \frac{1}{3}\left(\frac{1}{2}\right)\left[0 + 4(4e^{-4}) + e^{-1}\right]$$

$$\approx 0.1101549$$

$$S_4 = \frac{1}{3}\left(\frac{1}{4}\right)\left[0 + 4(16e^{-16}) + 2(4e^{-4}) \right.$$

$$\left. + 4\left(\frac{16}{9}e^{-16/9}\right) + e^{-1}\right]$$

$$\approx 0.1430237$$

$$S_8 = \frac{1}{3}\left(\frac{1}{8}\right)\left[0 + 4\left(64e^{-64} + \frac{64}{9}e^{-64/9} + \frac{64}{25}e^{-64/25} + \right.\right.$$

$$\left.\frac{64}{49}e^{-64/49}\right) + 2\left(16e^{-16} + 4e^{-4} + \frac{16}{9}e^{-16/9}\right) + e^{-1}\right]$$

$$\approx 0.1393877.$$

Hence, $I \approx 0.14$, accurate to 2 decimal places. These approximations do not converge very quickly, because the fourth derivative of e^{-1/t^2} has very large values for some values of t near 0. In fact, higher and higher derivatives behave more and more badly near 0, so higher order methods cannot be expected to work well either.

10. We are given that $\displaystyle\int_0^\infty e^{-x^2} \, dx = \frac{1}{2}\sqrt{\pi}$ and from the previous exercise $\displaystyle\int_0^1 e^{-x^2} \, dx = 0.74684$. Therefore,

$$\int_1^\infty e^{-x^2} \, dx = \int_0^\infty e^{-x^2} \, dx - \int_0^1 e^{-x^2} \, dx$$

$$= \frac{1}{2}\sqrt{\pi} - 0.74684$$

$$= 0.139 \quad \text{(to 3 decimal places)}.$$

12. For any function f we use the approximation

$$\int_{-1}^{1} f(x)\,dx \approx f(-1/\sqrt{3}) + f(1/\sqrt{3}).$$

We have

$$\int_{-1}^{1} x^4\,dx \approx \left(-\frac{1}{\sqrt{3}}\right)^4 + \left(\frac{1}{\sqrt{3}}\right)^4 = \frac{2}{9}$$

$$\text{Error } = \int_{-1}^{1} x^4\,dx - \frac{2}{9} = \frac{2}{5} - \frac{2}{9} \approx 0.17778$$

$$\int_{-1}^{1} \cos x\,dx \approx \cos\left(-\frac{1}{\sqrt{3}}\right) + \cos\left(\frac{1}{\sqrt{3}}\right) \approx 1.67582$$

$$\text{Error } = \int_{-1}^{1} \cos x\,dx - 1.67582 \approx 0.00712$$

$$\int_{-1}^{1} e^x\,dx \approx e^{-1/\sqrt{3}} + e^{1/\sqrt{3}} \approx 2.34270$$

$$\text{Error } = \int_{-1}^{1} e^x\,dx - 2.34270 \approx 0.00771.$$

14. For any function f we use the approximation

$$\int_{-1}^{1} f(x)\,dx \approx \frac{5}{9}\left[f(-\sqrt{3/5}) + f(\sqrt{3/5})\right] + \frac{8}{9}f(0).$$

We have

$$\int_{-1}^{1} x^6\,dx \approx \frac{5}{9}\left[\left(-\sqrt{\frac{3}{5}}\right)^6 + \left(\sqrt{\frac{3}{5}}\right)^6\right] + 0 = 0.24000$$

$$\text{Error } = \int_{-1}^{1} x^6\,dx - 0.24000 \approx 0.04571$$

$$\int_{-1}^{1} \cos x\,dx \approx \frac{5}{9}\left[\cos\left(-\sqrt{\frac{3}{5}}\right) + \cos\left(\sqrt{\frac{3}{5}}\right)\right] + \frac{8}{9}$$

$$\approx 1.68300$$

$$\text{Error } = \int_{-1}^{1} \cos x\,dx - 1.68300 \approx 0.00006$$

$$\int_{-1}^{1} e^x\,dx \approx e^{-\sqrt{3/5}} + e^{\sqrt{3/5}} \approx 2.35034$$

$$\text{Error } = \int_{-1}^{1} e^x\,dx - 2.35034 \approx 0.00006.$$

16. From Exercise 9 in Section 7.6, for $I = \int_{0}^{1.6} f(x)\,dx$,

$$T_0^0 = T_1 = 1.9196$$
$$T_1^0 = T_2 = 2.00188$$
$$T_2^0 = T_4 = 2.02622$$
$$T_3^0 = T_8 = 2.02929.$$

Hence,

$$R_1 = T_1^1 = \frac{4T_1^0 - T_0^0}{3} = 2.0346684$$

$$T_2^1 = \frac{4T_2^0 - T_1^0}{3} = 2.0343333 = S_4$$

$$R_2 = T_2^2 = \frac{16T_2^1 - T_1^1}{15} = 2.0346684$$

$$T_3^1 = \frac{4T_3^0 - T_2^0}{3} = 2.0303133 = S_8$$

$$T_3^2 = \frac{16T_3^1 - T_2^1}{15} = 2.0300453$$

$$R_3 = T_3^3 = \frac{64T_3^2 - T_2^2}{63} = 2.0299719.$$

18. Let

$$I = \int_{\pi}^{\infty} \frac{\sin x}{1 + x^2}\,dx \quad \text{Let } x = \frac{1}{t}$$
$$dx = -\frac{dt}{t^2}$$

$$= \int_{1/\pi}^{0} \frac{\sin\left(\dfrac{1}{t}\right)}{1 + \left(\dfrac{1}{t^2}\right)}\left(-\frac{1}{t^2}\right)\,dt$$

$$= \int_{0}^{1/\pi} \frac{\sin\left(\dfrac{1}{t}\right)}{1 + t^2}\,dt.$$

The transformation is not suitable because the derivative of $\sin\left(\dfrac{1}{t}\right)$ is $-\dfrac{1}{t^2}\cos\left(\dfrac{1}{t}\right)$, which has very large values at some points close to 0.

In order to approximate the integral I to an desired degree of accuracy, say with error less than ϵ in absolute value, we have to divide the integral into two parts:

$$I = \int_{\pi}^{\infty} \frac{\sin x}{1 + x^2}\,dx$$
$$= \int_{\pi}^{t} \frac{\sin x}{1 + x^2}\,dx + \int_{t}^{\infty} \frac{\sin x}{1 + x^2}\,dx$$
$$= I_1 + I_2.$$

If $t \geq \tan\dfrac{\pi - \epsilon}{2}$, then

$$\int_{t}^{\infty} \frac{\sin x}{1 + x^2}\,dx < \int_{t}^{\infty} \frac{dx}{1 + x^2}$$
$$= \tan^{-1}(x)\Big|_{t}^{\infty} = \frac{\pi}{2} - \tan^{-1}(t) \leq \frac{\epsilon}{2}.$$

Now let A be a numerical approximation to the proper integral $\int_{\pi}^{t} \frac{\sin x}{1 + x^2}\, dx$, having error less than $\epsilon/2$ in absolute value. Then

$$|I - A| = |I_1 + I_2 - A|$$
$$\leq |I_1 - A| + |I_2|$$
$$\leq \frac{\epsilon}{2} + \frac{\epsilon}{2} = \epsilon.$$

Hence, A is an approximation to the integral I with the desired accuracy.

Review Exercises on Techniques of Integration (page 365)

2. $\displaystyle\int \frac{x}{(x-1)^3}\, dx$　Let $u = x - 1$
$$du = dx$$
$$= \int \frac{u+1}{u^3}\, du = \int \left(\frac{1}{u^2} + \frac{1}{u^3} \right) du$$
$$= -\frac{1}{u} - \frac{1}{2u^2} + C = -\frac{1}{x-1} - \frac{1}{2(x-1)^2} + C.$$

4. $\displaystyle\int \frac{(1+\sqrt{x})^{1/3}}{\sqrt{x}}\, dx$　Let $u = 1 + \sqrt{x}$
$$du = \frac{dx}{2\sqrt{x}}$$
$$= 2\int u^{1/3}\, du = 2(\tfrac{3}{4})u^{4/3} + C$$
$$= \frac{3}{2}(1 + \sqrt{x})^{4/3} + C.$$

6. $\displaystyle\int (x^2 + x - 2)\sin 3x\, dx$
$$U = x^2 + x - 2 \qquad dV = \sin 3x$$
$$dU = (2x + 1)\, dx \qquad V = -\tfrac{1}{3}\cos 3x$$
$$= -\tfrac{1}{3}(x^2 + x - 2)\cos 3x + \tfrac{1}{3}\int (2x + 1)\cos 3x\, dx$$

$$U = 2x + 1 \qquad dV = \cos 3x\, dx$$
$$dU = 2\, dx \qquad V = \tfrac{1}{3}\sin 3x$$
$$= -\tfrac{1}{3}(x^2 + x - 2)\cos 3x + \tfrac{1}{9}(2x + 1)\sin 3x$$
$$\qquad - \tfrac{2}{9}\int \sin 3x\, dx$$
$$= -\tfrac{1}{3}(x^2 + x - 2)\cos 3x + \tfrac{1}{9}(2x + 1)\sin 3x$$
$$\qquad + \tfrac{2}{27}\cos 3x + C.$$

8. $\displaystyle\int x^3 \cos(x^2)\, dx$　Let $w = x^2$
$$dw = 2x\, dx$$
$$= \frac{1}{2}\int w\cos w\, dw$$
$$U = w \qquad dV = \cos w\, dw$$
$$dU = dw \qquad V = \sin w$$
$$= \frac{1}{2}w\sin w - \frac{1}{2}\int \sin w\, dw$$
$$= \frac{1}{2}x^2 \sin(x^2) + \frac{1}{2}\cos(x^2) + C.$$

10. $\displaystyle\frac{1}{x^2 + 2x - 15} = \frac{A}{x-3} + \frac{B}{x+5} = \frac{(A+B)x + (5A - 3B)}{x^2 + 2x - 15}$
$$\Rightarrow \begin{cases} A + B = 0 \\ 5A - 3B = 1 \end{cases} \Rightarrow A = \frac{1}{8},\ B = -\frac{1}{8}.$$
$$\int \frac{dx}{x^2 + 2x - 15} = \frac{1}{8}\int \frac{dx}{x-3} - \frac{1}{8}\int \frac{dx}{x+5}$$
$$= \frac{1}{8}\ln\left| \frac{x-3}{x+5} \right| + C.$$

12. $\displaystyle\int (\sin x + \cos x)^2\, dx = \int (1 + \sin 2x)\, dx$
$$= x - \tfrac{1}{2}\cos 2x + C.$$

14. $\displaystyle\int \frac{\cos x}{1 + \sin^2 x}\, dx$　Let $u = \sin x$
$$du = \cos x\, dx$$
$$= \int \frac{du}{1 + u^2} = \tan^{-1} u + C$$
$$= \tan^{-1}(\sin x) + C.$$

16. We have
$$\int \frac{x^2\, dx}{(3 + 5x^2)^{3/2}} \qquad \text{Let } x = \sqrt{\tfrac{3}{5}}\tan u$$
$$dx = \sqrt{\tfrac{3}{5}}\sec^2 u\, du$$
$$= \int \frac{(\tfrac{3}{5}\tan^2 u)(\sqrt{\tfrac{3}{5}}\sec^2 u)\, du}{(3)^{3/2}\sec^3 u}$$
$$= \frac{1}{5\sqrt{5}}\int (\sec u - \cos u)\, du$$
$$= \frac{1}{5\sqrt{5}}(\ln|\sec u + \tan u| - \sin u) + C$$
$$= \frac{1}{5\sqrt{5}}\left(\ln\left| \frac{\sqrt{5x^2 + 3}}{\sqrt{3}} + \frac{\sqrt{5x}}{\sqrt{3}} \right| - \frac{\sqrt{5x}}{\sqrt{5x^2 + 3}} \right) + C$$
$$= \frac{1}{5\sqrt{5}}\ln\left(\sqrt{5x^2 + 3} + \sqrt{5}x \right) - \frac{x}{5\sqrt{5x^2 + 3}} + C_0,$$
where $C_0 = C - \dfrac{1}{5\sqrt{5}}\ln\sqrt{3}$.

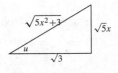

Fig. RT.16

18.
$$I = \int \frac{2x^2 + 4x - 3}{x^2 + 5x} \, dx = \int \frac{2x^2 + 10x - 6x - 3}{x^2 + 5x} \, dx$$

$$= \int \left[2 - \frac{6x + 3}{x(x+5)} \right] dx$$

$$\frac{6x + 3}{x(x+5)} = \frac{A}{x} + \frac{B}{x+5} = \frac{(A+B)x + 5A}{x(x+5)}$$

$$\Rightarrow \begin{cases} A + B = 6 \\ 5A = 3 \end{cases} \Rightarrow A = \frac{3}{5}, \ B = \frac{27}{5}.$$

$$I = \int 2 \, dx - \frac{3}{5} \int \frac{dx}{x} - \frac{27}{5} \int \frac{dx}{x+5}$$

$$= 2x - \frac{3}{5} \ln |x| - \frac{27}{5} \ln |x+5| + C.$$

20.
$$\frac{1}{4x^3 + x} = \frac{A}{x} + \frac{Bx + C}{4x^2 + 1}$$

$$= \frac{A(4x^2 + 1) + Bx^2 + Cx}{4x^3 + x}$$

$$\Rightarrow \begin{cases} 4A + B = 0 \\ C = 0, \ A = 1 \end{cases} \Rightarrow B = -4.$$

$$\int \frac{1}{4x^3 + x} \, dx = \int \frac{dx}{x} - 4 \int \frac{x \, dx}{4x^2 + 1}$$

$$= \ln |x| - \frac{1}{2} \ln(4x^2 + 1) + C.$$

22.
$$\int \sin^2 x \cos^4 x \, dx$$

$$= \int \frac{1}{2}(1 - \cos 2x)[\frac{1}{2}(1 + \cos 2x)]^2 \, dx$$

$$= \frac{1}{8} \int (1 + \cos 2x - \cos^2 2x - \cos^3 2x) \, dx$$

$$= \frac{1}{8}x + \frac{1}{16} \sin 2x - \frac{1}{16} \int (1 + \cos 4x) \, dx$$

$$- \frac{1}{8} \int (1 - \sin^2 2x) \cos 2x \, dx$$

$$= \frac{x}{8} + \frac{1}{16} \sin 2x - \frac{x}{16} - \frac{1}{64} \sin 4x - \frac{1}{16} \sin 2x$$

$$+ \frac{1}{48} \sin^3 2x + C$$

$$= \frac{x}{16} - \frac{\sin 4x}{64} + \frac{\sin^3 2x}{48} + C.$$

24. We have

$$I = \int \tan^4 x \sec x \, dx$$

$$U = \tan^3 x \qquad\qquad dV = \tan x \sec x \, dx$$
$$dU = 3 \tan^2 x \sec^2 x \, dx \qquad V = \sec x$$

$$= \tan^3 x \sec x - 3 \int \tan^2 x \sec^3 x \, dx$$

$$= \tan^3 x \sec x - 3 \int \tan^2 x (\tan^2 x + 1) \sec x \, dx$$

$$= \tan^3 x \sec x - 3I - 3J \text{ where}$$

$$J = \int \tan^2 x \sec x \, dx$$

$$U = \tan x \qquad\qquad dV = \tan x \sec x \, dx$$
$$dU = \sec^2 x \, dx \qquad V = \sec x$$

$$= \tan x \sec x - \int \sec^3 x \, dx$$

$$= \tan x \sec x - \int (\tan^2 x + 1) \sec x \, dx$$

$$= \tan x \sec x - J - \ln |\sec x + \tan x| + C$$

$$J = \frac{1}{2} \tan x \sec x - \frac{1}{2} \ln |\sec x + \tan x| + C.$$

$$I = \frac{1}{4} \tan^3 x \sec x - \frac{3}{8} \tan x \sec x$$

$$+ \frac{3}{8} \ln |\sec x + \tan x| + C.$$

26. We have

$$\int x \sin^{-1} \left(\frac{x}{2} \right) dx$$

$$U = \sin^{-1} \left(\frac{x}{2} \right) \quad dV = x \, dx$$

$$dU = \frac{dx}{\sqrt{4 - x^2}} \qquad V = \frac{x^2}{2}$$

$$= \frac{x^2}{2} \sin^{-1} \left(\frac{x}{2} \right) - \frac{1}{2} \int \frac{x^2 \, dx}{\sqrt{4 - x^2}} \quad \begin{array}{l} \text{Let } x = 2 \sin u \\ dx = 2 \cos u \, du \end{array}$$

$$= \frac{x^2}{2} \sin^{-1} \left(\frac{x}{2} \right) - 2 \int \sin^2 u \, du$$

$$= \frac{x^2}{2} \sin^{-1} \left(\frac{x}{2} \right) - \int (1 - \cos 2u) \, du$$

$$= \frac{x^2}{2} \sin^{-1} \left(\frac{x}{2} \right) - u + \sin u \cos u + C$$

$$= \left(\frac{x^2}{2} - 1 \right) \sin^{-1} \left(\frac{x}{2} \right) + \frac{1}{4} x \sqrt{4 - x^2} + C.$$

28. We have

$$I = \int \frac{dx}{x^5 - 2x^3 + x} = \int \frac{x \, dx}{x^6 - 2x^4 + x^2} \quad \begin{array}{l} \text{Let } u = x^2 \\ du = 2x \, dx \end{array}$$

$$= \frac{1}{2} \int \frac{du}{u^3 - 2u^2 + u} = \frac{1}{2} \int \frac{du}{u(u-1)^2}$$

$$\frac{1}{u(u-1)^2} = \frac{A}{u} + \frac{B}{u-1} + \frac{C}{(u-1)^2}$$

$$= \frac{A(u^2 - 2u + 1) + B(u^2 - u) + Cu}{u^3 - 2u^2 + u}$$

$$\Rightarrow \begin{cases} A + B = 0 \\ -2A - B + C = 0 \Rightarrow A = 1, \ B = -1, \ C = 1. \\ A = 1 \end{cases}$$

$$\frac{1}{2} \int \frac{du}{u^3 - 2u^2 + u} = \frac{1}{2} \int \frac{du}{u} - \frac{1}{2} \int \frac{du}{u-1}$$

$$+ \frac{1}{2} \int \frac{du}{(u-1)^2}$$

$$= \frac{1}{2} \ln |u| - \frac{1}{2} \ln |u - 1| - \frac{1}{2} \frac{1}{u-1} + K$$

$$= \frac{1}{2} \ln \frac{x^2}{|x^2 - 1|} - \frac{1}{2(x^2 - 1)} + K.$$

30.   Let

$$I_n = \int x^n 3^x \, dx$$

$$U = x^n \qquad\qquad dV = 3^x \, dx$$
$$dU = nx^{n-1} \, dx \qquad V = \frac{3^x}{\ln 3}$$

$$= \frac{x^n 3^x}{\ln 3} - \frac{n}{\ln 3} I_{n-1}.$$

$$I_0 = \int 3^x \, dx = \frac{3^x}{\ln 3} + C.$$

Hence,

$$I_3 = \int x^3 3^x \, dx$$

$$= \frac{x^3 3^x}{\ln 3} - \frac{3}{\ln 3}\left[\frac{x^2 3^x}{\ln 3} - \frac{2}{\ln 3}\left(\frac{x 3^x}{\ln 3} - \frac{1}{\ln 3} I_0\right)\right] + C_1$$

$$= 3^x\left[\frac{x^3}{\ln 3} - \frac{3x^2}{(\ln 3)^2} + \frac{6x}{(\ln 3)^3} - \frac{6}{(\ln 3)^4}\right] + C_1.$$

32.   We have

$$\int \frac{x^2 + 1}{x^2 + 2x + 2} \, dx = \int \left(1 - \frac{2x + 1}{x^2 + 2x + 2}\right) dx$$

$$= x - \int \frac{2x + 1}{(x + 1)^2 + 1} \, dx \qquad \text{Let } u = x + 1$$
$$\qquad\qquad\qquad\qquad\qquad\qquad du = dx$$

$$= x - \int \frac{2u - 1}{u^2 + 1} \, du$$

$$= x - \ln|u^2 + 1| + \tan^{-1} u + C$$

$$= x - \ln(x^2 + 2x + 2) + \tan^{-1}(x + 1) + C.$$

34.   We have

$$\int x^3 (\ln x)^2 \, dx$$

$$U = (\ln x)^2 \qquad dV = x^3 \, dx$$
$$dU = \frac{2}{x} \ln x \, dx \qquad V = \frac{1}{4} x^4$$

$$= \frac{1}{4} x^4 (\ln x)^2 - \frac{1}{2} \int x^3 \ln x \, dx$$

$$U = \ln x \qquad dV = x^3 \, dx$$
$$dU = \frac{1}{x} \, dx \qquad V = \frac{1}{4} x^4$$

$$= \frac{1}{4} x^4 (\ln x)^2 - \frac{1}{8} x^4 \ln x + \frac{1}{8} \int x^3 \, dx$$

$$= \frac{x^4}{4}\left[(\ln x)^2 - \frac{1}{2} \ln x + \frac{1}{8}\right] + C.$$

36. $\displaystyle\int \frac{e^{1/x}}{x^2} \, dx$ Let $u = \dfrac{1}{x}$

$$du = -\frac{1}{x^2} \, dx$$

$$= -\int e^u \, du = -e^u + C = -e^{1/x} + C.$$

38. $\displaystyle\int e^{(x^{1/3})}$ Let $x = u^3$
$$dx = 3u^2 \, du$$

$$= 3 \int u^2 e^u \, du = 3I_2$$

See solution to #16 of Section 6.6 for
$$I_n = \int u^n e^u \, du = u^n e^u - nI_{n-1}.$$

$$= 3[u^2 e^u - 2(u e^u - e^u)] + C$$

$$= e^{(x^{1/3})}(3x^{2/3} - 6x^{1/3} + 6) + C.$$

40. $\displaystyle\int \frac{10^{\sqrt{x+2}} \, dx}{\sqrt{x + 2}}$ Let $u = \sqrt{x + 2}$
$$du = \frac{dx}{2\sqrt{x + 2}}$$

$$= 2 \int 10^u \, du = \frac{2}{\ln 10} 10^u + C = \frac{2}{\ln 10} 10^{\sqrt{x+2}} + C.$$

42.   Assume that $x \geq 1$ and let $x = \sec u$ and $dx = \sec u \tan u \, du$. Then

$$\int \frac{x^2 \, dx}{\sqrt{x^2 - 1}}$$

$$= \int \frac{\sec^3 u \tan u \, du}{\tan u} = \int \sec^3 u \, du$$

$$= \frac{1}{2} \sec u \tan u + \frac{1}{2} \ln|\sec u + \tan u| + C$$

$$= \frac{1}{2} x\sqrt{x^2 - 1} + \frac{1}{2} \ln|x + \sqrt{x^2 - 1}| + C.$$

Differentiation shows that this solution is valid for $x \leq -1$ also.

44. $\displaystyle\int \frac{2x - 3}{\sqrt{4 - 3x + x^2}} \, dx$ Let $u = 4 - 3x + x^2$
$$du = (-3 + 2x) \, dx$$

$$= \int \frac{du}{\sqrt{u}} = 2\sqrt{u} + C = 2\sqrt{4 - 3x + x^2} + C.$$

46.   Let $\sqrt{3}x = \sec u$ and $\sqrt{3} \, dx = \sec u \tan u \, du$. Then

$$\int \frac{\sqrt{3x^2 - 1}}{x} \, dx$$

$$= \int \frac{\tan u \dfrac{1}{\sqrt{3}} \sec u \tan u \, du}{\dfrac{1}{\sqrt{3}} \sec u}$$

$$= \int \tan^2 u \, du = \int (\sec^2 u - 1) \, du$$

$$= \tan u - u + C = \sqrt{3x^2 - 1} - \sec^{-1}(\sqrt{3}x) + C$$

$$= \sqrt{3x^2 - 1} + \sin^{-1}\left(\frac{1}{\sqrt{3}x}\right) + C_1.$$

48. $\displaystyle\int \sqrt{x - x^2}\, dx$

$\displaystyle= \int \sqrt{\tfrac{1}{4} - (x - \tfrac{1}{2})^2}\, dx$ Let $x - \tfrac{1}{2} = \tfrac{1}{2}\sin u$

$\qquad\qquad\qquad\qquad\qquad dx = \tfrac{1}{2}\cos u\, du$

$\displaystyle= \tfrac{1}{4}\int \cos^2 u\, du = \tfrac{1}{8}u + \tfrac{1}{8}\sin u\,\cos u + C$

$\displaystyle= \tfrac{1}{8}\sin^{-1}(2x - 1) + \tfrac{1}{4}(2x - 1)\sqrt{x - x^2} + C.$

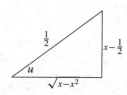

Fig. RT.48

50. $\displaystyle\int x \tan^{-1}\left(\frac{x}{3}\right) dx$

$\qquad U = \tan^{-1}\left(\dfrac{x}{3}\right)\qquad dV = x\, dx$

$\qquad dU = \dfrac{3\, dx}{9 + x^2}\qquad V = \dfrac{x^2}{2}$

$\displaystyle= \frac{x^2}{2}\tan^{-1}\left(\frac{x}{3}\right) - \frac{3}{2}\int \frac{x^2}{9 + x^2}\, dx$

$\displaystyle= \frac{x^2}{2}\tan^{-1}\left(\frac{x}{3}\right) - \frac{3}{2}\int \left(1 - \frac{9}{9 + x^2}\right) dx$

$\displaystyle= \frac{x^2}{2}\tan^{-1}\left(\frac{x}{3}\right) - \frac{3x}{2} + \frac{9}{2}\tan^{-1}\left(\frac{x}{3}\right) + C.$

52. Let $u = x^2$ and $du = 2x\, dx$; then we have

$\displaystyle I = \int \frac{dx}{x(x^2 + 4)^2} = \int \frac{x\, dx}{x^2(x^2 + 4)^2} = \frac{1}{2}\int \frac{du}{u(u + 4)^2}.$

Since

$\displaystyle\frac{1}{u(u + 4)^2} = \frac{A}{u} + \frac{B}{u + 4} + \frac{C}{(u + 4)^2}$

$\displaystyle= \frac{A(u^2 + 8u + 16) + B(u^2 + 4u) + Cu}{u(u + 4)^2}$

$\Rightarrow \begin{cases} A + B = 0 \\ 8A + 4B + C = 0 \\ 16A = 1 \end{cases} \Rightarrow A = \dfrac{1}{16},\ B = -\dfrac{1}{16},\ C = -\dfrac{1}{4},$

therefore

$\displaystyle I = \frac{1}{32}\int \frac{du}{u} - \frac{1}{32}\int \frac{du}{u + 4} - \frac{1}{8}\int \frac{du}{(u + 4)^2}$

$\displaystyle= \frac{1}{32}\ln\left|\frac{u}{u + 4}\right| + \frac{1}{8}\frac{1}{u + 4} + C$

$\displaystyle= \frac{1}{32}\ln\left|\frac{x^2}{x^2 + 4}\right| + \frac{1}{8(x^2 + 4)} + C.$

54. Since

$\displaystyle I = \int \frac{\sin(\ln x)}{x^2}\, dx$

$\qquad U = \sin(\ln x)\qquad\qquad dV = \dfrac{dx}{x^2}$

$\qquad dU = \dfrac{\cos(\ln x)}{x}\, dx\qquad V = -\dfrac{1}{x}$

$\displaystyle= -\frac{\sin(\ln x)}{x} + \int \frac{\cos(\ln x)}{x^2}\, dx$

$\qquad U = \cos(\ln x)\qquad\qquad dV = \dfrac{dx}{x^2}$

$\qquad dU = -\dfrac{\sin(\ln x)}{x}\, dx\qquad V = \dfrac{-1}{x}$

$\displaystyle= -\frac{\sin(\ln x)}{x} - \frac{\cos(\ln x)}{x} - I,$

therefore

$\displaystyle I = -\frac{1}{2x}\Big[\sin(\ln x) + \cos(\ln x)\Big] + C.$

56. We have

$\displaystyle I = \int \frac{x^3 + x - 2}{x^2 - 7}\, dx = \int \frac{x^3 - 7x + 8x - 2}{x^2 - 7}\, dx$

$\displaystyle= \int \left(x + \frac{8x - 2}{x^2 - 7}\right) dx.$

Since

$\displaystyle\frac{8x - 2}{x^2 - 7} = \frac{A}{x + \sqrt{7}} + \frac{B}{x - \sqrt{7}} = \frac{(A + B)x + (B - A)\sqrt{7}}{x^2 - 7}$

$\Rightarrow \begin{cases} A + B = 8 \\ B - A = -\dfrac{2}{\sqrt{7}} \end{cases} \Rightarrow A = 4 + \dfrac{1}{\sqrt{7}},\ B = 4 - \dfrac{1}{\sqrt{7}},$

therefore

$\displaystyle I = \int \left(x + \frac{8x - 2}{x^2 - 7}\right) dx$

$\displaystyle= \frac{x^2}{2} + \left(4 + \frac{1}{\sqrt{7}}\right)\int \frac{dx}{x + \sqrt{7}} + \left(4 - \frac{1}{\sqrt{7}}\right)\int \frac{dx}{x - \sqrt{7}}$

$\displaystyle= \frac{x^2}{2} + \left(4 + \frac{1}{\sqrt{7}}\right)\ln|x + \sqrt{7}| + \left(4 - \frac{1}{\sqrt{7}}\right)\ln|x - \sqrt{7}| + C.$

58. $\displaystyle\int \cos^7 x\, dx = \int (1 - \sin^2 x)^3 \cos x\, dx$ Let $u = \sin x$

$\qquad\qquad\qquad\qquad\qquad\qquad\qquad\qquad\qquad du = \cos x\, dx$

$\displaystyle= \int (1 - u^2)^3\, du = \int (1 - 3u^2 + 3u^4 - u^6)\, du$

$\displaystyle= u - u^3 + \tfrac{3}{5}u^5 - \tfrac{1}{7}u^7 + C$

$\displaystyle= \sin x - \sin^3 x + \tfrac{3}{5}\sin^5 x - \tfrac{1}{7}\sin^7 x + C.$

60. We have

$$\int \tan^4(\pi x)\, dx = \int \tan^2(\pi x)[\sec^2(\pi x) - 1]\, dx$$

$$= \int \tan^2(\pi x)\sec^2(\pi x)\, dx - \int [\sec^2(\pi x) - 1]\, dx$$

$$= \frac{1}{3\pi}\tan^3(\pi x) - \frac{1}{\pi}\tan(\pi x) + x + C.$$

62. $\displaystyle\int e^x(1 - e^{2x})^{5/2}\, dx$ Let $e^x = \sin u$

$$e^x\, dx = \cos u\, du$$

$$= \int \cos^6 u\, du = \left(\frac{1}{2}\right)^3 \int (1 + \cos 2u)^3\, du$$

$$= \frac{1}{8}\int (1 + 3\cos 2u + 3\cos^2 2u + \cos^3 2u)\, du$$

$$= \frac{u}{8} + \frac{3}{16}\sin 2u + \frac{3}{16}\int (1 + \cos 4u)\, du +$$

$$\frac{1}{8}\int (1 - \sin^2 2u)\cos 2u\, du$$

$$= \frac{5u}{16} + \frac{3}{16}\sin 2u + \frac{3}{64}\sin 4u + \frac{\sin 2u}{16}$$

$$- \frac{1}{48}\sin^3 2u + C$$

$$= \frac{5}{16}\sin^{-1}(e^x) + \frac{1}{4}\sin[2\sin^{-1}(e^x)] +$$

$$\frac{3}{64}\sin[4\sin^{-1}(e^x)] - \frac{1}{48}\sin^3[2\sin^{-1}(e^x)] + C$$

$$= \frac{5}{16}\sin^{-1}(e^x) + \frac{1}{2}e^x\sqrt{1 - e^{2x}}$$

$$+ \frac{3}{16}e^x\sqrt{1 - e^{2x}}\left(1 - 2e^{2x}\right)$$

$$- \frac{1}{6}e^{3x}\left(1 - e^{2x}\right)^{3/2} + C.$$

64. $\displaystyle\int \frac{x^2}{2x^2 - 3}\, dx = \frac{1}{2}\int \left(1 + \frac{3}{2x^2 - 3}\right) dx$

$$= \frac{x}{2} + \frac{\sqrt{3}}{4}\int \left(\frac{1}{\sqrt{2}x - \sqrt{3}} - \frac{1}{\sqrt{2}x + \sqrt{3}}\right) dx$$

$$= \frac{x}{2} + \frac{\sqrt{3}}{4\sqrt{2}}\ln\left|\frac{\sqrt{2}x - \sqrt{3}}{\sqrt{2}x + \sqrt{3}}\right| + C.$$

66. We have

$$\int \frac{dx}{x(x^2 + x + 1)^{1/2}}$$

$$= \int \frac{dx}{x[(x + \frac{1}{2})^2 + \frac{3}{4}]^{1/2}}$$ Let $x + \dfrac{1}{2} = \dfrac{\sqrt{3}}{2}\tan\theta$

$$dx = \frac{\sqrt{3}}{2}\sec^2\theta\, d\theta$$

$$= \int \frac{\frac{\sqrt{3}}{2}\sec^2\theta\, d\theta}{\left(\frac{\sqrt{3}}{2}\tan\theta - \frac{1}{2}\right)\left(\frac{\sqrt{3}}{2}\sec\theta\right)}$$

$$= \int \frac{2\sec\theta\, d\theta}{\sqrt{3}\tan\theta - 1} = 2\int \frac{d\theta}{\sqrt{3}\sin\theta - \cos\theta}$$

$$= 2\int \frac{\sqrt{3}\sin\theta + \cos\theta}{3\sin^2\theta - \cos^2\theta}\, d\theta$$

$$= 2\sqrt{3}\int \frac{\sin\theta\, d\theta}{3\sin^2\theta - \cos^2\theta} + 2\int \frac{\cos\theta\, d\theta}{3\sin^2\theta - \cos^2\theta}$$

$$= 2\sqrt{3}\int \frac{\sin\theta\, d\theta}{3 - 4\cos^2\theta} + 2\int \frac{\cos\theta\, d\theta}{4\sin^2\theta - 1}$$

Let $u = \cos\theta$, $du = -\sin\theta\, d\theta$ in the first integral;
let $v = \sin\theta$, $dv = \cos\theta\, d\theta$ in the second integral.

$$= -2\sqrt{3}\int \frac{du}{3 - 4u^2} + 2\int \frac{dv}{4v^2 - 1}$$

$$= -\frac{\sqrt{3}}{2}\int \frac{du}{\frac{3}{4} - u^2} - \frac{1}{2}\int \frac{du}{\frac{1}{4} - v^2}$$

$$= -\frac{\sqrt{3}}{2}\left(\frac{1}{2}\right)\left(\frac{2}{\sqrt{3}}\right)\ln\left|\frac{\cos\theta + \frac{\sqrt{3}}{2}}{\cos\theta - \frac{\sqrt{3}}{2}}\right|$$

$$- \frac{1}{2}\left(\frac{1}{2}\right)(2)\ln\left|\frac{\sin\theta + \frac{1}{2}}{\sin\theta - \frac{1}{2}}\right| + C$$

$$= \frac{1}{2}\ln\left|\frac{\left(\cos\theta - \frac{\sqrt{3}}{2}\right)\left(\sin\theta - \frac{1}{2}\right)}{\left(\cos\theta + \frac{\sqrt{3}}{2}\right)\left(\sin\theta + \frac{1}{2}\right)}\right| + C.$$

Since $\sin\theta = \dfrac{2x + 1}{2\sqrt{x^2 + x + 1}}$ and $\cos\theta = \dfrac{\sqrt{3}}{2\sqrt{x^2 + x + 1}}$,
therefore

$$\int \frac{dx}{x(x^2 + x + 1)^{1/2}} = \frac{1}{2}\ln\left|\frac{(x + 2) - 2\sqrt{x^2 + x + 1}}{(x + 2) + 2\sqrt{x^2 + x + 1}}\right| + C.$$

68. $\displaystyle\int \frac{x\, dx}{4x^4 + 4x^2 + 5}$ Let $u = x^2$

$$du = 2x\, dx$$

$$= \frac{1}{2}\int \frac{du}{4u^2 + 4u + 5}$$

$$= \frac{1}{2}\int \frac{du}{(2u + 1)^2 + 4}$$ Let $w = 2u + 1$

$$dw = 2\, du$$

$$= \frac{1}{4}\int \frac{dw}{w^2 + 4} = \frac{1}{8}\tan^{-1}\left(\frac{w}{2}\right) + C$$

$$= \frac{1}{8}\tan^{-1}\left(x^2 + \frac{1}{2}\right) + C.$$

70. Use the partial fraction decomposition

$$\frac{1}{x^3 + x^2 + x} = \frac{A}{x} + \frac{Bx + C}{x^2 + x + 1}$$
$$= \frac{A(x^2 + x + 1) + Bx^2 + Cx}{x^3 + x^2 + x}$$
$$\Rightarrow \begin{cases} A + B = 0 \\ A + C = 0 \\ A = 1 \end{cases} \Rightarrow A = 1, \ B = -1, \ C = -1.$$

Therefore,

$$\int \frac{dx}{x^3 + x^2 + x}$$
$$= \int \frac{dx}{x} - \int \frac{x + 1}{x^2 + x + 1} \, dx \quad \text{Let } u = x + \tfrac{1}{2}$$
$$\phantom{= \int \frac{dx}{x} - \int \frac{x + 1}{x^2 + x + 1} \, dx} \quad du = dx$$
$$= \ln|x| - \int \frac{u + \tfrac{1}{2}}{u^2 + \tfrac{3}{4}} \, du$$
$$= \ln|x| - \frac{1}{2} \ln\left(x^2 + x + 1\right) - \frac{1}{\sqrt{3}} \tan^{-1}\left(\frac{2x + 1}{\sqrt{3}}\right) + C.$$

72. $\displaystyle \int e^x \sec(e^x) \, dx \quad \text{Let } u = e^x$
$$ \quad du = e^x \, dx$$
$$= \int \sec u \, du = \ln|\sec u + \tan u| + C$$
$$= \ln|\sec(e^x) + \tan(e^x)| + C.$$

74. $\displaystyle \int \frac{dx}{x^{1/3} - 1} \quad \text{Let } x = (u + 1)^3$
$$\phantom{\int \frac{dx}{x^{1/3} - 1}} \quad dx = 3(u + 1)^2 \, du$$
$$= 3 \int \frac{(u + 1)^2}{u} \, du = 3 \int \left(u + 2 + \frac{1}{u}\right) du$$
$$= 3\left(\frac{u^2}{2} + 2u + \ln|u|\right) + C$$
$$= \frac{3}{2}(x^{1/3} - 1)^2 + 6(x^{1/3} - 1) + 3 \ln|x^{1/3} - 1| + C.$$

76. $\displaystyle \int \frac{x \, dx}{\sqrt{3 - 4x - 4x^2}} = \int \frac{x \, dx}{\sqrt{4 - (2x + 1)^2}} \quad \text{Let } u = 2x + 1$
$$\phantom{\int \frac{x \, dx}{\sqrt{3 - 4x - 4x^2}}} \quad du = 2 \, dx$$
$$= \frac{1}{4} \int \frac{u - 1}{\sqrt{4 - u^2}} \, du$$
$$= -\frac{1}{4} \sqrt{4 - u^2} - \frac{1}{4} \sin^{-1}\left(\frac{u}{2}\right) + C$$
$$= -\frac{1}{4} \sqrt{3 - 4x - 4x^2} - \frac{1}{4} \sin^{-1}\left(x + \frac{1}{2}\right) + C.$$

78. $\displaystyle \int \sqrt{1 + e^x} \, dx \quad \text{Let } u^2 = 1 + e^x$
$$\phantom{\int \sqrt{1 + e^x} \, dx} \quad 2u \, du = e^x \, dx$$
$$= \int \frac{2u^2 \, du}{u^2 - 1} = \int \left(2 + \frac{2}{u^2 - 1}\right) du$$
$$= \int \left(2 + \frac{1}{u - 1} - \frac{1}{u + 1}\right) du$$
$$= 2u + \ln\left|\frac{u - 1}{u + 1}\right| + C$$
$$= 2\sqrt{1 + e^x} + \ln\left|\frac{\sqrt{1 + e^x} - 1}{\sqrt{1 + e^x} + 1}\right| + C.$$

80. By the procedure used in Example 4 of Section 7.1,

$$\int e^x \cos x \, dx = \tfrac{1}{2} e^x (\sin x + \cos x) + C,$$
$$\int e^x \sin x \, dx = \tfrac{1}{2} e^x (\sin x - \cos x) + C.$$

Now

$$\int x e^x \cos x \, dx$$
$$\quad U = x \qquad dV = e^x \cos x \, dx$$
$$\quad dU = dx \qquad V = \tfrac{1}{2} e^x (\sin x + \cos x)$$
$$= \tfrac{1}{2} x e^x (\sin + \cos x) - \tfrac{1}{2} \int e^x (\sin x + \cos x) \, dx$$
$$= \tfrac{1}{2} x e^x (\sin + \cos x)$$
$$\quad - \tfrac{1}{4} e^x (\sin x - \cos x + \sin x + \cos x) + C$$
$$= \tfrac{1}{2} x e^x (\sin x + \cos x) - \tfrac{1}{2} e^x \sin x + C.$$

Other Review Exercises 6   (page 366)

2. $\displaystyle \int_0^\infty x^r e^{-x} \, dx$
$$= \lim_{\substack{c \to 0+ \\ R \to \infty}} \int_c^R x^r e^{-x} \, dx$$
$$\quad U = x^r \qquad dV = e^{-x} \, dx$$
$$\quad dU = r x^{r-1} \, dr \qquad V = -e^{-x}$$
$$= \lim_{\substack{c \to 0+ \\ R \to \infty}} -x^r e^{-x} \Big|_c^R + r \int_0^\infty x^{r-1} e^{-x} \, dx$$
$$= \lim_{c \to 0+} c^r e^{-c} + r \int_0^\infty x^{r-1} e^{-x} \, dx$$

because $\lim_{R \to \infty} R^r e^{-R} = 0$ for any r. In order to ensure that $\lim_{c \to 0+} c^r e^{-c} = 0$ we must have $\lim_{c \to 0+} c^r = 0$, so we need $r > 0$.

4. $\displaystyle\int_1^\infty \frac{dx}{x+x^3} = \lim_{R\to\infty}\int_1^R \left(\frac{1}{x} - \frac{x}{1+x^2}\right)dx$

$\displaystyle = \lim_{R\to\infty}\left(\ln|x| - \frac{1}{2}\ln(1+x^2)\right)\bigg|_1^R$

$\displaystyle = \lim_{R\to\infty}\frac{1}{2}\left(\ln\frac{R^2}{1+R^2} + \ln 2\right) = \frac{\ln 2}{2}$

6. $\displaystyle\int_0^1 \frac{dx}{x\sqrt{1-x^2}} > \int_0^1 \frac{dx}{x} = \infty$ (diverges)

Therefore $\displaystyle\int_{-1}^1 \frac{dx}{x\sqrt{1-x^2}}$ diverges.

8. Volume $= \int_0^{60} A(x)\,dx$. The approximation is

$$T_6 = \frac{10}{2}\Big[10,200 + 2(9,200 + 8,000 + 7,100$$
$$+ 4,500 + 2,400) + 100\Big]$$
$$\approx 364,000 \text{ m}^3.$$

10. $\displaystyle I = \int_0^1 \sqrt{2+\sin(\pi x)}\,dx$

$T_4 = \dfrac{1}{8}\Big[\sqrt{2} + 2(\sqrt{2+\sin(\pi/4)} + \sqrt{2+\sin(\pi/2)}$

$\qquad + \sqrt{2+\sin(3\pi/4)} + \sqrt{2}\Big]$

≈ 1.609230

$M_4 = \dfrac{1}{4}\Big[\sqrt{2+\sin(\pi/8)} + \sqrt{2+\sin(3\pi/8)}$

$\qquad \sqrt{2+\sin(5\pi/8)} + \sqrt{2+\sin(7\pi/8)}\Big]$

≈ 1.626765

$I \approx 1.6$

12. $\displaystyle I = \int_{1/2}^\infty \frac{x^2}{x^5+x^3+1}\,dx$ Let $x = 1/t$

$\qquad\qquad\qquad\qquad\qquad dx = -(1/t^2)\,dt$

$\displaystyle = \int_0^2 \frac{(1/t^4)\,dt}{(1/t^5)+(1/t^3)+1} = \int_0^2 \frac{t\,dt}{t^5+t^2+1}$

$T_4 \approx 0.4444 \qquad M_4 \approx 0.4799$

$T_8 \approx 0.4622 \qquad M_8 \approx 0.4708$

$S_8 \approx 0.4681 \qquad S_{16} \approx 0.4680$

$I \approx 0.468$ to 3 decimal places

Challenging Problems 6   (page 367)

2. a) $\displaystyle I_n = \int (1-x^2)^n\,dx$

$\qquad U = (1-x^2)^n \qquad\qquad dV = dx$

$\qquad dU = -2nx(1-x^2)^{n-1}\,dx \qquad V = x$

$\displaystyle = x(1-x^2)^n + 2n\int x^2(1-x^2)^{n-1}\,dx$

$\displaystyle = x(1-x^2)^n - 2n\int (1-x^2-1)(1-x^2)^{n-1}\,dx$

$= x(1-x^2)^n - 2nI_n + 2nI_{n-1}, \quad$ so

$\displaystyle I_n = \frac{1}{2n+1}x(1-x^2)^n + \frac{2n}{2n+1}I_{n-1}.$

b) Let $\displaystyle J_n = \int_0^1 (1-x^2)^n\,dx$. Observe that $J_0 = 1$. By (a), if $n > 0$, then we have

$$J_n = \frac{x(1-x^2)^n}{2n+1}\bigg|_0^1 + \frac{2n}{2n+1}J_{n-1} = \frac{2n}{2n+1}J_{n-1}.$$

Therefore,

$$J_n = \frac{2n}{2n+1}\cdot\frac{2n-2}{2n-1}\cdots\frac{4}{5}\cdot\frac{2}{3}J_0$$
$$= \frac{[(2n)(2n-2)\cdots(4)(2)]^2}{(2n+1)!} = \frac{2^{2n}(n!)^2}{(2n+1)!}.$$

c) From (a):

$$I_{n-1} = \frac{2n+1}{2n}I_n - \frac{1}{2n}x(1-x^2)^n.$$

Thus

$$\int (1-x^2)^{-3/2}\,dx = I_{-3/2}$$
$$= \frac{2(-1/2)+1}{-1}I_{-1/2} - \frac{1}{-1}x(1-x^2)^{-1/2}$$
$$= \frac{x}{\sqrt{1-x^2}}.$$

4. $I_{m,n} = \displaystyle\int_0^1 x^m (\ln x)^n \, dx$ Let $x = e^{-t}$
$$dx = -e^{-t} \, dt$$

$$= \int_0^\infty e^{-mt} (-t)^n e^{-t} \, dt$$

$$= (-1)^n \int_0^\infty t^n e^{-(m+1)t} \, dt \quad \text{Let } u = (m+1)t$$
$$du = (m+1) \, dt$$

$$= \frac{(-1)^n}{(m+1)^n} \int_0^\infty u^n e^{-u} \, du$$

$$= \frac{(-1)^n}{(m+1)^n} \Gamma(n+1) \quad \text{(see \#50 in Section 7.5)}$$

$$= \frac{(-1)^n n!}{(m+1)^n}.$$

6. $I = \displaystyle\int_0^1 e^{-Kx} \, dx = \left.\frac{e^{-Kx}}{-K}\right|_0^1 = \frac{1}{K}\left(1 - \frac{1}{e^K}\right).$

For very large K, the value of I is very small ($I < 1/K$). However,

$$T_{100} = \frac{1}{100}(1 + \cdots) > \frac{1}{100}$$

$$S_{100} = \frac{1}{300}(1 + \cdots) > \frac{1}{300}$$

$$M_{100} = \frac{1}{100}(e^{-K/200} + \cdots) < \frac{1}{100}.$$

In each case the $\cdots$ represent terms much less than the first term (shown) in the sum. Evidently M_{100} is smallest if k is much greater than 100, and is therefore the best approximation. T_{100} appears to be the worst.

8.   a) $f'(x) < 0$ on $[1, \infty)$, and $\lim_{x \to \infty} f(x) = 0$. Therefore

$$\int_1^\infty |f'(x)| \, dx = -\int_1^\infty f'(x) \, dx$$

$$= -\lim_{R \to \infty} \int_1^R f'(x) \, dx$$

$$= \lim_{R \to \infty} (f(1) - f(R)) = f(1).$$

Thus

$$\left|\int_R^\infty f'(x) \cos x \, dx\right| \le \int_R^\infty |f'(x)| \, dx \to 0 \text{ as } R \to \infty.$$

Thus $\lim_{R \to \infty} \displaystyle\int_1^R f'(x) \cos x \, dx$ exists.

b) $\displaystyle\int_1^\infty f(x) \sin x \, dx$

$$U = f(x) \qquad dV = \sin x \, dx$$
$$dU = f'(x) \, dx \qquad V = -\cos x$$

$$= \lim_{R \to \infty} \left. f(x) \cos x \right|_1^R + \int_1^\infty f'(x) \cos x \, dx$$

$$= -f(1) \cos(1) + \int_1^\infty f'(x) \cos x \, dx;$$

the integral converges.

c) $f(x) = 1/x$ satisfies the conditions of part (a), so

$$\int_1^\infty \frac{\sin x}{x} \, dx \quad \text{converges}$$

by part (b). Similarly, it can be shown that

$$\int_1^\infty \frac{\cos(2x)}{x} \, dx \quad \text{converges}.$$

But since $|\sin x| \ge \sin^2 x = \frac{1}{2}(1 - \cos(2x))$, we have

$$\int_1^\infty \frac{|\sin x|}{x} \, dx \ge \int_1^\infty \frac{1 - \cos(2x)}{2x}.$$

The latter integral diverges because $\int_1^\infty (1/x) \, dx$ diverges to infinity while $\int_1^\infty (\cos(2x))/(2x) \, dx$ converges. Therefore

$$\int_1^\infty \frac{|\sin x|}{x} \, dx \quad \text{diverges to infinity}.$$

CHAPTER 7. APPLICATIONS OF INTEGRATION

Section 7.1   Volumes of Solids of Revolution (page 376)

2. Slicing:

$$V = \pi \int_0^1 (1 - y)\, dy$$

$$= \pi \left(y - \frac{1}{2} y^2 \right) \Big|_0^1 = \frac{\pi}{2} \text{ cu. units.}$$

Shells:

$$V = 2\pi \int_0^1 x^3\, dx$$

$$= 2\pi \left(\frac{x^4}{4} \right) \Big|_0^1 = \frac{\pi}{2} \text{ cu. units.}$$

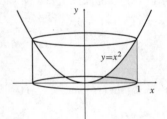

Fig. 7.1.2

4. Slicing:

$$V = \pi \int_0^1 (y - y^4)\, dy$$

$$= \pi \left(\frac{1}{2} y^2 - \frac{1}{5} y^5 \right) \Big|_0^1 = \frac{3\pi}{10} \text{ cu. units.}$$

Shells:

$$V = 2\pi \int_0^1 x(x^{1/2} - x^2)\, dx$$

$$= 2\pi \left(\frac{2}{5} x^{5/2} - \frac{1}{4} x^4 \right) \Big|_0^1 = \frac{3\pi}{10} \text{ cu. units.}$$

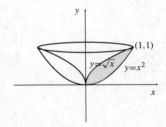

Fig. 7.1.4

6. Rotate about

a) the x-axis

$$V = \pi \int_0^1 (x^2 - x^4)\, dx$$

$$= \pi \left(\frac{1}{3} x^3 - \frac{1}{5} x^5 \right) \Big|_0^1 = \frac{2\pi}{15} \text{ cu. units.}$$

b) the y-axis

$$V = 2\pi \int_0^1 x(x - x^2)\, dx$$

$$= 2\pi \left(\frac{1}{3} x^3 - \frac{1}{4} x^4 \right) \Big|_0^1 = \frac{\pi}{6} \text{ cu. units.}$$

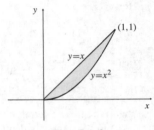

Fig. 7.1.6

8. Rotate about

a) the x-axis

$$V = \pi \int_0^\pi [(1 + \sin x)^2 - 1]\, dx$$

$$= \pi \int_0^\pi (2 \sin x + \sin^2 x)\, dx$$

$$= \left(-2\pi \cos x + \frac{\pi}{2} x - \frac{\pi}{4} \sin 2x \right) \Big|_0^\pi$$

$$= 4\pi + \frac{1}{2} \pi^2 \text{ cu. units.}$$

b) the y-axis

$$V = 2\pi \int_0^\pi x \sin x\, dx$$

$$U = x \qquad dV = \sin x\, dx$$

$$dU = dx \qquad V = -\cos x$$

$$= 2\pi \left[-x \cos x \Big|_0^\pi + \int_0^\pi \cos x\, dx \right]$$

$$= 2\pi^2 \text{ cu. units.}$$

10. By symmetry, rotation about the x-axis gives the same volume as rotation about the y-axis, namely

$$V = 2\pi \int_{1/3}^{3} x\left(\frac{10}{3} - x - \frac{1}{x}\right) dx$$

$$= 2\pi \left(\frac{5}{3}x^2 - \frac{1}{3}x^3 - x\right)\Big|_{1/3}^{3}$$

$$= \frac{512\pi}{81} \text{ cu. units.}$$

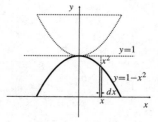

Fig. 7.1.10

12. $V = \pi \int_{-1}^{1} [(1)^2 - (x^2)^2] \, dx$

$$= \pi \left(x - \frac{1}{5}x^5\right)\Big|_{-1}^{1}$$

$$= \frac{8\pi}{5} \text{ cu. units.}$$

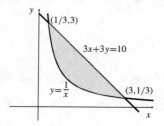

Fig. 7.1.12

14. The radius of the hole is $\sqrt{R^2 - \frac{1}{4}L^2}$. Thus, by slicing, the remaining volume is

$$V = \pi \int_{-L/2}^{L/2} \left[\left(R^2 - x^2\right) - \left(R^2 - \frac{L^2}{4}\right)\right] dx$$

$$= 2\pi \left(\frac{L^2}{4}x - \frac{1}{3}x^3\right)\Big|_{0}^{L/2}$$

$$= \frac{\pi}{6}L^3 \text{ cu. units (independent of } R).$$

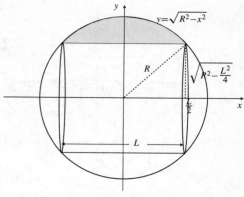

Fig. 7.1.14

16. Let a circular disk with radius a have centre at point $(a, 0)$. Then the disk is rotated about the y-axis which is one of its tangent lines. The volume is:

$$V = 2 \times 2\pi \int_{0}^{2a} x\sqrt{a^2 - (x-a)^2}\, dx \quad \text{Let } u = x - a$$
$$\qquad\qquad\qquad\qquad\qquad\qquad\qquad du = dx$$

$$= 4\pi \int_{-a}^{a} (u + a)\sqrt{a^2 - u^2}\, du$$

$$= 4\pi \int_{-a}^{a} u\sqrt{a^2 - u^2}\, du + 4\pi a \int_{-a}^{a} \sqrt{a^2 - u^2}\, du$$

$$= 0 + 4\pi a\left(\frac{1}{2}\pi a^2\right) = 2\pi^2 a^3 \text{ cu. units.}$$

(Note that the first integral is zero because the integrand is odd and the interval is symmetric about zero; the second integral is the area of a semicircle.)

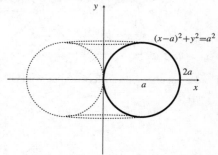

Fig. 7.1.16

18. Let the centre of the bowl be at $(0, 30)$. Then the volume of the water in the bowl is

$$V = \pi \int_{0}^{20} \left[30^2 - (y - 30)^2\right] dy$$

$$= \pi \int_{0}^{20} 60y - y^2 \, dy$$

$$= \pi \left[30y^2 - \frac{1}{3}y^3\right]\Big|_{0}^{20}$$

$$\approx 29322 \text{ cm}^3.$$

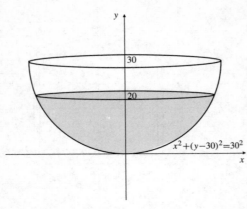

Fig. 7.1.18

20. The cross-section at height y is an annulus (ring) having inner radius $b - \sqrt{a^2 - y^2}$ and outer radius $b + \sqrt{a^2 - y^2}$. Thus the volume of the torus is

$$V = \pi \int_{-a}^{a} \left[(b + \sqrt{a^2 - y^2})^2 - (b - \sqrt{a^2 - y^2})^2 \right] dy$$

$$= 2\pi \int_{0}^{a} 4b\sqrt{a^2 - y^2}\, dy$$

$$= 8\pi b \frac{\pi a^2}{4} = 2\pi^2 a^2 b \text{ cu. units..}$$

We used the area of a quarter-circle of radius a to evaluate the last integral.

22. The volume is

$$V = \pi \int_{1}^{\infty} x^{-2k}\, dx = \pi \lim_{R \to \infty} \frac{x^{1-2k}}{1 - 2k} \Big|_{1}^{R}$$

$$= \pi \lim_{R \to \infty} \frac{R^{1-2k}}{1 - 2k} + \frac{\pi}{2k - 1}.$$

In order for the solid to have finite volume we need

$$1 - 2k < 0, \quad \text{that is,} \quad k > \frac{1}{2}.$$

24. A solid consisting of points on parallel line segments between parallel planes will certainly have congruent cross-sections in planes parallel to and lying between the two base planes, any solid satisfying the new definition will certainly satisfy the old one. But not vice versa; congruent cross-sections does not imply a family of parallel line segments giving all the points in a solid. For a counterexample, see the next exercise. Thus the earlier, incorrect definition defines a larger class of solids than does the current definition. However, the formula $V = Ah$ for the volume of such a solid is still valid, as all congruent cross-sections still have the same area, A, as the base region.

26. Using heights $f(x)$ estimated from the given graph, we obtain

$$V = \pi \int_{1}^{9} \left(f(x) \right)^2 dx$$

$$\approx \frac{\pi}{3} \Big[3^2 + 4(3.8)^2 + 2(5)^2 + 4(6.7)^2 + 2(8)^2 \\ + 4(8)^2 + 2(7)^2 + 4(5.2)^2 + 3^2 \Big] \approx 938 \text{ cu. units.}$$

28. Using heights $f(x)$ estimated from the given graph, we obtain

$$V = 2\pi \int_{1}^{9} (x + 1) f(x)\, dx$$

$$\approx \frac{2\pi}{3} \Big[2(3) + 4(3)(3.8) + 2(4)(5) + 4(5)(6.7) + 2(6)(8) \\ + 4(7)(8) + 2(8)(7) + 4(9)(5.2) + 10(3) \Big] \approx 1832 \text{ cu. unit}$$

30. The volume of the ball is $\frac{4}{3}\pi R^3$. Expressing this volume as the "sum" (i.e., integral) of volume elements that are concentric spherical shells of radius r and thickness dr, and therefore surface area kr^2 and volume $kr^2\, dr$, we obtain

$$\frac{4}{3}\pi R^3 = \int_{0}^{R} kr^2\, dr = \frac{k}{3} R^3.$$

Thus $k = 4\pi$.

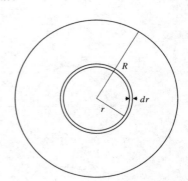

Fig. 7.1.30

32. Let P be the point $(t, \frac{5}{2} - t)$. The line through P perpendicular to AB has equation $y = x + \frac{5}{2} - 2t$, and meets the curve $xy = 1$ at point Q with x-coordinate s equal to the positive root of $s^2 + (\frac{5}{2} - 2t)s = 1$. Thus,

$$s = \frac{1}{2} \left[2t - \frac{5}{2} + \sqrt{\left(\frac{5}{2} - 2t \right)^2 + 4} \right].$$

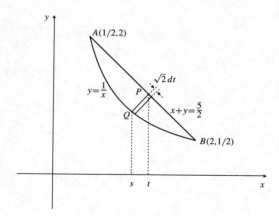

Fig. 7.1.32

The volume element at P has radius

$$PQ = \sqrt{2}(t - s)$$
$$= \sqrt{2}\left[\frac{5}{4} - \frac{1}{2}\sqrt{\left(\frac{5}{2} - 2t\right)^2 + 4}\right]$$

and thickness $\sqrt{2}\,dt$. Hence, the volume of the solid is

$$V = \pi \int_{1/2}^{2}\left[\sqrt{2}\left(\frac{5}{4} - \frac{1}{2}\sqrt{\left(\frac{5}{2} - 2t\right)^2 + 4}\right)\right]^2 \sqrt{2}\,dt$$

$$= 2\sqrt{2}\pi \int_{1/2}^{2}\left[\frac{25}{16} - \frac{5}{4}\left(\sqrt{\left(\frac{5}{2} - 2t\right)^2 + 4}\right) + \right.$$

$$\left. \frac{1}{4}\left[\left(\frac{5}{2} - 2t\right)^2 + 4\right]\right]dt \quad \text{Let } u = 2t - \frac{5}{2}$$
$$\hspace{6cm} du = 2\,dt$$

$$= \sqrt{2}\pi \int_{-3/2}^{3/2}\left(\frac{41}{16} - \frac{5}{4}\sqrt{u^2 + 4} + \frac{u^2}{4}\right)du$$

$$= \sqrt{2}\pi\left(\frac{41}{16}u + \frac{1}{12}u^3\right)\Big|_{-3/2}^{3/2} -$$

$$\frac{5\sqrt{2}\pi}{4}\int_{-3/2}^{3/2}\sqrt{u^2 + 4}\,du \quad \text{Let } u = 2\tan v$$
$$\hspace{6cm} du = 2\sec^2 v\,dv$$

$$= \frac{33\sqrt{2}\pi}{4} - 5\sqrt{2}\pi\int_{\tan^{-1}(-3/4)}^{\tan^{-1}(3/4)}\sec^3 v\,dv$$

$$= \frac{33\sqrt{2}\pi}{4} - 10\sqrt{2}\pi\int_{0}^{\tan^{-1}(3/4)}\sec^3 v\,dv$$

$$= \frac{33\sqrt{2}\pi}{4} - 5\sqrt{2}\pi\Big(\sec v\tan v +$$

$$\ln|\sec v + \tan v|\Big)\Big|_{0}^{\tan^{-1}(3/4)}$$

$$= \sqrt{2}\pi\left[\frac{33}{4} - 5\left(\frac{15}{16} + \ln 2 - 0 - \ln 1\right)\right]$$

$$= \sqrt{2}\pi\left(\frac{57}{16} - 5\ln 2\right) \text{ cu. units.}$$

Section 7.2   Other Volumes by Slicing (page 380)

2. A horizontal slice of thickness dz at height a has volume $dV = z(h - z)\,dz$. Thus the volume of the solid is

$$V = \int_{0}^{h}(z(h - z)\,dz = \left(\frac{hz^2}{2} - \frac{z^3}{3}\right)\Big|_{0}^{h} = \frac{h^3}{6} \text{ units}^3.$$

4. $V = \int_{1}^{3}x^2\,dx = \frac{x^3}{3}\Big|_{1}^{3} = \frac{26}{3}$ cu. units

6. The area of an equilateral triangle of edge $\sqrt{x}$ is $A(x) = \frac{1}{2}\sqrt{x}\left(\frac{\sqrt{3}}{2}\sqrt{x}\right) = \frac{\sqrt{3}}{4}x$ sq. units. The volume of the solid is

$$V = \int_{1}^{4}\frac{\sqrt{3}}{4}x\,dx = \frac{\sqrt{3}}{8}x^2\Big|_{1}^{4} = \frac{15\sqrt{3}}{8} \text{ cu. units.}$$

8. Since $V = 4$, we have

$$4 = \int_{0}^{2}kx^3\,dx = k\,\frac{x^4}{4}\Big|_{0}^{2} = 4k.$$

Thus $k = 1$.

10. This is similar to Exercise 7. We have $4z = \int_{0}^{z}A(t)\,dt$, so $A(z) = 4$. Thus the square cross-section at height z has side 2 units.

12. The area of an equilateral triangle of base $2y$ is $\frac{1}{2}(2y)(\sqrt{3}y) = \sqrt{3}y^2$. Hence, the solid has volume

$$V = 2\int_{0}^{r}\sqrt{3}(r^2 - x^2)\,dx$$
$$= 2\sqrt{3}\left(r^2x - \frac{1}{3}x^3\right)\Big|_{0}^{r}$$
$$= \frac{4}{\sqrt{3}}r^3 \text{ cu. units.}$$

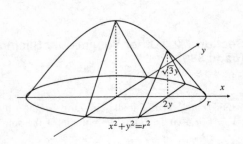

Fig. 7.2.12

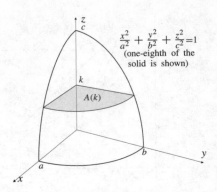

$$\frac{x^2}{a^2} + \frac{y^2}{b^2} + \frac{z^2}{c^2} = 1$$
(one-eighth of the solid is shown)

Fig. 7.2.16

14. The volume of a solid of given height h and given cross-sectional area $A(z)$ at height z above the base is given by

$$V = \int_0^h A(z)\, dz.$$

If two solids have the same height h and the same area function $A(z)$, then they must necessarily have the same volume.

18. The solution is similar to that of Exercise 15 except that the legs of the right-triangular cross-sections are $y - 10$ instead of y, and x goes from $-10\sqrt{3}$ to $10\sqrt{3}$ instead of -20 to 20. The volume of the notch is

$$V = 2\int_0^{10\sqrt{3}} \frac{1}{2}(\sqrt{400 - x^2} - 10)^2\, dx$$

$$= \int_0^{10\sqrt{3}} \left(500 - x^2 - 20\sqrt{400 - x^2}\right) dx$$

$$= 3,000\sqrt{3} - \frac{4,000\pi}{3} \approx 1,007 \text{ cm}^3.$$

16. The plane $z = k$ meets the ellipsoid in the ellipse

$$\left(\frac{x}{a}\right)^2 + \left(\frac{y}{b}\right)^2 = 1 - \left(\frac{k}{c}\right)^2$$

that is, $\dfrac{x^2}{a^2\left[1 - \left(\frac{k}{c}\right)^2\right]} + \dfrac{y^2}{b^2\left[1 - \left(\frac{k}{c}\right)^2\right]} = 1$

which has area

$$A(k) = \pi ab\left[1 - \left(\frac{k}{c}\right)^2\right].$$

The volume of the ellipsoid is found by summing volume elements of thickness dk:

$$V = \int_{-c}^{c} \pi ab\left[1 - \left(\frac{k}{c}\right)^2\right] dk$$

$$= \pi ab\left[k - \frac{1}{3c^2}k^3\right]\Bigg|_{-c}^{c}$$

$$= \frac{4}{3}\pi abc \text{ cu. units.}$$

20. One eighth of the region lying inside both cylinders is shown in the figure. If the region is sliced by a horizontal plane at height z, then the intersection is a rectangle with area

$$A(z) = \sqrt{b^2 - z^2}\sqrt{a^2 - z^2}.$$

The volume of the whole region is

$$V = 8\int_0^b \sqrt{b^2 - z^2}\sqrt{a^2 - z^2}\, dz.$$

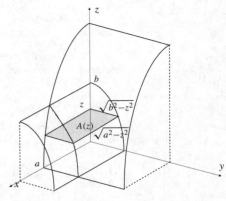

Fig. 7.2.20

Section 7.3   Arc Length and Surface Area (page 387)

2. $y = ax + b$, $A \le x \le B$, $y' = a$. The length is

$$L = \int_A^B \sqrt{1 + a^2}\, dx = \sqrt{1 + a^2}(B - A) \text{ units.}$$

4. $y^2 = (x - 1)^3$, $y = (x - 1)^{3/2}$, $y' = \frac{3}{2}\sqrt{x - 1}$

$$L = \int_1^2 \sqrt{1 + \frac{9}{4}(x - 1)}\, dx = \frac{1}{2}\int_1^2 \sqrt{9x - 5}\, dx$$

$$= \frac{1}{27}(9x - 5)^{3/2}\Big|_1^2 = \frac{13^{3/2} - 8}{27} \text{ units.}$$

6. $2(x + 1)^3 = 3(y - 1)^2$, $y = 1 + \sqrt{\frac{2}{3}}(x + 1)^{3/2}$

$y' = \sqrt{\frac{3}{2}}(x + 1)^{1/2}$,

$$ds = \sqrt{1 + \frac{3x + 3}{2}}\, dx = \sqrt{\frac{3x + 5}{2}}\, dx$$

$$L = \frac{1}{\sqrt{2}}\int_{-1}^0 \sqrt{3x + 5}\, dx = \frac{\sqrt{2}}{9}(3x + 5)^{3/2}\Big|_{-1}^0$$

$$= \frac{\sqrt{2}}{9}\left(5^{3/2} - 2^{3/2}\right) \text{ units.}$$

8. $y = \frac{x^3}{3} + \frac{1}{4x}$, $y' = x^2 - \frac{1}{4x^2}$

$$ds = \sqrt{1 + \left(x^2 - \frac{1}{4x^2}\right)^2}\, dx = \left(x^2 + \frac{1}{4x^2}\right) dx$$

$$L = \int_1^2 \left(x^2 + \frac{1}{4x^2}\right) dx = \left(\frac{x^3}{3} - \frac{1}{4x}\right)\Big|_1^2 = \frac{59}{24} \text{ units.}$$

10. If $y = x^2 - \frac{\ln x}{8}$ then $y' = 2x - \frac{1}{8x}$ and

$$1 + (y')^2 = \left(2x + \frac{1}{8x}\right)^2.$$

Thus the arc length is given by

$$s = \int_1^2 \sqrt{1 + \left(2x - \frac{1}{8x}\right)^2}\, dx$$

$$= \int_1^2 \left(2x + \frac{1}{8x}\right) dx$$

$$= \left(x^2 + \frac{1}{8}\ln x\right)\Big|_1^2 = 3 + \frac{1}{8}\ln 2 \text{ units.}$$

12.
$$s = \int_{\pi/6}^{\pi/4} \sqrt{1 + \tan^2 x}\, dx$$

$$= \int_{\pi/6}^{\pi/4} \sec x\, dx = \ln|\sec x + \tan x|\Big|_{\pi/6}^{\pi/4}$$

$$= \ln(\sqrt{2} + 1) - \ln\left(\frac{2}{\sqrt{3}} + \frac{1}{\sqrt{3}}\right)$$

$$= \ln\frac{\sqrt{2} + 1}{\sqrt{3}} \text{ units.}$$

14. $y = \ln\frac{e^x - 1}{e^x + 1}$, $2 \le x \le 4$

$y' = \frac{e^x + 1}{e^x - 1} \cdot \frac{(e^x + 1)e^x - (e^x - 1)e^x}{(e^x + 1)^2}$

$= \frac{2e^x}{e^{2x} - 1}$.

The length of the curve is

$$L = \int_2^4 \sqrt{1 + \frac{4e^{2x}}{(e^{2x} - 1)^2}}\, dx$$

$$= \int_2^4 \frac{e^{2x} + 1}{e^{2x} - 1}\, dx$$

$$= \int_2^4 \frac{e^x + e^{-x}}{e^x - e^{-x}}\, dx = \ln\left|e^x - e^{-x}\right|\Big|_2^4$$

$$= \ln\left(e^4 - \frac{1}{e^4}\right) - \ln\left(e^2 - \frac{1}{e^2}\right)$$

$$= \ln\left(\frac{e^8 - 1}{e^4}\,\frac{e^2}{e^4 - 1}\right) = \ln\frac{e^4 + 1}{e^2} \text{ units.}$$

16. The required length is

$$L = \int_0^1 \sqrt{1 + (4x^3)^2}\, dx = \int_0^1 \sqrt{1 + 16x^6}\, dx.$$

Using a calculator we calculate some Simpson's Rule approximations as described in Section 7.2:

$$S_2 \approx 1.59921 \qquad S_4 \approx 1.60110$$
$$S_8 \approx 1.60025 \qquad S_{16} \approx 1.60023.$$

To four decimal places the length is 1.6002 units.

18. For the ellipse $3x^2 + y^2 = 3$, we have $6x + 2yy' = 0$, so $y' = -3x/y$. Thus

$$ds = \sqrt{1 + \frac{9x^2}{3 - 3x^2}}\, dx = \sqrt{\frac{3 + 6x^2}{3 - 3x^2}}\, dx.$$

The circumference of the ellipse is

$$4\int_0^1 \sqrt{\frac{3 + 6x^2}{3 - 3x^2}}\, dx \approx 8.73775 \text{ units}$$

(with a little help from Maple's numerical integration routine.)

20. $S = 2\pi \int_0^2 |x|\sqrt{1+4x^2}\,dx$ Let $u = 1 + 4x^2$
$$du = 8x\,dx$$

$$= \frac{\pi}{4}\int_1^{17}\sqrt{u}\,du = \frac{\pi}{4}\left(\frac{2}{3}u^{3/2}\right)\Big|_1^{17}$$

$$= \frac{\pi}{6}(17\sqrt{17}-1) \text{ sq. units.}$$

22. $y = x^{3/2}$, $0 \le x \le 1$. $ds = \sqrt{1+\frac{9}{4}x}\,dx$.

The area of the surface of rotation about the x-axis is

$$S = 2\pi \int_0^1 x^{3/2}\sqrt{1+\frac{9x}{4}}\,dx \quad \text{Let } 9x = 4u^2$$
$$9\,dx = 8u\,du$$

$$= \frac{128\pi}{243}\int_0^{3/2} u^4\sqrt{1+u^2}\,du \quad \text{Let } u = \tan v$$
$$du = \sec^2 v\,dv$$

$$= \frac{128\pi}{243}\int_0^{\tan^{-1}(3/2)} \tan^4 v \sec^3 v\,dv$$

$$= \frac{128\pi}{243}\int_0^{\tan^{-1}(3/2)} (\sec^7 v - 2\sec^5 v + \sec^3 v)\,dv.$$

At this stage it is convenient to use the reduction formula

$$\int \sec^n v\,dv = \frac{1}{n-1}\sec^{n-2} v \tan v + \frac{n-2}{n-1}\int \sec^{n-2} v\,dv$$

(see Exercise 36 of Section 7.1) to reduce the powers of secant down to 3, and then use

$$\int_0^a \sec^3 v\,dv = \frac{1}{2}(\sec a \tan a + \ln|\sec a + \tan a|.$$

We have

$$I = \int_0^a (\sec^7 v - 2\sec^5 v + \sec^3 v)\,dv$$

$$= \frac{\sec^5 v \tan v}{6}\Big|_0^a + \left(\frac{5}{6}-2\right)\int_0^a \sec^5 v\,dv + \int_0^a \sec^3 v\,dv$$

$$= \frac{\sec^5 a \tan a}{6} - \frac{7}{6}\left[\frac{\sec^3 v \tan v}{4}\Big|_0^a + \frac{3}{4}\int_0^a \sec^3 v\,dv\right]$$
$$+ \int_0^a \sec^3 v\,dv$$

$$= \frac{\sec^5 a \tan a}{6} - \frac{7\sec^3 a \tan a}{24} + \frac{1}{8}\int_0^a \sec^3 v\,dv$$

$$= \frac{\sec^5 a \tan a}{6} - \frac{7\sec^3 a \tan a}{24}$$
$$+ \frac{\sec a \tan a + \ln|\sec a + \tan a|}{16}.$$

Substituting $a = \arctan(3/2)$ now gives the following value for the surface area:

$$S = \frac{28\sqrt{13}\pi}{81} + \frac{8\pi}{243}\ln\left(\frac{3+\sqrt{13}}{2}\right) \text{ sq. units.}$$

24. We have

$$S = 2\pi \int_0^1 e^x\sqrt{1+e^{2x}}\,dx \quad \text{Let } e^x = \tan\theta$$
$$e^x\,dx = \sec^2\theta\,d\theta$$

$$= 2\pi \int_{x=0}^{x=1}\sqrt{1+\tan^2\theta}\,\sec^2\theta\,d\theta = 2\pi\int_{x=0}^{x=1}\sec^3\theta\,d\theta$$

$$= \pi\left[\sec\theta\tan\theta + \ln|\sec\theta+\tan\theta|\right]\Big|_{x=0}^{x=1}.$$

Since

$$x = 1 \Rightarrow \tan\theta = e, \ \sec\theta = \sqrt{1+e^2},$$
$$x = 0 \Rightarrow \tan\theta = 1, \ \sec\theta = \sqrt{2},$$

therefore

$$S = \pi\left[e\sqrt{1+e^2} + \ln|\sqrt{1+e^2}+e| - \sqrt{2} - \ln|\sqrt{2}+1|\right]$$

$$= \pi\left[e\sqrt{1+e^2} - \sqrt{2} + \ln\frac{\sqrt{1+e^2}+e}{\sqrt{2}+1}\right] \text{ sq. units.}$$

26. $1 + (y')^2 = 1 + \left(\frac{x^2}{4} - \frac{1}{x^2}\right)^2 = \left(\frac{x^2}{4}+\frac{1}{x^2}\right)^2$

$$S = 2\pi\int_1^4\left(\frac{x^3}{12}+\frac{1}{x}\right)\left(\frac{x^2}{4}+\frac{1}{x^2}\right)dx$$

$$= 2\pi\int_1^4\left(\frac{x^5}{48}+\frac{x}{3}+\frac{1}{x^3}\right)dx$$

$$= 2\pi\left(\frac{x^6}{288}+\frac{x^2}{6}-\frac{1}{2x^2}\right)\Big|_1^4$$

$$= \frac{275}{8}\pi \text{ sq. units.}$$

28. The area of the cone obtained by rotating the line $y = (h/r)x$, $0 \le x \le r$, about the y-axis is

$$S = 2\pi\int_0^r x\sqrt{1+(h/r)^2}\,dx = 2\pi\frac{\sqrt{r^2+h^2}}{r}\frac{x^2}{2}\Big|_0^r$$

$$= \pi r\sqrt{r^2+h^2} \text{ sq. units.}$$

30. The top half of $x^2 + 4y^2 = 4$ is $y = \dfrac{1}{2}\sqrt{4 - x^2}$, so

$\dfrac{dy}{dx} = \dfrac{-x}{2\sqrt{4 - x^2}}$, and

$$S = 2 \times 2\pi \int_0^2 \frac{\sqrt{4 - x^2}}{2}\sqrt{1 + \left(\frac{x}{2\sqrt{4 - x^2}}\right)^2}\,dx$$

$$= \pi \int_0^2 \sqrt{16 - 3x^2}\,dx \quad \text{Let } x = \sqrt{\frac{16}{3}}\sin\theta$$

$$dx = \sqrt{\frac{16}{3}}\cos\theta\,d\theta$$

$$= \pi \int_0^{\pi/3} (4\cos\theta)\frac{4}{\sqrt{3}}\cos\theta\,d\theta$$

$$= \frac{16\pi}{\sqrt{3}} \int_0^{\pi/3} \cos^2\theta\,d\theta$$

$$= \frac{8\pi}{\sqrt{3}}\left(\theta + \sin\theta\cos\theta\right)\Big|_0^{\pi/3}$$

$$= \frac{2\pi(4\pi + 3\sqrt{3})}{3\sqrt{3}} \text{ sq. units.}$$

32. As in Example 4, the arc length element for the ellipse is

$$ds = \sqrt{1 + \left(\frac{dy}{dx}\right)^2}\,dx = \sqrt{\frac{a^2 - \dfrac{a^2 - b^2}{a^2}x^2}{a^2 - x^2}}\,dx.$$

To get the area of the ellipsoid, we must rotate both the upper and lower semi-ellipses (see the figure for Exercise 20 of Section 8.1):

$$S = 2 \times 2\pi \int_0^a \left[\left(c - b\sqrt{1 - \left(\frac{x}{a}\right)^2}\right) + \left(c + b\sqrt{1 - \left(\frac{x}{a}\right)^2}\right)\right]ds$$

$$= 8\pi c \int_0^a \sqrt{\frac{a^2 - \dfrac{a^2 - b^2}{a^2}x^2}{a^2 - x^2}}\,dx$$

$$= 8\pi c \left[\frac{1}{4} \text{ of the circumference of the ellipse}\right]$$

$$= 8\pi c a E(\varepsilon)$$

where $\varepsilon = \dfrac{\sqrt{a^2 - b^2}}{a}$ and $E(\varepsilon) = \int_0^{\pi/2} \sqrt{1 - \varepsilon^2 \sin t}\,dt$ as defined in Example 4.

34. Let the equation of the sphere be $x^2 + y^2 = R^2$. Then the surface area between planes $x = a$ and $x = b$ $(-R \le a < b \le R)$ is

$$S = 2\pi \int_a^b \sqrt{R^2 - x^2}\sqrt{1 + \left(\frac{dy}{dx}\right)^2}\,dx$$

$$= 2\pi \int_a^b \sqrt{R^2 - x^2}\frac{R}{\sqrt{R^2 - x^2}}\,dx$$

$$= 2\pi R \int_a^b dx = 2\pi R(b - a) \text{ sq. units.}$$

Thus, the surface area depends only on the radius R of the sphere, and the distance $(b - a)$ between the parellel planes.

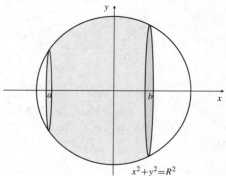

Fig. 7.3.34

36.
$$S = 2\pi \int_0^1 |x|\sqrt{1 + \frac{1}{x^2}}\,dx$$

$$= 2\pi \int_0^1 \sqrt{x^2 + 1}\,dx \quad \text{Let } x = \tan\theta$$

$$dx = \sec^2\theta\,d\theta$$

$$= 2\pi \int_0^{\pi/4} \sec^3\theta\,d\theta$$

$$= \pi\left(\sec\theta\tan\theta + \ln|\sec\theta + \tan\theta|\right)\Big|_0^{\pi/4}$$

$$= \pi[\sqrt{2} + \ln(\sqrt{2} + 1)] \text{ sq. units.}$$

Section 7.4 Mass, Moments, and Centre of Mass (page 394)

2. A slice of the wire of width dx at x has volume $dV = \pi(a + bx)^2\,dx$. Therefore the mass of the whole wire is

$$m = \int_0^L \delta_0\pi(a + bx)^2\,dx$$

$$= \delta_0\pi \int_0^L (a^2 + 2abx + b^2x^2)\,dx$$

$$= \delta_0\pi\left(a^2 L + abL^2 + \frac{1}{3}b^2 L^3\right).$$

Its moment about $x = 0$ is

$$M_{x=0} = \int_0^L x\delta_0\pi(a+bx)^2\,dx$$

$$= \delta_0\pi \int_0^L (a^2x + 2abx^2 + b^2x^3)\,dx$$

$$= \delta_0\pi\left(\frac{1}{2}a^2L^2 + \frac{2}{3}abL^3 + \frac{1}{4}b^2L^4\right).$$

Thus, the centre of mass is

$$\bar{x} = \frac{\delta_0\pi\left(\dfrac{1}{2}a^2L^2 + \dfrac{2}{3}abL^3 + \dfrac{1}{4}b^2L^4\right)}{\delta_0\pi\left(a^2L + abL^2 + \dfrac{1}{3}b^2L^3\right)}$$

$$= \frac{L\left(\dfrac{1}{2}a^2 + \dfrac{2}{3}abL + \dfrac{1}{4}b^2L^2\right)}{a^2 + abL + \dfrac{1}{3}b^2L^2}.$$

4. A vertical strip has area $dA = \sqrt{a^2 - x^2}\,dx$. Therefore, the mass of the quarter-circular plate is

$$m = \int_0^a (\delta_0 x)\sqrt{a^2 - x^2}\,dx \quad \text{Let } u = a^2 - x^2$$
$$du = -2x\,dx$$

$$= \frac{1}{2}\delta_0 \int_0^{a^2} \sqrt{u}\,du = \frac{1}{2}\delta_0\left(\frac{2}{3}u^{3/2}\right)\Big|_0^{a^2} = \frac{1}{3}\delta_0 a^3.$$

The moment about $x = 0$ is

$$M_{x=0} = \int_0^a \delta_0 x^2\sqrt{a^2 - x^2}\,dx \quad \text{Let } x = a\sin\theta$$
$$dx = a\cos\theta\,d\theta$$

$$= \delta_0 a^4 \int_0^{\pi/2} \sin^2\theta\cos^2\theta\,d\theta$$

$$= \frac{\delta_0 a^4}{4} \int_0^{\pi/2} \sin^2 2\theta\,d\theta$$

$$= \frac{\delta_0 a^4}{8} \int_0^{\pi/2} (1 - \cos 4\theta)\,d\theta = \frac{\pi\delta_0 a^4}{16}.$$

The moment about $y = 0$ is

$$M_{y=0} = \frac{1}{2}\delta_0 \int_0^a x(a^2 - x^2)\,dx$$

$$= \frac{1}{2}\delta_0\left(\frac{a^2x^2}{2} - \frac{x^4}{4}\right)\Big|_0^a = \frac{1}{8}a^4\delta_0.$$

Thus, $\bar{x} = \dfrac{3}{16}\pi a$ and $\bar{y} = \dfrac{3}{8}a$. Hence, the centre of mass is located at $(\dfrac{3}{16}\pi a, \dfrac{3}{8}a)$.

6. A vertical strip at h has area $dA = (2 - \frac{2}{3}h)\,dh$. Thus, the mass of the plate is

$$m = \int_0^3 (5h)\left(2 - \frac{2}{3}h\right)dh = 10\int_0^3\left(h - \frac{h^2}{3}\right)dh$$

$$= 10\left(\frac{h^2}{2} - \frac{h^3}{9}\right)\Big|_0^3 = 15 \text{ kg.}$$

The moment about $x = 0$ is

$$M_{x=0} = 10\int_0^3\left(h^2 - \frac{h^3}{3}\right)dh$$

$$= 10\left(\frac{h^3}{3} - \frac{h^4}{12}\right)\Big|_0^3 = \frac{45}{2} \text{ kg-m.}$$

The moment about $y = 0$ is

$$M_{y=0} = 10\int_0^3 \frac{1}{2}\left(2 - \frac{2}{3}h\right)\left(h - \frac{1}{3}h^2\right)dh$$

$$= 10\int_0^3\left(h - \frac{2}{3}h^2 + \frac{1}{9}h^3\right)dh$$

$$= 10\left(\frac{h^2}{2} - \frac{2h^3}{9} + \frac{h^4}{36}\right)\Big|_0^3 = \frac{15}{2} \text{ kg-m.}$$

Thus, $\bar{x} = \dfrac{\left(\dfrac{45}{2}\right)}{15} = \dfrac{3}{2}$ and $\bar{y} = \dfrac{\left(\dfrac{15}{2}\right)}{15} = \dfrac{1}{2}$. The centre of mass is located at $(\frac{3}{2}, \frac{1}{2})$.

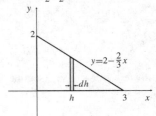

Fig. 7.4.6

8. A vertical strip has area $dA = 2\left(\dfrac{a}{\sqrt{2}} - r\right)dr$. Thus, the mass is

$$m = 2\int_0^{a/\sqrt{2}} kr\left[2\left(\frac{a}{\sqrt{2}} - r\right)\right]dr$$

$$= 4k\int_0^{a/\sqrt{2}}\left(\frac{a}{\sqrt{2}}r - r^2\right)dr = \frac{k}{3\sqrt{2}}a^3 \text{ g.}$$

Since the mass is symmetric about the y-axis, and the plate is symmetric about both the x- and y-axis, therefore the centre of mass must be located at the centre of the square.

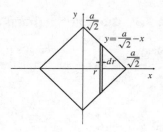

Fig. 7.4.8

10. The slice of the brick shown in the figure has volume $dV = 50\,dx$. Thus, the mass of the brick is

$$m = \int_0^{20} kx50\,dx = 25kx^2 \Big|_0^{20} = 10000k \text{ g}.$$

The moment about $x = 0$, i.e., the yz-plane, is

$$M_{x=0} = 50k \int_0^{20} x^2\,dx = \frac{50}{3}kx^3 \Big|_0^{20}$$
$$= \frac{50}{3}(8000)k \text{ g-cm}.$$

Thus, $\bar{x} = \dfrac{\dfrac{50}{3}(8000)k}{10000k} = \dfrac{40}{3}$. Since the density is independent of y and z, $\bar{y} = \dfrac{5}{2}$ and $\bar{z} = 5$. Hence, the centre of mass is located on the 20 cm long central axis of the brick, two-thirds of the way from the least dense 10×5 face to the most dense such face.

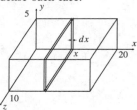

Fig. 7.4.10

12. A slice at height z has volume $dV = \pi y^2\,dz$ and density kz g/cm^3. Thus, the mass of the cone is

$$m = \int_0^b kz\pi y^2\,dz$$
$$= \pi ka^2 \int_0^b z\left(1 - \frac{z}{b}\right)^2 dz$$
$$= \pi ka^2 \left(\frac{z^2}{2} - \frac{2z^3}{3b} + \frac{z^4}{4b^2}\right)\Big|_0^b$$
$$= \frac{1}{12}\pi ka^2 b^2 \text{ g}.$$

The moment about $z = 0$ is

$$M_{z=0} = \pi ka^2 \int_0^b z^2 \left(1 - \frac{z}{b}\right)^2 dz = \frac{1}{30}\pi ka^2 b^3 \text{ g-cm}.$$

Thus, $\bar{z} = \dfrac{2b}{5}$. Hence, the centre of mass is on the axis of the cone at height $2b/5$ cm above the base.

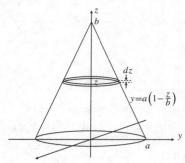

Fig. 7.4.12

14. Assume the cone has its base in the xy-plane and its vertex at height b on the z-axis. By symmetry, the centre of mass lies on the z-axis. A cylindrical shell of thickness dx and radius x about the z-axis has height $z = b(1 - (x/a))$. Since it's density is constant kx, its mass is

$$dm = 2\pi bkx^2 \left(1 - \frac{x}{a}\right) dx.$$

Also its centre of mass is at half its height,

$$\bar{y}_{\text{shell}} = \frac{b}{2}\left(1 - \frac{x}{a}\right).$$

Thus its moment about $z = 0$ is

$$dM_{z=0} = \bar{y}_{\text{shell}}\,dm = \pi bkx^2 \left(1 - \frac{x}{a}\right)^2 dx.$$

Hence

$$m = \int_0^a 2\pi bkx^2 \left(1 - \frac{x}{a}\right) dx = \frac{\pi kba^3}{6}$$
$$M_{z=0} = \int_0^a \pi bkx^2 \left(1 - \frac{x}{a}\right)^2 dx = \frac{\pi kb^2 a^3}{30}$$

and $\bar{z} = M_{z=0}/m = b/5$. The centre of mass is on the axis of the cone at height $b/5$ cm above the base.

16.

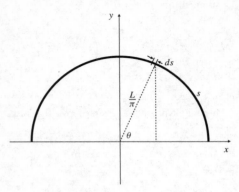

Fig. 7.4.16

The radius of the semicircle is $\dfrac{L}{\pi}$. Let s measure the distance along the wire from the point where it leaves the positive x-axis. Thus, the density at position s is $\delta\delta(s) = \sin\left(\dfrac{\pi s}{L}\right)$ g/cm. The mass of the wire is

$$m = \int_0^L \sin\frac{\pi s}{L}\, ds = -\frac{L}{\pi}\cos\frac{\pi s}{L}\bigg|_0^L = \frac{2L}{\pi}\ \text{g}.$$

Since an arc element ds at position s is at height $y = \dfrac{L}{\pi}\sin\theta = \dfrac{L}{\pi}\sin\dfrac{\pi s}{L}$, the moment of the wire about $y = 0$ is

$$M_{y=0} = \int_0^L \frac{L}{\pi}\sin^2\frac{\pi s}{L}\, ds \quad \text{Let } \theta = \pi s/L$$
$$\hspace{4cm} d\theta = \pi ds/L$$
$$= \left(\frac{L}{\pi}\right)^2 \int_0^\pi \sin^2\theta\, d\theta$$
$$= \frac{L^2}{2\pi^2}\Big(\theta - \sin\theta\cos\theta\Big)\bigg|_0^\pi = \frac{L^2}{2\pi}\ \text{g-cm}.$$

Since the wire and the density function are both symmetric about the y-axis, we have $M_{x=0} = 0$.

Hence, the centre of mass is located at $\left(0, \dfrac{L}{4}\right)$.

18.
$$\bar{r} = \frac{1}{m}\int_0^\infty rCe^{-kr^2}(4\pi r^2)\, dr$$
$$= \frac{4\pi C}{C\pi^{3/2}k^{-3/2}}\int_0^\infty r^3 e^{-kr^2}\, dr \quad \text{Let } u = kr^2$$
$$\hspace{5cm} du = 2kr\, dr$$
$$= \frac{4k^{3/2}}{\sqrt{\pi}}\frac{1}{2k^2}\int_0^\infty u e^{-u}\, du$$
$$\quad U = u \qquad dV = e^{-u}\, du$$
$$\quad dU = du \qquad V = -e^{-u}$$
$$= \frac{2}{\sqrt{\pi k}}\lim_{R\to\infty}\left(-u e^{-u}\bigg|_0^R + \int_0^R e^{-u}\, du\right)$$
$$= \frac{2}{\sqrt{\pi k}}\Big(0 + \lim_{R\to\infty}(e^0 - e^{-R})\Big) = \frac{2}{\sqrt{\pi k}}.$$

Section 7.5   Centroids   (page 399)

2. By symmetry, $\bar{x} = 0$. A horizontal strip at y has mass $dm = 2\sqrt{9-y}\, dy$ and moment $dM_{y=0} = 2y\sqrt{9-y}\, dy$ about $y = 0$. Thus,

$$m = 2\int_0^9 \sqrt{9-y}\, dy = -2\left(\frac{2}{3}\right)(9-y)^{3/2}\bigg|_0^9 = 36$$

and

$$M_{y=0} = 2\int_0^9 y\sqrt{9-y}\, dy \quad \text{Let } u^2 = 9-y$$
$$\hspace{4cm} 2u\, du = -dy$$
$$= 4\int_0^3 (9u^2 - u^4)\, du = 4(3u^3 - \tfrac{1}{5}u^5)\bigg|_0^3 = \frac{648}{5}.$$

Thus, $\bar{y} = \dfrac{648}{5\times 36} = \dfrac{18}{5}$. Hence, the centroid is at $\left(0, \dfrac{18}{5}\right)$.

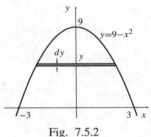

Fig. 7.5.2

4. The area of the sector is $A = \tfrac{1}{8}\pi r^2$. Its moment about $x = 0$ is

$$M_{x=0} = \int_0^{r/\sqrt{2}} x^2\, dx + \int_{r/\sqrt{2}}^r x\sqrt{r^2 - x^2}\, dx$$
$$= \frac{r^3}{6\sqrt{2}} - \frac{1}{3}(r^2 - x^2)^{3/2}\bigg|_{r/\sqrt{2}}^r = \frac{r^3}{3\sqrt{2}}.$$

Thus, $\bar{x} = \dfrac{r^3}{3\sqrt{2}}\times\dfrac{8}{\pi r^2} = \dfrac{8r}{3\sqrt{2}\pi}$. By symmetry, the centroid must lie on the line $y = x\left(\tan\dfrac{\pi}{8}\right) = x(\sqrt{2}-1)$.

Thus, $\bar{y} = \dfrac{8r(\sqrt{2}-1)}{3\sqrt{2}\pi}$.

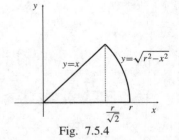

Fig. 7.5.4

6. By symmetry, $\bar{x} = 0$. The area is $A = \frac{1}{2}\pi ab$. The moment about $y = 0$ is

$$M_{y=0} = \frac{1}{2}\int_{-a}^{a} b^2\left[1 - \left(\frac{x}{a}\right)^2\right]dx = b^2\int_0^a 1 - \frac{x^2}{a^2}\,dx$$

$$= b^2\left(x - \frac{x^3}{3a^2}\right)\Big|_0^a = \frac{2}{3}ab^2.$$

Thus, $\bar{y} = \frac{2ab^2}{3} \times \frac{2}{\pi ab} = \frac{4b}{3\pi}$.

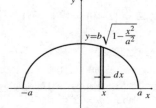

Fig. 7.5.6

8. The region is the union of a half-disk and a triangle. The centroid of the half-disk is known to be at $\left(1, \frac{4}{3\pi}\right)$ and that of the triangle is at $\left(\frac{2}{3}, -\frac{2}{3}\right)$. The area of the semi-circle is $\frac{\pi}{2}$ and the triangle is 2. Hence,

$$M_{x=0} = \left(\frac{\pi}{2}\right)(1) + (2)\left(\frac{2}{3}\right) = \frac{3\pi + 8}{6};$$

$$M_{y=0} = \left(\frac{\pi}{2}\right)\left(\frac{4}{3\pi}\right) + (2)\left(-\frac{2}{3}\right) = -\frac{2}{3}.$$

Since the area of the whole region is $\frac{\pi}{2} + 2$, then

$$\bar{x} = \frac{3\pi + 8}{3(\pi + 4)} \text{ and } \bar{y} = -\frac{4}{3(\pi + 4)}.$$

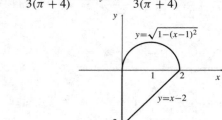

Fig. 7.5.8

10. By symmetry, $\bar{x} = \bar{y} = 0$. The volume is $V = \frac{2}{3}\pi r^3$. A thin slice of the solid at height z will have volume $dV = \pi y^2\,dz = \pi(r^2 - z^2)\,dz$. Thus, the moment about $z = 0$ is

$$M_{z=0} = \int_0^r z\pi(r^2 - z^2)\,dz$$

$$= \pi\left(\frac{r^2z^2}{2} - \frac{z^4}{4}\right)\Big|_0^r = \frac{\pi r^4}{4}.$$

Thus, $\bar{z} = \frac{\pi r^4}{4} \times \frac{3}{2\pi r^3} = \frac{3r}{8}$. Hence, the centroid is on the axis of the hemisphere at distance $3r/8$ from the base.

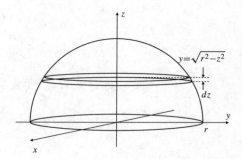

Fig. 7.5.10

12. A band at height z with vertical width dz has radius $y = r\left(1 - \frac{z}{h}\right)$, and has actual (slant) width

$$ds = \sqrt{1 + \left(\frac{dy}{dz}\right)^2}\,dz = \sqrt{1 + \frac{r^2}{h^2}}\,dz.$$

Its area is

$$dA = 2\pi r\left(1 - \frac{z}{h}\right)\sqrt{1 + \frac{r^2}{h^2}}\,dz.$$

Thus the area of the conical surface is

$$A = 2\pi r\sqrt{1 + \frac{r^2}{h^2}}\int_0^h\left(1 - \frac{z}{h}\right)dz = \pi r\sqrt{r^2 + h^2}.$$

The moment about $z = 0$ is

$$M_{z=0} = 2\pi r\sqrt{1 + \frac{r^2}{h^2}}\int_0^h z\left(1 - \frac{z}{h}\right)dz$$

$$= 2\pi r\sqrt{1 + \frac{r^2}{h^2}}\left(\frac{z^2}{2} - \frac{z^3}{3h}\right)\Big|_0^h = \frac{1}{3}\pi rh\sqrt{r^2 + h^2}.$$

Thus, $\bar{z} = \frac{\pi rh\sqrt{r^2 + h^2}}{3} \times \frac{1}{\pi r\sqrt{r^2 + h^2}} = \frac{h}{3}$. By symmetry, $\bar{x} = \bar{y} = 0$. Hence, the centroid is on the axis of the conical surface, at distance $h/3$ from the base.

14. The area of the region is

$$A = \int_0^{\pi/2}\cos x\,dx = \sin x\Big|_0^{\pi/2} = 1.$$

The moment about $x = 0$ is

$$M_{x=0} = \int_0^{\pi/2} x \cos x \, dx$$

$$U = x \qquad dV = \cos x \, dx$$
$$dU = dx \qquad V = \sin x$$

$$= x \sin x \Big|_0^{\pi/2} - \int_0^{\pi/2} \sin x \, dx = \frac{\pi}{2} - 1.$$

Thus, $\bar{x} = \frac{\pi}{2} - 1$. The moment about $y = 0$ is

$$M_{y=0} = \frac{1}{2} \int_0^{\pi/2} \cos^2 x \, dx$$

$$= \frac{1}{4} \left(x + \frac{1}{2} \sin 2x \right) \Big|_0^{\pi/2} = \frac{\pi}{8}.$$

Thus, $\bar{y} = \frac{\pi}{8}$. The centroid is $\left(\frac{\pi}{2} - 1, \frac{\pi}{8} \right)$.

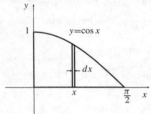

Fig. 7.5.14

16. The solid S in question consists of a solid cone $\mathcal{C}$ with vertex at the origin, height 1, and top a circular disk of radius 2, and a solid cylinder D of radius 2 and height 1 sitting on top of the cone. These solids have volumes $V_C = 4\pi/3$, $V_D = 4\pi$, and $V_S = V_C + V_D = 16\pi/3$.

By symmetry, the centroid of the solid lies on its vertical axis of symmetry; let us continue to call this the y-axis. We need only determine $\bar{y}_S$. Since D lies between $y = 1$ and $y = 2$, its centroid satisfies $\bar{y}_D = 3/2$. Also, by Exercise 11, the centroid of the solid cone satisfies $\bar{y}_C = 3/4$. Thus $\mathcal{C}$ and D have moments about $y = 0$:

$$M_{C,y=0} = \left(\frac{4\pi}{3} \right) \left(\frac{3}{4} \right) = \pi, \quad M_{D,y=0} = (4\pi) \left(\frac{3}{2} \right) = 6\pi.$$

Thus $M_{S,y=0} = \pi + 6\pi = 7\pi$, and $\bar{z}_S = 7\pi/(16\pi/3) = 21/16$. The centroid of the solid S is on its vertical axis of symmetry at height 21/16 above the vertex of the conical part.

18. The region in figure (b) is the union of a square of area $(\sqrt{2})^2 = 2$ and centroid $(0, 0)$ and a triangle of area $1/2$ and centroid $(2/3, 2/3)$. Therefore its area is $5/2$ and its centroid is $(\bar{x}, \bar{y})$, where

$$\frac{5}{2} \bar{x} = 2(0) + \frac{1}{2} \left(\frac{2}{3} \right) = \frac{1}{3}.$$

Therefore, $\bar{x} = \bar{y} = 2/15$, and the centroid is $(2/15, 2/15)$.

20. The region in figure (d) is the union of three half-disks, one with area $\pi/2$ and centroid $(0, 4/(3\pi))$, and two with areas $\pi/8$ and centroids $(-1/2, -2/(3\pi))$ and $(1/2, -2/(3\pi))$. Therefore its area is $3\pi/4$ and its centroid is $(\bar{x}, \bar{y})$, where

$$\frac{3\pi}{4}(\bar{x}) = \frac{\pi}{2}(0) + \frac{\pi}{8} \left(\frac{-1}{2} \right) + \frac{\pi}{8} \left(\frac{1}{2} \right) = 0$$

$$\frac{3\pi}{4}(\bar{y}) = \frac{\pi}{2} \left(\frac{4}{3\pi} \right) + \frac{\pi}{8} \left(\frac{-2}{3\pi} \right) + \frac{\pi}{8} \left(\frac{-2}{3\pi} \right) = \frac{1}{2}.$$

Therefore, the centroid is $(0, 2/(3\pi))$.

22. The line segment from $(1, 0)$ to $(0, 1)$ has centroid $(\frac{1}{2}, \frac{1}{2})$ and length $\sqrt{2}$. By Pappus's Theorem, the surface area of revolution about $x = 2$ is

$$A = 2\pi \left(2 - \frac{1}{2} \right) \sqrt{2} = 3\pi \sqrt{2} \text{ sq. units.}$$

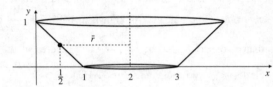

Fig. 7.5.22

24. The altitude h of the triangle is $\frac{s\sqrt{3}}{2}$. Its centroid is at height $\frac{h}{3} = \frac{s}{2\sqrt{3}}$ above the base side. Thus, by Pappus's Theorem, the volume of revolution is

$$V = 2\pi \left(\frac{s}{2\sqrt{3}} \right) \left(\frac{s}{2} \times \frac{\sqrt{3}s}{2} \right) = \frac{\pi s^3}{4} \text{ cu. units.}$$

The centroid of one side is $\frac{h}{2} = \frac{s\sqrt{3}}{4}$ above the base. Thus, the surface area of revolution is

$$S = 2 \times 2\pi \left(\frac{\sqrt{3}s}{4} \right) (s) = s^2 \pi \sqrt{3} \text{ sq. units.}$$

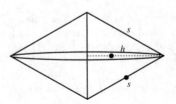

Fig. 7.5.24

26. The region bounded by $y = 0$ and $y = \ln(\sin x)$ between $x = 0$ and $x = \pi/2$ lies below the x-axis, so

$$A = -\int_0^{\pi/2} \ln(\sin x)\, dx \approx 1.088793$$

$$\bar{x} = \frac{-1}{A}\int_0^{\pi/2} x\ln(\sin x)\, dx \approx 0.30239$$

$$\bar{y} = \frac{-1}{2A}\int_0^{\pi/2} \big(\ln(\sin x)\big)^2\, dx \approx -0.93986.$$

28. The surface area is given by
$S = 2\pi\displaystyle\int_{-\infty}^{\infty} e^{-x^2}\sqrt{1 + 4x^2 e^{-2x^2}}\, dx.$ Since
$\displaystyle\lim_{x\to\pm\infty} 1 + 4x^2 e^{-2x^2} = 1$, this expression must be bounded for all x, that is, $1 \le 1 + 4x^2 e^{-2x^2} \le K^2$ for some constant K. Thus, $S \le 2\pi K\displaystyle\int_{-\infty}^{\infty} e^{-x^2}\, dx = 2K\pi\sqrt{\pi}$. The integral converges and the surface area is finite. Since the whole curve $y = e^{-x^2}$ lies above the x-axis, its centroid would have to satisfy $\bar{y} > 0$. However, Pappus's Theorem would then imply that the surface of revolution would have infinite area: $S = 2\pi\bar{y} \times (\text{length of curve}) = \infty$. The curve cannot, therefore, have any centroid.

30. Let us take L to be the y-axis and suppose that a plane curve $\mathcal{C}$ lies between $x = a$ and $x = b$ where $0 < a < b$. Thus, $\bar{r} = \bar{x}$, the x-coordinate of the centroid of $\mathcal{C}$. Let ds denote an arc length element of $\mathcal{C}$ at position x. This arc length element generates, on rotation about L, a circular band of surface area $dS = 2\pi x\, ds$, so the surface area of the surface of revolution is

$$S = 2\pi\int_{x=a}^{x=b} x\, ds = 2\pi M_{x=0} = 2\pi\bar{r}s.$$

Section 7.6   Other Physical Applications (page 406)

2. A vertical slice of water at position y with thickness dy is in contact with the botttom over an area $8\sec\theta\, dy = \frac{4}{5}\sqrt{101}\, dy$ m², which is at depth $x = \frac{1}{10}y + 1$ m. The force exerted on this area is then $dF = \rho g(\frac{1}{10}y + 1)\frac{4}{5}\sqrt{101}\, dy$. Hence, the total force exerted on the bottom is

$$F = \frac{4}{5}\sqrt{101}\,\rho g\int_0^{20}\left(\frac{1}{10}y + 1\right) dy$$

$$= \frac{4}{5}\sqrt{101}\,(1000)(9.8)\left(\frac{y^2}{20} + y\right)\Big|_0^{20}$$

$$\approx 3.1516 \times 10^6 \text{ N}.$$

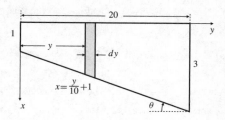

Fig. 7.6.2

4. The height of each triangular face is $2\sqrt{3}$ m and the height of the pyramid is $2\sqrt{2}$ m. Let the angle between the triangular face and the base be θ, then $\sin\theta = \sqrt{\dfrac{2}{3}}$ and $\cos\theta = \dfrac{1}{\sqrt{3}}$.

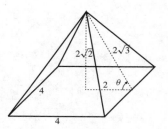

Fig. 7.6.4

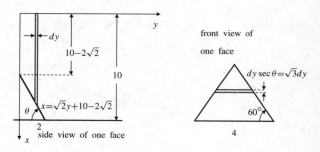

Fig. 7.6.4

A vertical slice of water with thickness dy at a distance y from the vertex of the pyramid exerts a force on the shaded strip shown in the front view, which has area $2\sqrt{3}y\, dy$ m² and which is at depth $\sqrt{2}y + 10 - 2\sqrt{2}$ m. Hence, the force exerted on the triangular face is

$$F = \rho g\int_0^2 (\sqrt{2}y + 10 - 2\sqrt{2})2\sqrt{3}y\, dy$$

$$= 2\sqrt{3}(9800)\left[\frac{\sqrt{2}}{3}y^3 + (5 - \sqrt{2})y^2\right]\Big|_0^2$$

$$\approx 6.1495 \times 10^5 \text{ N}.$$

6. The spring force is $F(x) = kx$, where x is the amount of compression. The work done to compress the spring 3 cm is

$$100 \text{ N·cm} = W = \int_0^3 kx \, dx = \frac{1}{2}kx^2 \Big|_0^3 = \frac{9}{2}k.$$

Hence, $k = \dfrac{200}{9}$ N/cm. The work necessary to compress the spring a further 1 cm is

$$W = \int_3^4 kx \, dx = \left(\frac{200}{9}\right)\frac{1}{2}x^2 \Big|_3^4 = \frac{700}{9} \text{ N·cm.}$$

8. The horizontal cross-sectional area of the pool at depth h is

$$A(h) = \begin{cases} 160, & \text{if } 0 \le h \le 1; \\ 240 - 80h, & \text{if } 1 < h \le 3. \end{cases}$$

The work done to empty the pool is

$$W = \rho g \int_0^3 h A(h) \, dh$$

$$= \rho g \left[\int_0^1 160h \, dh + \int_1^3 240h - 80h^2 \, dh \right]$$

$$= 9800 \left[80h^2 \Big|_0^1 + \left(120h^2 - \frac{80}{3}h^3\right)\Big|_1^3 \right]$$

$$= 3.3973 \times 10^6 \text{ N·m.}$$

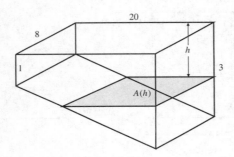

Fig. 7.6.8

10. Let the time required to raise the bucket to height h m be t minutes. Given that the velocity is 2 m/min, then $t = \dfrac{h}{2}$. The weight of the bucket at time t is

$$16 \text{ kg} - (1 \text{ kg/min})(t \text{ min}) = 16 - \frac{h}{2} \text{ kg. Therefore,}$$

the work done required to move the bucket to a height of 10 m is

$$W = g \int_0^{10} \left(16 - \frac{h}{2}\right) dh$$

$$= 9.8 \left(16h - \frac{h^2}{4}\right)\Big|_0^{10} = 1323 \text{ N·m.}$$

Section 7.7 Applications in Business, Finance, and Ecology   (page 409)

2. The number of chips sold in the first year was

$$1,000 \int_0^{52} te^{-t/10} \, dt = 100,000 - 620,000e^{-26/5}$$

that is, about 96,580.

4. The price per kg at time t (years) is $\$10 + 5t$. Thus the revenue per year at time t is $400(10 + 5t)/(1 + 0.1t)$ \$/year. The total revenue over the year is

$$\int_0^1 \frac{400(10 + 5t)}{1 + 0.1t} \, dt \approx \$4,750.37.$$

6. The present value of continuous payments of $\$1,000$ per year for 10 years at a discount rate of 5% is

$$V = \int_0^{10} 1,000e^{-0.05t} \, dt = \frac{1,000}{-0.05} e^{-0.05t} \Big|_0^{10} = \$7,869.39.$$

8. The present value of continuous payments of $\$1,000$ per year for 25 years beginning 10 years from now at a discount rate of 5% is

$$V = \int_{10}^{35} 1,000e^{-0.05t} \, dt = \frac{1,000}{-0.05} e^{-0.05t} \Big|_{10}^{35} = \$8,655.13.$$

10. The present value of continuous payments of $\$1,000$ per year beginning 10 years from now and continuing for all future time at a discount rate of 5% is

$$V = \int_{10}^{\infty} 1,000e^{-0.05t} \, dt = \frac{1,000}{-0.05} e^{-0.5} = \$12,130.61.$$

12. After t years, money is flowing at $\$1,000(1.1)^t$ per year. The present value of 10 years of payments discounted at 5% is

$$V = 1,000 \int_0^{10} e^{t \ln(1.1)} e^{-0.05t} \, dt$$

$$= \frac{1,000}{\ln(1.1) - 0.05} e^{t(\ln(1.1) - 0.05)} \Big|_0^{10} = \$12,650.23.$$

14. Let T be the time required for the account balance to reach $\$1,000,000$. The $\$5,000(1.1)^t\,dt$ deposited in the time interval $[t, t+dt]$ grows for $T-t$ years, so the balance after T years is

$$\int_0^T 5,000(1.1)^t(1.06)^{T-t}\,dt = 1,000,000$$

$$(1.06)^T \int_0^T \left(\frac{1.1}{1.06}\right)^t dt = \frac{1,000,000}{5,000} = 200$$

$$\frac{(1.06)^T}{\ln(1.1/1.06)}\left[\left(\frac{1.1}{1.06}\right)^T - 1\right] = 200$$

$$(1.1)^T - (1.06)^T = 200\ln\frac{1.1}{1.06}.$$

This equation can be solved by Newton's method or using a calculator "solve" routine. The solution is $T \approx 26.05$ years.

16. The analysis carried out in the text for the logistic growth model showed that the total present value of future harvests could be maximized by holding the population size x at a value that maximizes the quadratic expression

$$Q(x) = kx\left(1 - \frac{x}{L}\right) - \delta x.$$

If the logistic model $dx/dt = kx(1-(x/L))$ is replaced with a more general growth model $dx/dt = F(x)$, exactly the same analysis leads us to maximize

$$Q(x) = F(x) - \delta x.$$

For realistic growth functions, the maximum will occur where $Q'(x) = 0$, that is, where $F'(x) = \delta$.

18. We are given that $k = 0.02$, $L = 150,000$, $p = \$10,000$. The growth rate at population level x is

$$\frac{dx}{dt} = 0.02x\left(1 - \frac{x}{150,000}\right).$$

a) The maximum sustainable annual harvest is

$$\left.\frac{dx}{dt}\right|_{x=L/2} = 0.02(75,000)(0.5) = 750 \text{ whales.}$$

b) The resulting annual revenue is $\$750p = \$7,500,000$.

c) If the whole population of 75,000 is harvested and the proceeds invested at 2%, the annual interest will be

$$75,000(\$10,000)(0.02) = \$15,000,000.$$

d) At 5%, the interest would be $(5/2)(\$15,000) = \$37,500,000$.

e) The total present value of all future harvesting revenue if the population level is maintained at 75,000 and $\delta = 0.05$ is

$$\int_0^\infty e^{-0.05t}7,500,000\,dt = \frac{7,500,000}{0.05} = \$150,000,000.$$

Section 7.8 Probability (page 421)

2. (a) We need $\sum_{n=1}^6 Kn = 1$. Thus $21K = 1$, and $K = 1/21$.
(b) $\Pr(X \le 3) = (1/21)(1 + 2 + 3) = 2/7$.

4. Since $\Pr(X = n) = n/21$, we have

$$\mu = \sum_{n=1}^6 n\Pr(X = n) = \frac{1\times1 + 2\times2 + \cdots + 6\times6}{21} = \frac{13}{3} \approx 4.33$$

$$\sigma^2 = \sum_{n=1}^6 n^2\Pr(X = n) - \mu^2 = \frac{1^2 + 2^3 + \cdots + 6^3}{21} - \mu^2$$

$$= 21 - \frac{169}{9} = \frac{20}{9} \approx 2.22$$

$$\sigma = \frac{\sqrt{20}}{3} \approx 1.49.$$

6. (a) Calculating as we did to construct the probability function in Example 2, but using the different values for the probabilities of "1" and "6", we obtain

$$f(2) = \frac{9}{60} \times \frac{9}{60} \approx 0.0225$$

$$f(3) = 2 \times \frac{9}{60} \times 16 = 0.0500$$

$$f(4) = 2 \times \frac{9}{60} \times 16 + \frac{1}{36} = 0.0778$$

$$f(5) = 2 \times \frac{9}{60} \times 16 + \frac{2}{36} = 0.1056$$

$$f(6) = 2 \times \frac{9}{60} \times 16 + \frac{3}{36} = 0.1333$$

$$f(7) = 2 \times \frac{9}{60} \times 1160 + \frac{4}{36} = 0.1661$$

$$f(8) = 2 \times \frac{11}{60} \times 16 + \frac{3}{36} = 0.1444$$

$$f(9) = 2 \times \frac{11}{60} \times 16 + \frac{2}{36} = 0.1167$$

$$f(10) = 2 \times \frac{11}{60} \times 16 + \frac{1}{36} = 0.0889$$

$$f(11) = 2 \times \frac{11}{60} \times 16 = 0.0611$$

$$f(12) = \frac{11}{60} \times 1160 = 0.0336.$$

(b) Multiplying each value $f(n)$ by n and summing, we get

$$\mu = \sum_{n=2}^{12} n f(n) \approx 7.1665.$$

Similarly,

$$E(X^2) = \sum_{n=2}^{12} n^2 f(n) \approx 57.1783,$$

so the standard deviation of X is

$$\sigma = \sqrt{E(X^2) - \mu^2} \approx 2.4124.$$

The mean is somewhat larger than the value (7) obtained for the unweighted dice, because the weighting favours more 6s than 1s showing if the roll is repeated many times. The standard deviation is just a tiny bit smaller than that found for the unweighted dice (2.4152); the distribution of probability is just slightly more concentrated around the mean here.

8. The number of red balls in the sack must be $0.6 \times 20 = 12$. Thus there are 8 blue balls.

(a) The probability of pulling out one blue ball is 8/20. If you got a blue ball, then there would be only 7 blue balls left among the 19 balls remaining in the sack, so the probability of pulling out a second blue ball is 7/19. Thus the probability of pulling out two blue balls is $\dfrac{8}{20} \times \dfrac{7}{19} = \dfrac{14}{95}$.

(b) The sample space for the three ball selection consists of all eight triples of the form (x, y, z), where each of x, y, z is either R(ed) or B(lue). Let X be the number of red balls among the three balls pulled out. Arguing in the same way as in (a), we calculate

$$\Pr(X = 0) = \Pr(B, B, B) = \frac{8}{20} \times \frac{7}{19} \times \frac{6}{18} = \frac{14}{285}$$
$$\approx 0.0491$$
$$\Pr(X = 1) = \Pr(R, B, B) + \Pr(B, R, B) + \Pr(B, B, R)$$
$$= 3 \times \frac{12}{20} \times \frac{8}{19} \times \frac{7}{18} = \frac{28}{95} \approx 0.2947$$
$$\Pr(X = 2) = \Pr(R, R, B) + \Pr(R, B, R) + \Pr(B, R, R)$$
$$= 3 \times \frac{12}{20} \times \frac{11}{19} \times \frac{8}{18} = \frac{44}{95} \approx 0.4632$$
$$\Pr(X = 3) = \Pr(R, R, R) = \frac{12}{20} \times \frac{11}{19} \times \frac{10}{18} = \frac{11}{57}$$
$$\approx 0.1930$$

Thus the expected value of X is

$$E(X) = 0 \times \frac{14}{285} + 1 \times \frac{28}{95} + 2 \times \frac{44}{95} + 3 \times \frac{11}{57}$$
$$= \frac{9}{5} = 1.8.$$

10. We have $f(x) = Cx$ on $[1, 2]$.

a) To find C, we have

$$1 = \int_1^2 Cx\, dx = \frac{C}{2} x^2 \Big|_1^2 = \frac{3}{2} C.$$

Hence, $C = \dfrac{2}{3}$.

b) The mean is

$$\mu = E(X) = \frac{2}{3} \int_1^2 x^2\, dx = \frac{2}{9} x^3 \Big|_1^2 = \frac{14}{9} \approx 1.556.$$

Since $E(X^2) = \dfrac{2}{3} \displaystyle\int_1^2 x^3\, dx = \dfrac{1}{6} x^4 \Big|_1^2 = \dfrac{5}{2}$, the variance is

$$\sigma^2 = E(X^2) - \mu^2 = \frac{5}{2} - \frac{196}{81} = \frac{13}{162}$$

and the standard deviation is

$$\sigma = \sqrt{\frac{13}{162}} \approx 0.283.$$

c) We have

$$\Pr(\mu - \sigma \le X \le \mu + \sigma) = \frac{2}{3} \int_{\mu - \sigma}^{\mu + \sigma} x\, dx$$
$$= \frac{(\mu + \sigma)^2 - (\mu - \sigma)^2}{3} = \frac{4\mu\sigma}{3} \approx 0.5875.$$

12. We have $f(x) = C \sin x$ on $[0, \pi]$.

a) To find C, we calculate

$$1 = \int_0^\pi C \sin x\, dx = -C \cos x \Big|_0^\pi = 2C.$$

Hence, $C = \dfrac{1}{2}$.

b) The mean is

$$\mu = E(X) = \frac{1}{2} \int_0^\pi x \sin x\, dx$$
$$\qquad U = x \qquad dV = \sin x\, dx$$
$$\qquad dU = dx \qquad V = -\cos x$$
$$= \frac{1}{2} \left[-x \cos x \Big|_0^\pi + \int_0^\pi \cos x\, dx \right]$$
$$= \frac{\pi}{2} = 1.571.$$

Since

$$E(X^2) = \frac{1}{2}\int_0^\pi x^2 \sin x \, dx$$

$$U = x^2 \qquad dV = \sin x \, dx$$
$$dU = 2x \, dx \qquad V = -\cos x$$

$$= \frac{1}{2}\left[-x^2\cos x\Big|_0^\pi + 2\int_0^\pi x\cos x\,dx\right]$$

$$U = x \qquad dV = \cos x \, dx$$
$$dU = dx \qquad V = \sin x$$

$$= \frac{1}{2}\left[\pi^2 + 2\left(x\sin x\Big|_0^\pi - \int_0^\pi \sin x\,dx\right)\right]$$

$$= \frac{1}{2}(\pi^2 - 4).$$

Hence, the variance is

$$\sigma^2 = E(X^2) - \mu^2 = \frac{\pi^2-4}{2} - \frac{\pi^2}{4} = \frac{\pi^2-8}{4} \approx 0.467$$

and the standard deviation is

$$\sigma = \sqrt{\frac{\pi^2-8}{4}} \approx 0.684.$$

c) Then

$$\Pr(\mu - \sigma \le X \le \mu + \sigma) = \frac{1}{2}\int_{\mu-\sigma}^{\mu+\sigma}\sin x\,dx$$

$$= -\frac{1}{2}\left[\cos(\mu+\sigma) - \cos(\mu-\sigma)\right]$$

$$= \sin\mu\sin\sigma = \sin\sigma \approx 0.632.$$

14. It was shown in Section 6.1 (p. 349) that

$$\int x^n e^{-x}\,dx = -x^n e^{-x} + n\int x^{n-1}e^{-x}\,dx.$$

If $I_n = \int_0^\infty x^n e^{-x}\,dx$, then

$$I_n = \lim_{R\to\infty} -R^n e^{-R} + nI_{n-1} = nI_{n-1} \qquad \text{if } n \ge 1.$$

Since $I_0 = \int_0^\infty e^{-x}\,dx = 1$, therefore $I_n = n!$ for $n \ge 1$.
Let $u = kx$; then

$$\int_0^\infty x^n e^{-kx}\,dx = \frac{1}{k^{n+1}}\int_0^\infty u^n e^{-u}\,du = \frac{1}{k^{n+1}}I_n = \frac{n!}{k^{n+1}}.$$

Now let $f(x) = Cxe^{-kx}$ on $[0,\infty)$.

a) To find C, observe that

$$1 = C\int_0^\infty xe^{-kx}\,dx = \frac{C}{k^2}.$$

Hence, $C = k^2$.

b) The mean is

$$\mu = E(X) = k^2\int_0^\infty x^2 e^{-kx}\,dx = k^2\left(\frac{2}{k^3}\right) = \frac{2}{k}.$$

Since $E(X^2) = k^2\int_0^\infty x^3 e^{-kx}\,dx = k^2\left(\frac{6}{k^4}\right) = \frac{6}{k^2}$, then the variance is

$$\sigma^2 = E(X^2) - \mu^2 = \frac{6}{k^2} - \frac{4}{k^2} = \frac{2}{k^2}$$

and the standard deviation is $\sigma = \dfrac{\sqrt{2}}{k}$.

c) Finally,

$$\Pr(\mu - \sigma \le X \le \mu + \sigma)$$

$$= k^2\int_{\mu-\sigma}^{\mu+\sigma} xe^{-kx}\,dx \quad \text{Let } u = kx$$
$$\qquad\qquad\qquad\qquad\qquad du = k\,dx$$

$$= \int_{k(\mu-\sigma)}^{k(\mu+\sigma)} ue^{-u}\,du$$

$$= -ue^{-u}\Big|_{k(\mu-\sigma)}^{k(\mu+\sigma)} + \int_{k(\mu-\sigma)}^{k(\mu+\sigma)} e^{-u}\,du$$

$$= -(2+\sqrt{2})e^{-(2+\sqrt{2})} + (2-\sqrt{2})e^{-(2-\sqrt{2})}$$
$$\quad - e^{-(2+\sqrt{2})} + e^{-(2-\sqrt{2})}$$

$$\approx 0.738.$$

16. No. The identity $\int_{-\infty}^\infty C\,dx = 1$ is not satisfied for any constant C.

18. Since $f(x) = \dfrac{2}{\pi(1+x^2)} > 0$ on $[0,\infty)$ and

$$\frac{2}{\pi}\int_0^\infty \frac{dx}{1+x^2} = \lim_{R\to\infty}\frac{2}{\pi}\tan^{-1}(R) = \frac{2}{\pi}\left(\frac{\pi}{2}\right) = 1,$$

therefore $f(x)$ is a probability density function on $[0,\infty)$. The expectation of X is

$$\mu = E(X) = \frac{2}{\pi}\int_0^\infty \frac{x\,dx}{1+x^2}$$

$$= \lim_{R\to\infty}\frac{1}{\pi}\ln(1+R^2) = \infty.$$

No matter what the cost per game, you should be willing to play (if you have an adequate bankroll). Your expected winnings per game in the long term is infinite.

20. The density function for T is $f(t) = ke^{-kt}$ on $[0, \infty)$, where $k = \dfrac{1}{\mu} = \dfrac{1}{20}$ (see Example 6). Then

$$\Pr(T \geq 12) = \frac{1}{20} \int_{12}^{\infty} e^{-t/20}\, dt = 1 - \frac{1}{20} \int_0^{12} e^{-t/20}\, dt$$

$$= 1 + e^{-t/20}\Big|_0^{12} = e^{-12/20} \approx 0.549.$$

The probability that the system will last at least 12 hours is about 0.549.

22. If X is the random variable giving the spinner's value, then $\Pr(X = 1/4) = 1/2$ and the density function for the other values of X is $f(x) = 1/2$. Thus the mean of X is

$$\mu = E(X) = \frac{1}{4}\Pr\left(X = \frac{1}{4}\right) + \int_0^1 x\, f(x)\, dx = \frac{1}{8} + \frac{1}{4} = \frac{3}{8}.$$

Also,

$$E(X^2) = \frac{1}{16}\Pr\left(X = \frac{1}{4}\right) + \int_0^1 x^2\, f(x)\, dx = \frac{1}{32} + \frac{1}{6} = \frac{19}{96}$$

$$\sigma^2 = E(X^2) - \mu^2 = \frac{19}{96} - \frac{9}{64} = \frac{11}{192}.$$

Thus $\sigma = \sqrt{11/192}$.

Section 7.9 First-Order Differential Equations (page 429)

2.
$$\frac{dy}{dx} = \frac{3y - 1}{x}$$
$$\int \frac{dy}{3y - 1} = \int \frac{dx}{x}$$
$$\frac{1}{3}\ln|3y - 1| = \ln|x| + \frac{1}{3}\ln C$$
$$\frac{3y - 1}{x^3} = C$$
$$\Rightarrow \quad y = \frac{1}{3}(1 + Cx^3).$$

4.
$$\frac{dy}{dx} = x^2 y^2$$
$$\int \frac{dy}{y^2} = \int x^2\, dx$$
$$-\frac{1}{y} = \frac{1}{3}x^3 + \frac{1}{3}C$$
$$\Rightarrow \quad y = -\frac{3}{x^3 + C}.$$

6.
$$\frac{dx}{dt} = e^x \sin t$$
$$\int e^{-x}\, dx = \int \sin t\, dt$$
$$-e^{-x} = -\cos t - C$$
$$\Rightarrow \quad x = -\ln(\cos t + C).$$

8.
$$\frac{dy}{dx} = 1 + y^2$$
$$\int \frac{dy}{1 + y^2} = \int dx$$
$$\tan^{-1} y = x + C$$
$$\Rightarrow \quad y = \tan(x + C).$$

10. We have

$$\frac{dy}{dx} = y^2(1 - y)$$
$$\int \frac{dy}{y^2(1 - y)} = \int dx = x + K.$$

Expand the left side in partial fractions:

$$\frac{1}{y^2(1 - y)} = \frac{A}{y} + \frac{B}{y^2} + \frac{C}{1 - y}$$
$$= \frac{A(y - y^2) + B(1 - y) + Cy^2}{y^2(1 - y)}$$
$$\Rightarrow \begin{cases} -A + C = 0; \\ A - B = 0; \quad \Rightarrow A = B = C = 1. \\ B = 1. \end{cases}$$

Hence,

$$\int \frac{dy}{y^2(1 - y)} = \int \left(\frac{1}{y} + \frac{1}{y^2} + \frac{1}{1 - y}\right) dy$$
$$= \ln|y| - \frac{1}{y} - \ln|1 - y|.$$

Therefore,

$$\ln\left|\frac{y}{1 - y}\right| - \frac{1}{y} = x + K.$$

12. We have $\dfrac{dy}{dx} + \dfrac{2y}{x} = \dfrac{1}{x^2}$. Let
$\mu = \displaystyle\int \frac{2}{x}\, dx = 2\ln x = \ln x^2$, then $e^\mu = x^2$, and

$$\frac{d}{dx}(x^2 y) = x^2 \frac{dy}{dx} + 2xy$$
$$= x^2\left(\frac{dy}{dx} + \frac{2y}{x}\right) = x^2\left(\frac{1}{x^2}\right) = 1$$
$$\Rightarrow \quad x^2 y = \int dx = x + C$$
$$\Rightarrow \quad y = \frac{1}{x} + \frac{C}{x^2}.$$

14. We have $\dfrac{dy}{dx} + y = e^x$. Let $\mu = \int dx = x$, then $e^\mu = e^x$, and

$$\frac{d}{dx}(e^x y) = e^x \frac{dy}{dx} + e^x y = e^x\left(\frac{dy}{dx} + y\right) = e^{2x}$$

$$\Rightarrow \quad e^x y = \int e^{2x}\, dx = \frac{1}{2}e^{2x} + C.$$

Hence, $y = \dfrac{1}{2}e^x + Ce^{-x}$.

16. We have $\dfrac{dy}{dx} + 2e^x y = e^x$. Let $\mu = \int 2e^x\, dx = 2e^x$, then

$$\frac{d}{dx}\left(e^{2e^x} y\right) = e^{2e^x}\frac{dy}{dx} + 2e^x e^{2e^x} y$$

$$= e^{2e^x}\left(\frac{dy}{dx} + 2e^x y\right) = e^{2e^x} e^x.$$

Therefore,

$$e^{2e^x} y = \int e^{2e^x} e^x\, dx \quad \text{Let } u = 2e^x$$
$$du = 2e^x\, dx$$
$$= \frac{1}{2}\int e^u\, du = \frac{1}{2}e^{2e^x} + C.$$

Hence, $y = \dfrac{1}{2} + Ce^{-2e^x}$.

18. $\dfrac{dy}{dx} + 3x^2 y = x^2, \qquad y(0) = 1$

$$\mu = \int 3x^2\, dx = x^3$$

$$\frac{d}{dx}(e^{x^3} y) = e^{x^3}\frac{dy}{dx} + 3x^2 e^{x^3} y = x^2 e^{x^3}$$

$$e^{x^3} y = \int x^2 e^{x^3}\, dx = \frac{1}{3}e^{x^3} + C$$

$$y(0) = 1 \;\Rightarrow\; 1 = \frac{1}{3} + C \;\Rightarrow\; C = \frac{2}{3}$$

$$y = \frac{1}{3} + \frac{2}{3}e^{-x^3}.$$

20. $y' + (\cos x)y = 2xe^{-\sin x}, \qquad y(\pi) = 0$

$$\mu = \int \cos x\, dx = \sin x$$

$$\frac{d}{dx}(e^{\sin x} y) = e^{\sin x}(y' + (\cos x)y) = 2x$$

$$e^{\sin x} y = \int 2x\, dx = x^2 + C$$

$$y(\pi) = 0 \;\Rightarrow\; 0 = \pi^2 + C \;\Rightarrow\; C = -\pi^2$$

$$y = (x^2 - \pi^2)e^{-\sin x}.$$

22. $y(x) = 1 + \displaystyle\int_0^x \frac{(y(t))^2}{1 + t^2}\, dt \implies y(0) = 1$

$$\frac{dy}{dx} = \frac{y^2}{1 + x^2}, \qquad \text{i.e. } dy/y^2 = dx/(1 + x^2)$$

$$-\frac{1}{y} = \tan^{-1} x + C$$

$$-1 = 0 + C \implies C = -1$$

$$y = 1/(1 - \tan^{-1} x).$$

24. $y(x) = 3 + \displaystyle\int_0^x e^{-y}\, dt \implies y(0) = 3$

$$\frac{dy}{dx} = e^{-y}, \qquad \text{i.e. } e^y\, dy = dx$$

$$e^y = x + C \implies y = \ln(x + C)$$

$$3 = y(0) = \ln C \implies C = e^3$$

$$y = \ln(x + e^3).$$

26. Since $b > a > 0$ and $k > 0$,

$$\lim_{t\to\infty} x(t) = \lim_{t\to\infty} \frac{ab\left(e^{(b-a)kt} - 1\right)}{be^{(b-a)kt} - a}$$

$$= \lim_{t\to\infty} \frac{ab\left(1 - e^{(a-b)kt}\right)}{b - ae^{(a-b)kt}}$$

$$= \frac{ab(1 - 0)}{b - 0} = a.$$

28. Given that $m\dfrac{dv}{dt} = mg - kv$, then

$$\int \frac{dv}{g - \dfrac{k}{m}v} = \int dt$$

$$-\frac{m}{k}\ln\left|g - \frac{k}{m}v\right| = t + C.$$

Since $v(0) = 0$, therefore $C = -\dfrac{m}{k}\ln g$. Also, $g - \dfrac{k}{m}v$ remains positive for all $t > 0$, so

$$\frac{m}{k}\ln\frac{g}{g - \dfrac{k}{m}v} = t$$

$$\frac{g - \dfrac{k}{m}v}{g} = e^{-kt/m}$$

$$\Rightarrow \quad v = v(t) = \frac{mg}{k}\left(1 - e^{-kt/m}\right).$$

Note that $\displaystyle\lim_{t\to\infty} v(t) = \frac{mg}{k}$. This limiting velocity can be obtained directly from the differential equation by setting $\dfrac{dv}{dt} = 0$.

30. The balance in the account after t years is $y(t)$ and $y(0) = 1000$. The balance must satisfy

$$\frac{dy}{dt} = 0.1y - \frac{y^2}{1,000,000}$$

$$\frac{dy}{dt} = \frac{10^5 y - y^2}{10^6}$$

$$\int \frac{dy}{10^5 y - y^2} = \int \frac{dt}{10^6}$$

$$\frac{1}{10^5} \int \left(\frac{1}{y} + \frac{1}{10^5 - y}\right) dy = \frac{t}{10^6} - \frac{C}{10^5}$$

$$\ln|y| - \ln|10^5 - y| = \frac{t}{10} - C$$

$$\frac{10^5 - y}{y} = e^{C - (t/10)}$$

$$y = \frac{10^5}{e^{C-(t/10)} + 1}.$$

Since $y(0) = 1000$, we have

$$1000 = y(0) = \frac{10^5}{e^C + 1} \quad \Rightarrow \quad C = \ln 99,$$

and

$$y = \frac{10^5}{99e^{-t/10} + 1}.$$

The balance after 1 year is

$$y = \frac{10^5}{99e^{-1/10} + 1} \approx \$1,104.01.$$

As $t \to \infty$, the balance can grow to

$$\lim_{t\to\infty} y(t) = \lim_{t\to\infty} \frac{10^5}{e^{(4.60-0.1t)} + 1} = \frac{10^5}{0+1} = \$100,000.$$

For the account to grow to \$50,000, t must satisfy

$$50,000 = y(t) = \frac{100,000}{99e^{-t/10} + 1}$$
$$\Rightarrow \quad 99e^{-t/10} + 1 = 2$$
$$\Rightarrow \quad t = 10 \ln 99 \approx 46 \text{ years.}$$

32. Let $x(t)$ be the number of kg of salt in the solution in the tank after t minutes. Thus, $x(0) = 50$. Salt is coming into the tank at a rate of 10 g/L $\times$ 12 L/min $= 0.12$ kg/min. Since the contents flow out at a rate of 10 L/min, the volume of the solution is increasing at 2 L/min and thus, at any time t, the volume of the solution is $1000 + 2t$ L. Therefore the concentration of salt is $\dfrac{x(t)}{1000 + 2t}$ L. Hence, salt is being removed at a rate

$$\frac{x(t)}{1000 + 2t} \text{ kg/L} \times 10 \text{ L/min} = \frac{5x(t)}{500 + t} \text{ kg/min.}$$

Therefore,

$$\frac{dx}{dt} = 0.12 - \frac{5x}{500 + t}$$

$$\frac{dx}{dt} + \frac{5}{500 + t}x = 0.12.$$

Let $\mu = \displaystyle\int \frac{5}{500 + t}\, dt = 5 \ln|500 + t| = \ln(500 + t)^5$ for $t > 0$. Then $e^\mu = (500 + t)^5$, and

$$\frac{d}{dt}\left[(500 + t)^5 x\right] = (500 + t)^5 \frac{dx}{dy} + 5(500 + t)^4 x$$

$$= (500 + t)^5 \left(\frac{dx}{dy} + \frac{5x}{500 + t}\right)$$

$$= 0.12(500 + t)^5.$$

Hence,

$$(500 + t)^5 x = 0.12 \int (500 + t)^5\, dt = 0.02(500 + t)^6 + C$$

$$\Rightarrow x = 0.02(500 + t) + C(500 + t)^{-5}.$$

Since $x(0) = 50$, we have $C = 1.25 \times 10^{15}$ and

$$x = 0.02(500 + t) + (1.25 \times 10^{15})(500 + t)^{-5}.$$

After 40 min, there will be

$$x = 0.02(540) + (1.25 \times 10^{15})(540)^{-5} = 38.023 \text{ kg}$$

of salt in the tank.

Review Exercises 7   (page 430)

2. Let $A(y)$ be the cross-sectional area of the bowl at height y above the bottom. When the depth of water in the bowl is Y, then the volume of water in the bowl is

$$V(Y) = \int_0^Y A(y)\, dy.$$

The water evaporates at a rate proportional to exposed surface area. Thus

$$\frac{dV}{dt} = kA(Y)$$

$$\frac{dV}{dY}\frac{dY}{dt} = kA(Y)$$

$$A(Y)\frac{dY}{dt} = kA(Y).$$

Hence $dY/dt = k$; the depth decreases at a constant rate.

4. A vertical slice parallel to the top ridge of the solid at distance x to the right of the centre is a rectangle of base $2\sqrt{100 - x^2}$ cm^and height $\sqrt{3}(10 - x)$ cm. Thus the solid has volume

$$V = 2\int_0^{10} \sqrt{3}(10 - x)2\sqrt{100 - x^2}\,dx$$

$$= 40\sqrt{3}\int_0^{10} \sqrt{100 - x^2}\,dx - 4\sqrt{3}\int_0^{10} x\sqrt{100 - x^2}\,dx$$

$$\text{Let } u = 100 - x^2$$

$$du = -2x\,dx$$

$$= 40\sqrt{3}\frac{100\pi}{4} - 2\sqrt{3}\int_0^{100} \sqrt{u}\,du$$

$$= 1,000\sqrt{3}\left(\pi - \frac{4}{3}\right)\ \text{cm}^3.$$

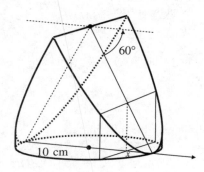

Fig. R-7.4

6. The area of revolution of $y = \sqrt{x}$, $(0 \le x \le 6)$, about the x-axis is

$$S = 2\pi\int_0^6 y\sqrt{1 + \left(\frac{dy}{dx}\right)^2}\,dx$$

$$= 2\pi\int_0^6 \sqrt{x}\sqrt{1 + \frac{1}{4x}}\,dx$$

$$= 2\pi\int_0^6 \sqrt{x + \frac{1}{4}}\,dx$$

$$= \frac{4\pi}{3}\left(x + \frac{1}{4}\right)^{3/2}\Big|_0^6 = \frac{4\pi}{3}\left[\frac{125}{8} - \frac{1}{8}\right] = \frac{62\pi}{3}\ \text{sq. units.}$$

8.

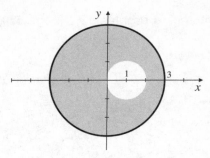

Fig. R-7.8

Let the disk have centre (and therefore centroid) at $(0, 0)$. Its area is 9π. Let the hole have centre (and therefore centroid) at $(1, 0)$. Its area is π. The remaining part has area 8π and centroid at $(\bar{x}, 0)$, where

$$(9\pi)(0) = (8\pi)\bar{x} + (\pi)(1).$$

Thus $\bar{x} = -1/8$. The centroid of the remaining part is $1/8$ ft from the centre of the disk on the side opposite the hole.

10. We are told that for any $a > 0$,

$$\pi\int_0^a \left[\left(f(x)\right)^2 - \left(g(x)\right)^2\right]dx = 2\pi\int_0^a x\left[f(x) - g(x)\right]dx.$$

Differentiating both sides of this equation with respect to a, we get

$$\left(f(a)\right)^2 - \left(g(a)\right)^2 = 2a\left[f(a) - g(a)\right],$$

or, equivalently, $f(a) + g(a) = 2a$. Thus f and g must satisfy

$$f(x) + g(x) = 2x \quad \text{for every } x > 0.$$

12. The ellipses $3x^2 + 4y^2 = C$ all satisfy the differential equation

$$6x + 8y\frac{dy}{dx} = 0, \quad \text{or} \quad \frac{dy}{dx} = -\frac{3x}{4y}.$$

A family of curves that intersect these ellipses at right angles must therefore have slopes given by $\frac{dy}{dx} = \frac{4y}{3x}$. Thus

$$3\int \frac{dy}{y} = 4\int \frac{dx}{x}$$

$$3\ln|y| = 4\ln|x| + \ln|C|.$$

The family is given by $y^3 = Cx^4$.

Challenging Problems 7   (page 430)

2.

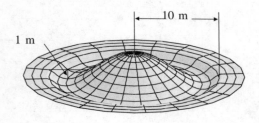

Fig. C-7.2

$h(r) = a(r^2 - 100)(r^2 - k^2),$ where $0 < k < 10$

$h'(r) = 2ar(r^2 - k^2) + 2ar(r^2 - 100) = 2ar(2r^2 - 100 - k^2).$

The deepest point occurs where $2r^2 = 100 + k^2$, i.e., $r^2 = 50 + (k^2/2)$. Since this depth must be 1 m, we require

$$a\left(\frac{k^2}{2} - 50\right)\left(50 - \frac{k^2}{2}\right) = -1,$$

or, equivalently, $a(100 - k^2)^2 = 4$. The volume of the pool is

$$V_P = 2\pi a \int_k^{10} r(100 - r^2)(r^2 - k^2)\, dr$$

$$= 2\pi a \left(\frac{250,000}{3} - 2,500k^2 + 25k^4 - \frac{1}{12}k^6\right).$$

The volume of the hill is

$$V_H = 2\pi a \int_0^k r(r^2 - 100)(r^2 - k^2)\, dr = 2\pi a \left(25k^4 - \frac{1}{12}k^6\right).$$

These two volumes must be equal, so $k^2 = 100/3$ and $k \approx 5.77$ m. Thus $a = 4/(100 - k^2)^2 = 0.0009$. The volume of earth to be moved is V_H with these values of a and k, namely

$$2\pi(0.0009)\left[25\left(\frac{100}{3}\right)^2 - \frac{1}{12}\left(\frac{100}{3}\right)^4\right] \approx 140 \text{ m}^3.$$

4.   a) If $f(x) = \begin{cases} a + bx + cx^2 & \text{for } 0 \le x \le 1 \\ p + qx + rx^2 & \text{for } 1 \le x \le 3 \end{cases}$, then

$f'(x) = \begin{cases} b + 2cx & \text{for } 0 < x < 1 \\ q + 2rx & \text{for } 1 < x < 3 \end{cases}$. We require that

$$
\begin{array}{ll}
a = 1 & p + 3q + 9r = 0 \\
a + b + c = 2 & p + q + r = 2 \\
b + 2c = m & q + 2r = m.
\end{array}
$$

The solutions of these systems are $a = 1$, $b = 2 - m$, $c = m - 1$, $p = \frac{3}{2}(1 - m)$, $q = 2m + 1$, and $r = -\frac{1}{2}(1 + m)$. $f(x, m)$ is $f(x)$ with these values of the six constants.

b) The length of the spline is

$$L(m) = \int_0^1 \sqrt{1 + (b + 2cx)^2}\, dx + \int_1^3 \sqrt{1 + (q + 2rx)^2}\, dx$$

with the values of b, c, q, and r determined above. A plot of the graph of $L(m)$ reveals a minimum value in the neighbourhood of $m = -0.3$. The derivative of $L(m)$ is a horrible expression, but Mathematica determined its zero to be about $m = -0.281326$, and the corresponding minimum value of L is about 4.41748. The polygonal line ABC has length $3\sqrt{2} \approx 4.24264$, which is only slightly shorter.

6.   Starting with $V_1(r) = 2r$, and using repeatedly the formula

$$V_n(r) = \int_{-r}^r V_{n-1}(\sqrt{r^2 - x^2})\, dx,$$

Maple gave the following results:

$$
\begin{array}{ll}
V_1(r) = 2r & V_2(r) = \pi r^2 \\[4pt]
V_3(r) = \frac{4}{3}\pi r^3 & V_4(r) = \frac{1}{2}\pi^2 r^4 \\[4pt]
V_5(r) = \frac{8}{15}\pi^2 r^5 & V_6(r) = \frac{1}{6}\pi^3 r^6 \\[4pt]
V_7(r) = \frac{16}{105}\pi^3 r^7 & V_8(r) = \frac{1}{24}\pi^4 r^8 \\[4pt]
V_9(r) = \frac{32}{945}\pi^4 r^9 & V_{10}(r) = \frac{1}{120}\pi^5 r^{10}
\end{array}
$$

It appears that

$$V_{2n}(r) = \frac{1}{n!}\pi^n r^{2n}, \quad \text{and}$$

$$V_{2n-1}(r) = \frac{2^n}{1 \cdot 3 \cdot 5 \cdots (2n - 1)}\pi^{n-1} r^{2n-1}$$

$$= \frac{2^{2n-1}(n - 1)!}{(2n - 1)!}\pi^{n-1} r^{2n-1}.$$

These formulas predict that

$$V_{11}(r) = \frac{2^{11} 5!}{11!}\pi^5 r^{11} \quad \text{and} \quad V_{12}(r) = \frac{1}{6!}\pi^6 r^{12},$$

both of which Maple is happy to confirm.

8.

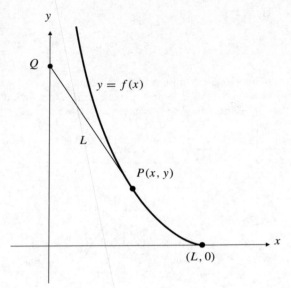

Fig. C-7.8

If $Q = (0, Y)$, then the slope of PQ is

$$\frac{y - Y}{x - 0} = f'(x) = \frac{dy}{dx}.$$

Since $|PQ| = L$, we have $(y - Y)^2 = L^2 - x^2$. Since the slope dy/dx is negative at P, $dy/dx = -\sqrt{L^2 - x^2}/x$. Thus

$$y = -\int \frac{\sqrt{L^2 - x^2}}{x}\, dx = L \ln\left(\frac{L + \sqrt{L^2 - x^2}}{x}\right) - \sqrt{L^2 - x^2} + C.$$

Since $y = 0$ when $x = L$, we have $C = 0$ and the equation of the tractrix is

$$y = L \ln\left(\frac{L + \sqrt{L^2 - x^2}}{x}\right) - \sqrt{L^2 - x^2}.$$

Note that the first term can be written in an alternate way:

$$y = L \ln\left(\frac{x}{L - \sqrt{L^2 - x^2}}\right) - \sqrt{L^2 - x^2}.$$

CHAPTER 8. CONICS, PARAMETRIC CURVES, AND POLAR CURVES

Section 8.1 Conics (page 443)

2. The ellipse with foci $(0, 1)$ and $(4, 1)$ has $c = 2$, centre $(2, 1)$, and major axis along $y = 1$. If $\epsilon = 1/2$, then $a = c/\epsilon = 4$ and $b^2 = 16 - 4 = 12$. The ellipse has equation

$$\frac{(x-2)^2}{16} + \frac{(y-1)^2}{12} = 1.$$

4. A parabola with focus at $(0, -1)$ and principal axis along $y = -1$ will have vertex at a point of the form $(v, -1)$. Its equation will then be of the form $(y + 1)^2 = \pm 4v(x - v)$. The origin lies on this curve if $1 = \pm 4(-v^2)$. Only the $-$ sign is possible, and in this case $v = \pm 1/2$. The possible equations for the parabola are $(y + 1)^2 = 1 \pm 2x$.

6. The hyperbola with foci at $(\pm 5, 1)$ and asymptotes $x = \pm(y - 1)$ is rectangular, has centre at $(0, 1)$ and has transverse axis along the line $y = 1$. Since $c = 5$ and $a = b$ (because the asymptotes are perpendicular to each other) we have $a^2 = b^2 = 25/2$. The equation of the hyperbola is

$$x^2 - (y - 1)^2 = \frac{25}{2}.$$

8. If $x^2 + 4y^2 - 4y = 0$, then

$$x^2 + 4\left(y^2 - y + \frac{1}{4}\right) = 1, \quad \text{or} \quad \frac{x^2}{1} + \frac{(y - \frac{1}{2})^2}{\frac{1}{4}} = 1.$$

This represents an ellipse with centre at $\left(0, \dfrac{1}{2}\right)$, semi-major axis 1, semi-minor axis $\dfrac{1}{2}$, and foci at $\left(\pm\dfrac{\sqrt{3}}{2}, \dfrac{1}{2}\right)$.

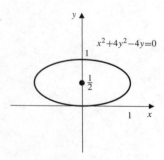

Fig. 8.1.8

10. If $4x^2 - y^2 - 4y = 0$, then

$$4x^2 - (y^2 + 4y + 4) = -4, \quad \text{or} \quad \frac{x^2}{1} - \frac{(y + 2)^2}{4} = -1.$$

This represents a hyperbola with centre at $(0, -2)$, semi-transverse axis 2, semi-conjugate axis 1, and foci at $(0, -2 \pm \sqrt{5})$. The asymptotes are $y = \pm 2x - 2$.

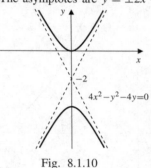

Fig. 8.1.10

12. If $x + 2y + 2y^2 = 1$, then

$$2\left(y^2 + y + \frac{1}{4}\right) = \frac{3}{2} - x$$

$$\Leftrightarrow \quad x = \frac{3}{2} - 2\left(y + \frac{1}{2}\right)^2.$$

This represents a parabola with vertex at $(\frac{3}{2}, -\frac{1}{2})$, focus at $(\frac{11}{8}, -\frac{1}{2})$ and directrix $x = \frac{13}{8}$.

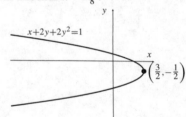

Fig. 8.1.12

14. If $9x^2 + 4y^2 - 18x + 8y = -13$, then

$$9(x^2 - 2x + 1) + 4(y^2 + 2y + 1) = 0$$
$$\Leftrightarrow 9(x - 1)^2 + 4(y + 1)^2 = 0.$$

This represents the single point $(1, -1)$.

16. The equation $(x - y)^2 - (x + y)^2 = 1$ simplifies to $4xy = -1$ and hence represents a rectangular hyperbola with centre at the origin, asymptotes along the coordinate axes, transverse axis along $y = -x$, conjugate axis along $y = x$, vertices at $(\frac{1}{2}, -\frac{1}{2})$ and $(-\frac{1}{2}, \frac{1}{2})$, semi-transverse and semi-conjugate axes equal to $1/\sqrt{2}$, semi-focal separation equal to $\sqrt{\frac{1}{2} + \frac{1}{2}} = 1$, and hence foci at the points $\left(\frac{1}{\sqrt{2}}, -\frac{1}{\sqrt{2}}\right)$ and $\left(-\frac{1}{\sqrt{2}}, \frac{1}{\sqrt{2}}\right)$. The eccentricity is $\sqrt{2}$.

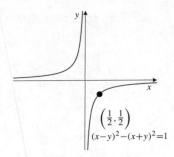

Fig. 8.1.16

18. The foci of the ellipse are $(0, 0)$ and $(3, 0)$, so the centre is $(3/2, 0)$ and $c = 3/2$. The semi-axes a and b must satisfy $a^2 - b^2 = 9/4$. Thus the possible equations of the ellipse are

$$\frac{(x - (3/2))^2}{(9/4) + b^2} + \frac{y^2}{b^2} = 1.$$

20. We have $x^2 + 2xy + y^2 = 4x - 4y + 4$ and $A = 1$, $B = 2$, $C = 1$, $D = -4$, $E = 4$ and $F = -4$. We rotate the axes through angle θ satisfying $\tan 2\theta = B/(A - C) = \infty \Rightarrow \theta = \frac{\pi}{4}$. Then $A' = 2$, $B' = 0$, $C' = 0$, $D' = 0$, $E' = 4\sqrt{2}$ and the transformed equation is

$$2u^2 + 4\sqrt{2}v - 4 = 0 \quad \Rightarrow \quad u^2 = -2\sqrt{2}\left(v - \frac{1}{\sqrt{2}}\right)$$

which represents a parabola with vertex at $(u, v) = \left(0, \frac{1}{\sqrt{2}}\right)$ and principal axis along $u = 0$. The distance a from the focus to the vertex is given by $4a = 2\sqrt{2}$, so $a = 1/\sqrt{2}$ and the focus is at $(0, 0)$. The directrix is $v = \sqrt{2}$.

Since $x = \frac{1}{\sqrt{2}}(u - v)$ and $y = \frac{1}{\sqrt{2}}(u + v)$, the vertex of the parabola in terms of xy-coordinates is $(-\frac{1}{2}, \frac{1}{2})$, and the focus is $(0, 0)$. The directrix is $x - y = 2$. The principal axis is $y = -x$.

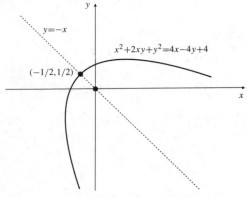

Fig. 8.1.20

22. We have $x^2 - 4xy + 4y^2 + 2x + y = 0$ and $A = 1$, $B = -4$, $C = 4$, $D = 2$, $E = 1$ and $F = 0$. We rotate the axes through angle θ satisfying $\tan 2\theta = B/(A - C) = \frac{4}{3}$. Then

$$\sec 2\theta = \sqrt{1 + \tan^2 2\theta} = \frac{5}{3} \quad \Rightarrow \quad \cos 2\theta = \frac{3}{5}$$

$$\Rightarrow \begin{cases} \cos\theta = \sqrt{\dfrac{1 + \cos 2\theta}{2}} = \sqrt{\dfrac{4}{5}} = \dfrac{2}{\sqrt{5}}; \\ \sin\theta = \sqrt{\dfrac{1 - \cos 2\theta}{2}} = \sqrt{\dfrac{1}{5}} = \dfrac{1}{\sqrt{5}}. \end{cases}$$

Then $A' = 0$, $B' = 0$, $C' = 5$, $D' = \sqrt{5}$, $E' = 0$ and the transformed equation is

$$5v^2 + \sqrt{5}u = 0 \quad \Rightarrow \quad v^2 = -\frac{1}{\sqrt{5}}u$$

which represents a parabola with vertex at $(u, v) = (0, 0)$, focus at $\left(-\frac{1}{4\sqrt{5}}, 0\right)$. The directrix is $u = \frac{1}{4\sqrt{5}}$ and the principal axis is $v = 0$. Since $x = \frac{2}{\sqrt{5}}u - \frac{1}{\sqrt{5}}v$ and $y = \frac{1}{\sqrt{5}}u + \frac{2}{\sqrt{5}}v$, in terms of the xy-coordinates, the vertex is at $(0, 0)$, the focus at $\left(-\frac{1}{10}, -\frac{1}{20}\right)$. The directrix is $2x + y = \frac{1}{4}$ and the principal axis is $2y - x = 0$.

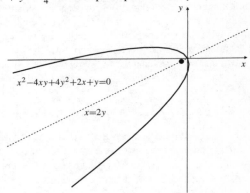

Fig. 8.1.22

24. Let the equation of the parabola be $y^2 = 4ax$. The focus F is at $(a, 0)$ and vertex at $(0, 0)$. Then the distance from the vertex to the focus is a. At $x = a$, $y = \sqrt{4a(a)} = \pm 2a$. Hence, $\ell = 2a$, which is twice the distance from the vertex to the focus.

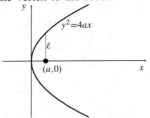

Fig. 8.1.24

26. Suppose the hyperbola has equation $\dfrac{x^2}{a^2} - \dfrac{y^2}{b^2} = 1$. The vertices are at $(\pm a, 0)$ and the foci are at $(\pm c, 0)$ where $c = \sqrt{a^2 + b^2}$. At $x = \sqrt{a^2 + b^2}$,

$$\frac{a^2 + b^2}{a^2} - \frac{y^2}{b^2} = 1$$
$$(a^2 + b^2)b^2 - a^2 y^2 = a^2 b^2$$
$$y = \pm \frac{b^2}{a}.$$

Hence, $\ell = \dfrac{b^2}{a}$.

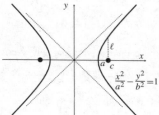

Fig. 8.1.26

28. Let F_1 and F_2 be the points where the plane is tangent to the spheres. Let P be an arbitrary point P on the hyperbola in which the plane intersects the cone. The spheres are tangent to the cone along two circles as shown in the figure. Let $PAVB$ be a generator of the cone (a straight line lying on the cone) intersecting these two circles at A and B as shown. (V is the vertex of the cone.) We have $PF_1 = PA$ because two tangents to a sphere from a point outside the sphere have equal lengths. Similarly, $PF_2 = PB$. Therefore

$$PF_2 - PF_1 = PB - PA = AB = \text{ constant},$$

since the distance between the two circles in which the spheres intersect the cone, measured along the generators of the cone, is the same for all generators. Hence, F_1 and F_2 are the foci of the hyperbola.

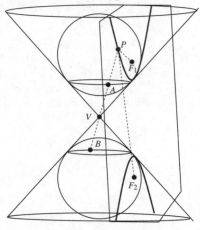

Fig. 8.1.28

Section 8.2 Parametric Curves   (page 449)

2. If $x = 2 - t$ and $y = t + 1$ for $0 \le t < \infty$, then $y = 2 - x + 1 = 3 - x$ for $-\infty < x \le 2$, which is a half line.

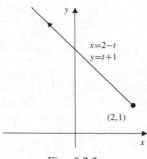

Fig. 8.2.2

4. If $x = \dfrac{1}{1 + t^2}$ and $y = \dfrac{t}{1 + t^2}$ for $-\infty < t < \infty$, then

$$x^2 + y^2 = \frac{1 + t^2}{(1 + t^2)^2} = \frac{1}{1 + t^2} = x$$
$$\Leftrightarrow \quad \left(x - \frac{1}{2}\right)^2 + y^2 = \frac{1}{4}.$$

This curve consists of all points of the circle with centre at $(\frac{1}{2}, 0)$ and radius $\frac{1}{2}$ except the origin $(0, 0)$.

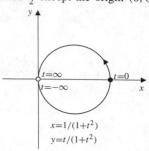

Fig. 8.2.4

6. If $x = a \sec t$ and $y = b \tan t$ for $-\dfrac{\pi}{2} < t < \dfrac{\pi}{2}$, then

$$\frac{x^2}{a^2} - \frac{y^2}{b^2} = \sec^2 t - \tan^2 t = 1.$$

The curve is one arch of this hyperbola.

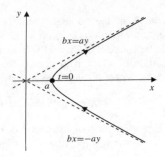

Fig. 8.2.6

8. If $x = \cos\sin s$ and $y = \sin\sin s$ for $-\infty < s < \infty$, then $x^2 + y^2 = 1$. The curve consists of the arc of this circle extending from $(a, -b)$ through $(1, 0)$ to (a, b) where $a = \cos(1)$ and $b = \sin(1)$, traversed infinitely often back and forth.

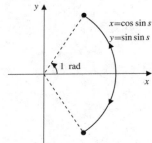

Fig. 8.2.8

10. If $x = 1 - \sqrt{4 - t^2}$ and $y = 2 + t$ for $-2 \le t \le 2$ then

$$(x - 1)^2 = 4 - t^2 = 4 - (y - 2)^2.$$

The parametric curve is the left half of the circle of radius 4 centred at $(1, 2)$, and is traced in the direction of increasing y.

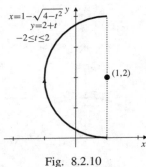

Fig. 8.2.10

12. $x = 2 - 3\cosh t$, $y = -1 + 2\sinh t$ represents the left half (branch) of the hyperbola

$$\frac{(x - 2)^2}{9} - \frac{(y + 1)^2}{4} = 1.$$

14. (i) If $x = \cos^4 t$ and $y = \sin^4 t$, then

$$\begin{aligned}
(x - y)^2 &= (\cos^4 t - \sin^4 t)^2 \\
&= \left[(\cos^2 t + \sin^2 t)(\cos^2 t - \sin^2 t) \right]^2 \\
&= (\cos^2 t - \sin^2 t)^2 \\
&= \cos^4 t + \sin^4 t - 2\cos^2 t \sin^2 t
\end{aligned}$$

and

$$1 = (\cos^2 t + \sin^2 t)^2 = \cos^4 t + \sin^4 t + 2\cos^2 t \sin^2 t.$$

Hence,

$$1 + (x - y)^2 = 2(\cos^4 t + \sin^4 t) = 2(x + y).$$

(ii) If $x = \sec^4 t$ and $y = \tan^4 t$, then

$$\begin{aligned}
(x - y)^2 &= (\sec^4 t - \tan^4 t)^2 \\
&= (\sec^2 t + \tan^2 t)^2 \\
&= \sec^4 t + \tan^4 t + 2\sec^2 t \tan^2 t
\end{aligned}$$

and

$$1 = (\sec^2 t - \tan^2 t)^2 = \sec^4 t + \tan^4 t - 2\sec^2 t \tan^2 t.$$

Hence,

$$1 + (x - y)^2 = 2(\sec^4 t + \tan^4 t) = 2(x + y).$$

(iii) Similarly, if $x = \tan^4 t$ and $y = \sec^4 t$, then

$$\begin{aligned}
1 + (x - y)^2 &= 1 + (y - x)^2 \\
&= (\sec^2 t - \tan^2 t)^2 + (\sec^4 t - \tan^4 t)^2 \\
&= 2(\tan^4 t + \sec^4 t) \\
&= 2(x + y).
\end{aligned}$$

These three parametric curves above correspond to different parts of the parabola $1 + (x - y)^2 = 2(x + y)$, as shown in the following diagram.

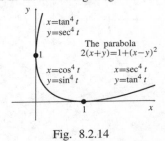

Fig. 8.2.14

16. If (x, y) is any point on the circle $x^2 + y^2 = R^2$ other than $(R, 0)$, then the line from (x, y) to $(R, 0)$ has slope $m = \dfrac{y}{x - R}$. Thus $y = m(x - R)$, and

$$x^2 + m^2(x - R)^2 = R^2$$
$$(m^2 + 1)x^2 - 2xRm^2 + (m^2 - 1)R^2 = 0$$
$$\left[(m^2 + 1)x - (m^2 - 1)R\right](x - R) = 0$$
$$\Rightarrow \quad x = \frac{(m^2 - 1)R}{m^2 + 1} \quad \text{or} \quad x = R.$$

The parametrization of the circle in terms of m is given by

$$x = \frac{(m^2 - 1)R}{m^2 + 1}$$
$$y = m\left[\frac{(m^2 - 1)R}{m^2 + 1} - R\right] = -\frac{2Rm}{m^2 + 1}$$

where $-\infty < m < \infty$. This parametrization gives every point on the circle except $(R, 0)$.

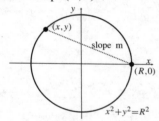

Fig. 8.2.16

18. The coordinates of P satisfy

$$x = a \sec t, \quad y = b \sin t.$$

The Cartesian equation is $\dfrac{y^2}{b^2} + \dfrac{a^2}{x^2} = 1$.

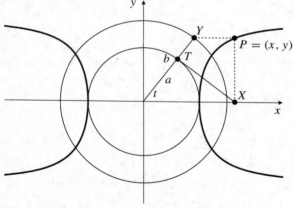

Fig. 8.2.18

20. Let C_0 and P_0 be the original positions of the centre of the wheel and a point at the bottom of the flange whose path is to be traced. The wheel is also shown in a subsequent position in which it makes contact with the rail at R. Since the wheel has been rotated by an angle θ,

$$OR = \text{arc } SR = a\theta.$$

Thus, the new position of the centre is $C = (a\theta, a)$. Let $P = (x, y)$ be the new position of the point; then

$$x = OR - PQ = a\theta - b\sin(\pi - \theta) = a\theta - b\sin\theta,$$
$$y = RC + CQ = a + b\cos(\pi - \theta) = a - b\cos\theta.$$

These are the parametric equations of the prolate cycloid.

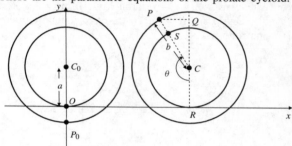

Fig. 8.2.20

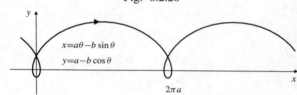

Fig. 8.2.20

22. a) From triangles in the figure,

$$x = |TX| = |OT|\tan t = \tan t$$
$$y = |OY| = \sin\left(\frac{\pi}{2} - t\right) = |OY|\cos t$$
$$= |OT|\cos t \cos t = \cos^2 t.$$

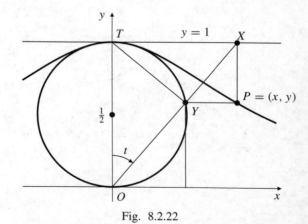

Fig. 8.2.22

b) $\frac{1}{y} = \sec^2 t = 1 + \tan^2 t = 1 + x^2$. Thus $y = \frac{1}{1+x^2}$.

24. $x = \sin t, \quad y = \sin(3t)$

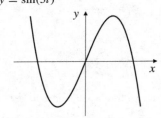

Fig. 8.2.24

26. $x = \sin(2t), \quad y = \sin(5t)$

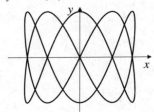

Fig. 8.2.26

28. $x = \left(1 + \frac{1}{n}\right)\cos t + \frac{1}{n}\cos((n-1)t)$

$y = \left(1 + \frac{1}{n}\right)\sin t - \frac{1}{n}\sin((n-1)t)$

represents a cycloid-like curve that is wound around the inside circle $x^2 + y^2 = \left(1 + (2/n)\right)^2$ and is externally tangent to $x^2 + y^2 = 1$. If $n \geq 2$ is an integer, the curve closes after one revolution and has n cusps. The figure shows the curve for $n = 7$. If n is a rational number but not an integer, the curve will wind around the circle more than once before it closes.

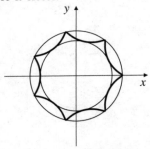

Fig. 8.2.28

Section 8.3 Smooth Parametric Curves and Their Slopes (page 453)

2. $x = t^2 - 2t \qquad y = t^2 + 2t$
$\frac{dx}{dt} = 2t - 2 \qquad \frac{dy}{dt} = 2t + 2$
Horizontal tangent at $t = -1$, i.e., at $(3, -1)$.
Vertical tangent at $t = 1$, i.e., at $(-1, 3)$.

4. $x = t^3 - 3t \qquad y = 2t^3 + 3t^2$
$\frac{dx}{dt} = 3(t^2 - 1) \qquad \frac{dy}{dt} = 6t(t+1)$
Horizontal tangent at $t = 0$, i.e., at $(0, 0)$.
Vertical tangent at $t = 1$, i.e., at $(-2, 5)$.
At $t = -1$ (i.e., at $(2, 1)$) both dx/dt and dy/dt change sign, so the curve is not smooth there. (It has a cusp.)

6. $x = \sin t \qquad y = \sin t - t\cos t$
$\frac{dx}{dt} = \cos t \qquad \frac{dy}{dt} = t\sin t$
Horizontal tangent at $t = n\pi$, i.e., at $(0, -(-1)^n n\pi)$ (for integers n).
Vertical tangent at $t = (n + \frac{1}{2})\pi$, i.e. at $(1, 1)$ and $(-1, -1)$.

8. $x = \frac{3t}{1 + t^3} \qquad y = \frac{3t^2}{1 + t^3}$
$\frac{dx}{dt} = \frac{3(1 - 2t^3)}{(1 + t^3)^2} \qquad \frac{dy}{dt} = \frac{3t(2 - t^3)}{(1 + t^3)^2}$
Horizontal tangent at $t = 0$ and $t = 2^{1/3}$, i.e., at $(0, 0)$ and $(2^{1/3}, 2^{2/3})$.
Vertical tangent at $t = 2^{-1/3}$, i.e., at $(2^{2/3}, 2^{1/3})$. The curve also approaches $(0, 0)$ vertically as $t \to \pm\infty$.

10. $x = t^4 - t^2 \qquad y = t^3 + 2t$
$\frac{dx}{dt} = 4t^3 - 2t \qquad \frac{dy}{dt} = 3t^2 + 2$
At $t = -1$; $\frac{dy}{dx} = \frac{3(-1)^2 + 2}{4(-1)^3 - 2(-1)} = -\frac{5}{2}$.

12. $x = e^{2t} \qquad y = te^{2t}$
$\frac{dx}{dt} = 2e^{2t} \qquad \frac{dy}{dt} = e^{2t}(1 + 2t)$
At $t = -2$; $\frac{dy}{dx} = \frac{e^{-4}(1 - 4)}{2e^{-4}} = -\frac{3}{2}$.

14.
$$x = t - \cos t = \frac{\pi}{4} - \frac{1}{\sqrt{2}}$$

$$\frac{dx}{dt} = 1 + \sin t = 1 + \frac{1}{\sqrt{2}}$$

$$y = 1 - \sin t = 1 - \frac{1}{\sqrt{2}} \quad \text{at } t = \frac{\pi}{4}$$

$$\frac{dy}{dt} = -\cos t = -\frac{1}{\sqrt{2}} \quad \text{at } t = \frac{\pi}{4}$$

Tangent line: $x = \frac{\pi}{4} - \frac{1}{\sqrt{2}} + \left(1 + \frac{1}{\sqrt{2}}\right)t,$

$$y = 1 - \frac{1}{\sqrt{2}} - \frac{t}{\sqrt{2}}.$$

16. $x = \sin t$, $y = \sin(2t)$ is at $(0,0)$ at $t = 0$ and $t = \pi$. Since

$$\frac{dy}{dx} = \frac{2\cos(2t)}{\cos t} = \begin{cases} 2 & \text{if } t = 0 \\ -2 & \text{if } t = \pi, \end{cases}$$

the tangents at $(0,0)$ at $t = 0$ and $t = \pi$ have slopes 2 and -2, respectively.

18. $x = (t-1)^4$

$$\frac{dx}{dt} = 4(t-1)^3$$

$$y = (t-1)^3$$

$$\frac{dy}{dt} = 3(t-1)^2 \quad \text{both vanish at } t = 1.$$

Since $\dfrac{dx}{dy} = \dfrac{4(t-1)}{3} \to 0$ as $t \to 1$, and dy/dt does not change sign at $t = 1$, the curve is smooth at $t = 1$ and therefore everywhere.

20. $x = t^3 \qquad y = t - \sin t$

$$\frac{dx}{dt} = 3t^2 \qquad \frac{dy}{dt} = 1 - \cos t \quad \text{both vanish at } t = 0.$$

$$\lim_{t\to 0}\frac{dx}{dy} = \lim_{t\to 0}\frac{3t^2}{1 - \cos t} = \lim_{t\to 0}\frac{6t}{\sin t} = 6 \text{ and } dy/dt \text{ does}$$
not change sign at $t = 0$. Thus the curve is smooth at $t = 0$, and hence everywhere.

22. If $x = f(t) = t^3$ and $y = g(t) = 3t^2 - 1$, then

$$f'(t) = 3t^2, \ f''(t) = 6t;$$
$$g'(t) = 6t, \ g''(t) = 6.$$

Both $f'(t)$ and $g'(t)$ vanish at $t = 0$. Observe that

$$\frac{dy}{dx} = \frac{6t}{3t^2} = \frac{2}{t}.$$

Thus,

$$\lim_{t\to 0+}\frac{dy}{dx} = \infty, \qquad \lim_{t\to 0-}\frac{dy}{dx} = -\infty$$

and the curve has a cusp at $t = 0$, i.e., at $(0, -1)$. Since

$$\frac{d^2y}{dx^2} = \frac{(3t^2)(6) - (6t)(6t)}{(3t^2)^3} = -\frac{2}{3t^4} < 0$$

for all t, the curve is concave down everywhere.

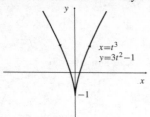

Fig. 8.3.22

24. If $x = f(t) = t^3 - 3t - 2$ and $y = g(t) = t^2 - t - 2$, then

$$f'(t) = 3t^2 - 3, \ f''(t) = 6t;$$
$$g'(t) = 2t - 1, \ g''(t) = 2.$$

The tangent is horizontal at $t = \frac{1}{2}$, i.e., at $\left(-\frac{27}{8}, -\frac{9}{4}\right)$. The tangent is vertical at $t = \pm 1$, i.e., $(-4, -2)$ and $(0, 0)$. Directional information is as follows:

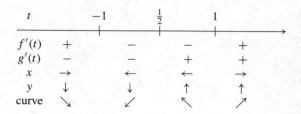

For concavity,

$$\frac{d^2y}{dx^2} = \frac{3(t^2-1)(2) - (2t-1)(6t)}{[3(t^2-1)]^3} = -\frac{2(t^2-t+1)}{9(t^2-1)^3}$$

which is undefined at $t = \pm 1$, therefore

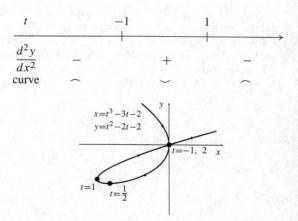

Fig. 8.3.24

Section 8.4   Arc Lengths and Areas for Parametric Curves   (page 458)

2. If $x = 1 + t^3$ and $y = 1 - t^2$ for $-1 \le t \le 2$, then the arc length is

$$s = \int_{-1}^{2} \sqrt{(3t^2)^2 + (-2t)^2}\, dt$$

$$= \int_{-1}^{2} |t| \sqrt{9t^2 + 4}\, dt$$

$$= \left(\int_0^1 + \int_0^2 \right) t\sqrt{9t^2 + 4}\, dt \quad \text{Let } u = 9t^2 + 4$$
$$\hspace{6cm} du = 18t\, dt$$

$$= \frac{1}{18} \left(\int_4^{13} + \int_4^{40} \right) \sqrt{u}\, du$$

$$= \frac{1}{27} \left(13\sqrt{13} + 40\sqrt{40} - 16 \right) \text{ units.}$$

4. If $x = \ln(1 + t^2)$ and $y = 2\tan^{-1} t$ for $0 \le t \le 1$, then

$$\frac{dx}{dt} = \frac{2t}{1 + t^2}; \qquad \frac{dy}{dt} = \frac{2}{1 + t^2}.$$

The arc length is

$$s = \int_0^1 \sqrt{\frac{4t^2 + 4}{(1 + t^2)^2}}\, dt$$

$$= 2\int_0^1 \frac{dt}{\sqrt{1 + t^2}} \quad \text{Let } t = \tan\theta$$
$$\hspace{4cm} dt = \sec^2\theta\, d\theta$$

$$= 2\int_0^{\pi/4} \sec\theta\, d\theta$$

$$= 2\ln|\sec\theta + \tan\theta|\Big|_0^{\pi/4} = 2\ln(1 + \sqrt{2}) \text{ units.}$$

6. $x = \cos t + t\sin t \qquad y = \sin t - t\cos t \quad (0 \le t \le 2\pi)$

$\dfrac{dx}{dt} = t\cos t \qquad\qquad \dfrac{dy}{dt} = t\sin t$

$$\text{Length} = \int_0^{2\pi} \sqrt{t^2\cos^2 t + t^2\sin^2 t}\, dt$$

$$= \int_0^{2\pi} t\, dt = \frac{t^2}{2}\Big|_0^{2\pi} = 2\pi^2 \text{ units.}$$

8. $x = \sin^2 t \qquad y = 2\cos t \quad (0 \le t \le \pi/2)$

$\dfrac{dx}{dt} = 2\sin t\cos t \qquad \dfrac{dy}{dt} = -2\sin t$

Length

$$= \int_0^{\pi/2} \sqrt{4\sin^2 t\cos^2 t + 4\sin^2 t}\, dt$$

$$= 2\int_0^{\pi/2} \sin t\sqrt{1 + \cos^2 t}\, dt \quad \text{Let } \cos t = \tan u$$
$$\hspace{5cm} -\sin t\, dt = \sec^2 u\, du$$

$$= 2\int_0^{\pi/4} \sec^3 u\, du$$

$$= \left(\sec u\tan u + \ln(\sec u + \tan u) \right)\Big|_0^{\pi/4}$$

$$= \sqrt{2} + \ln(1 + \sqrt{2}) \text{ units.}$$

10. If $x = at - a\sin t$ and $y = a - a\cos t$ for $0 \le t \le 2\pi$, then

$$\frac{dx}{dt} = a - a\cos t, \qquad \frac{dy}{dt} = a\sin t;$$

$$ds = \sqrt{(a - a\cos t)^2 + (a\sin t)^2}\, dt$$

$$= a\sqrt{2}\sqrt{1 - \cos t}\, dt = a\sqrt{2}\sqrt{2\sin^2\left(\frac{t}{2}\right)}\, dt$$

$$= 2a\sin\left(\frac{t}{2}\right) dt.$$

a) The surface area generated by rotating the arch about the x-axis is

$$S_x = 2\pi \int_0^{2\pi} |y|\, ds$$

$$= 4\pi \int_0^{\pi} (a - a\cos t)2a\sin\left(\frac{t}{2}\right) dt$$

$$= 16\pi a^2 \int_0^{\pi} \sin^3\left(\frac{t}{2}\right) dt$$

$$= 16\pi a^2 \int_0^{\pi} \left[1 - \cos^2\left(\frac{t}{2}\right) \right] \sin\left(\frac{t}{2}\right) dt$$

$$\text{Let } u = \cos\left(\frac{t}{2}\right)$$

$$du = -\frac{1}{2}\sin\left(\frac{t}{2}\right) dt$$

$$= -32\pi a^2 \int_1^0 (1 - u^2)\, du$$

$$= 32\pi a^2 \left[u - \frac{1}{3}u^3 \right]\Big|_0^1$$

$$= \frac{64}{3}\pi a^3 \text{ sq. units.}$$

b) The surface area generated by rotating the arch about the y-axis is

$$S_y = 2\pi \int_0^{2\pi} |x| \, ds$$

$$= 2\pi \int_0^{2\pi} (at - a \sin t) 2a \sin\left(\frac{t}{2}\right) dt$$

$$= 4\pi a^2 \int_0^{2\pi} \left[t - 2\sin\left(\frac{t}{2}\right)\cos\left(\frac{t}{2}\right)\right]\sin\left(\frac{t}{2}\right) dt$$

$$= 4\pi a^2 \int_0^{2\pi} t \sin\left(\frac{t}{2}\right) dt$$

$$\quad - 8\pi a^2 \int_0^{2\pi} \sin^2\left(\frac{t}{2}\right)\cos\left(\frac{t}{2}\right) dt$$

$$= 4\pi a^2 \left[-2t\cos\left(\frac{t}{2}\right)\Big|_0^{2\pi} + 2\int_0^{2\pi}\cos\left(\frac{t}{2}\right) dt\right] - 0$$

$$= 4\pi a^2 [4\pi + 0] = 16\pi^2 a^2 \text{ sq. units.}$$

12. The area of revolution of the curve in Exercise 11 about the y-axis is

$$\int_{t=0}^{t=\pi/2} 2\pi x \, ds = 2\sqrt{2}\pi \int_0^{\pi/2} e^{2t}\cos t \, dt$$

$$= 2\sqrt{2}\pi \frac{e^{2t}}{5}(2\cos t + \sin t)\Big|_0^{\pi/2}$$

$$= \frac{2\sqrt{2}\pi}{5}(e^\pi - 2) \text{ sq. units.}$$

14. The area of revolution of the curve of Exercise 13 about the x-axis is

$$\int_{t=0}^{t=1} 2\pi y \, ds = 24\pi \int_0^1 t^4\sqrt{1+t^2}\, dt \quad \text{Let } t = \tan u$$
$$\qquad\qquad\qquad\qquad\qquad\qquad\qquad dt = \sec^2 u \, du$$

$$= 24\pi \int_0^{\pi/4} \tan^4 u \sec^3 u \, du$$

$$= 24\pi \int_0^{\pi/4} (\sec^7 u - 2\sec^5 u + \sec^3 u)\, du$$

$$= \frac{\pi}{2}\left(7\sqrt{2} + 3\ln(1+\sqrt{2})\right) \text{ sq. units.}$$

We have omitted the details of evaluation of the final integral. See Exercise 24 of Section 8.3 for a similar evaluation.

16. Area of $R = 4 \times \int_{\pi/2}^0 (a\sin^3 t)(-3a\sin t\cos^2 t)\, dt$

$$= -12a^2 \int_{\pi/2}^0 \sin^4 t \cos^2 t \, dt$$

$$= 12a^2\left[\frac{t}{16} - \frac{\sin(4t)}{64} - \frac{\sin^3(2t)}{48}\right]\Big|_0^{\pi/2}$$

$$\text{(See Exercise 34 of Section 6.4.)}$$

$$= \frac{3}{8}\pi a^2 \text{ sq. units.}$$

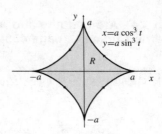

Fig. 8.4.16

18. If $x = \cos s \sin s = \frac{1}{2}\sin 2s$ and $y = \sin^2 s = \frac{1}{2} - \frac{1}{2}\cos 2s$ for $0 \le s \le \frac{1}{2}\pi$, then

$$x^2 + \left(y - \frac{1}{2}\right)^2 = \frac{1}{4}\sin^2 2s + \frac{1}{4}\cos^2 2s = \frac{1}{4}$$

which is the right half of the circle with radius $\frac{1}{2}$ and centre at $(0, \frac{1}{2})$. Hence, the area of R is

$$\frac{1}{2}\left[\pi\left(\frac{1}{2}\right)^2\right] = \frac{\pi}{8} \text{ sq. units.}$$

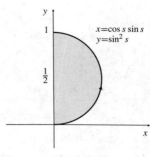

Fig. 8.4.18

20. To find the shaded area we subtract the area under the upper half of the hyperbola from that of a right triangle:

$$\text{Shaded area} = \text{Area } \triangle ABC - \text{Area sector } ABC$$

$$= \frac{1}{2}\sec t_0 \tan t_0 - \int_0^{t_0} \tan t(\sec t \tan t)\, dt$$

$$= \frac{1}{2}\sec t_0 \tan t_0 - \int_0^{t_0} (\sec^3 t - \sec t)\, dt$$

$$= \frac{1}{2}\sec t_0 \tan t_0 - \left[\frac{1}{2}\sec t \tan t +\right.$$

$$\left.\frac{1}{2}\ln|\sec t + \tan t| - \ln|\sec t + \tan t|\right]\Big|_0^{t_0}$$

$$= \frac{1}{2}\ln|\sec t_0 + \tan t_0| \text{ sq. units.}$$

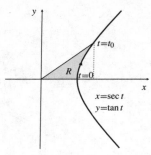

Fig. 8.4.20

22. If $x = f(t) = at - a\sin t$ and $y = g(t) = a - a\cos t$, then the volume of the solid obtained by rotating about the x-axis is

$$V = \int_{t=0}^{t=2\pi} \pi y^2\, dx = \pi \int_{t=0}^{t=2\pi} [g(t)]^2 f'(t)\, dt$$

$$= \pi \int_0^{2\pi} (a - a\cos t)^2 (a - a\cos t)\, dt$$

$$= \pi a^3 \int_0^{2\pi} (1 - \cos t)^3\, dt$$

$$= \pi a^3 \int_0^{2\pi} (1 - 3\cos t + 3\cos^2 t - \cos^3 t)\, dt$$

$$= \pi a^3 \left[2\pi - 0 + \frac{3}{2}\int_0^{2\pi}(1 + \cos 2t)\, dt - 0 \right]$$

$$= \pi a^3 \left[2\pi + \frac{3}{2}(2\pi) \right] = 5\pi^2 a^3 \text{ cu. units.}$$

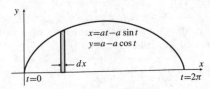

Fig. 8.4.22

Section 8.5 Polar Coordinates and Polar Curves (page 464)

$$r = -2\csc\theta \Rightarrow r\sin\theta = -2$$
$$\Leftrightarrow \quad y = -2 \quad \text{a horizontal line.}$$

$$r = \sin\theta + \cos\theta$$
$$r^2 = r\sin\theta + r\cos\theta$$
$$x^2 + y^2 = y + x$$
$$\left(x - \frac{1}{2}\right)^2 + \left(y - \frac{1}{2}\right)^2 = \frac{1}{2}$$
circle with centre $\left(\frac{1}{2}, \frac{1}{2}\right)$ and radius $\frac{1}{\sqrt{2}}$.

6. $r = \sec\theta\tan\theta \Rightarrow r\cos\theta = \dfrac{r\sin\theta}{r\cos\theta}$
$$x^2 = y \qquad \text{a parabola.}$$

8. $r = \dfrac{2}{\sqrt{\cos^2\theta + 4\sin^2\theta}}$
$$r^2\cos^2\theta + 4r^2\sin^2\theta = 4$$
$$x^2 + 4y^2 = 4 \qquad \text{an ellipse.}$$

10. $r = \dfrac{2}{2 - \cos\theta}$
$$2r - r\cos\theta = 2$$
$$4r^2 = (2 + x)^2$$
$$4x^2 + 4y^2 = 4 + 4x + x^2$$
$$3x^2 + 4y^2 - 4x = 4 \qquad \text{an ellipse.}$$

12. $r = \dfrac{2}{1 + \sin\theta}$
$$r + r\sin\theta = 2$$
$$r^2 = (2 - y)^2$$
$$x^2 + y^2 = 4 - 4y + y^2$$
$$x^2 = 4 - 4y \qquad \text{a parabola.}$$

14. If $r = 1 - \cos\left(\theta + \dfrac{\pi}{4}\right)$, then $r = 0$ at $\theta = -\dfrac{\pi}{4}$ and $\dfrac{7\pi}{4}$. This is a cardioid.

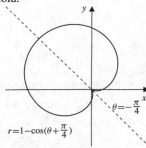

Fig. 8.5.14

16. If $r = 1 - 2\sin\theta$, then $r = 0$ at $\theta = \dfrac{\pi}{6}$ and $\dfrac{5\pi}{6}$.

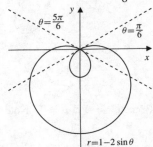

Fig. 8.5.16

18. If $r = 2\sin 2\theta$, then $r = 0$ at $\theta = 0$, $\pm\dfrac{\pi}{2}$ and π.

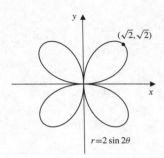

$(\sqrt{2}, \sqrt{2})$

$r = 2\sin 2\theta$

Fig. 8.5.18

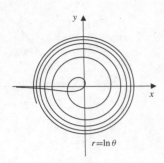

$r = \ln\theta$

Fig. 8.5.24

20. If $r = 2\cos 4\theta$, then $r = 0$ at $\theta = \pm\dfrac{\pi}{8}$, $\pm\dfrac{3\pi}{8}$, $\pm\dfrac{5\pi}{8}$ and $\pm\dfrac{7\pi}{8}$. (an eight leaf rosette)

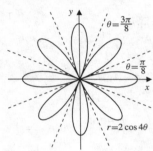

$\theta = \dfrac{3\pi}{8}$

$\theta = \dfrac{\pi}{8}$

$r = 2\cos 4\theta$

Fig. 8.5.20

22. If $r^2 = 4\cos 3\theta$, then $r = 0$ at $\theta = \pm\dfrac{\pi}{6}$, $\pm\dfrac{\pi}{2}$ and $\pm\dfrac{5\pi}{6}$. This equation defines two functions of r, namely $r = \pm 2\sqrt{\cos 3\theta}$. Each contributes 3 leaves to the graph.

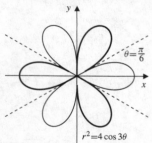

$\theta = \dfrac{\pi}{6}$

$r^2 = 4\cos 3\theta$

Fig. 8.5.22

24. If $r = \ln\theta$, then $r = 0$ at $\theta = 1$. Note that

$$y = r\sin\theta = \ln\theta \sin\theta = (\theta\ln\theta)\left(\frac{\sin\theta}{\theta}\right) \to 0$$

as $\theta \to 0+$. Therefore, the (negative) x-axis is an asymptote of the curve.

26. $r^2 = 2\cos(2\theta)$, $r = 1$.
$\cos(2\theta) = 1/2 \Rightarrow \theta = \pm\pi/6$ or $\theta = \pm 5\pi/6$.
Intersections: $[1, \pm\pi/6]$ and $[1, \pm 5\pi/6]$.

28. Let $r_1(\theta) = \theta$ and $r_2(\theta) = \theta + \pi$. Although the equation $r_1(\theta) = r_2(\theta)$ has no solutions, the curves $r = r_1(\theta)$ and $r = r_2(\theta)$ can still intersect if $r_1(\theta_1) = -r_2(\theta_2)$ for two angles θ_1 and θ_2 having the opposite directions in the polar plane. Observe that $\theta_1 = -n\pi$ and $\theta_2 = (n-1)\pi$ are two such angles provided n is any integer. Since

$$r_1(\theta_1) = -n\pi = -r_2((n-1)\pi),$$

the curves intersect at any point of the form $[n\pi, 0]$ or $[n\pi, \pi]$.

30. The graph of $r = \cos n\theta$ has $2n$ leaves if n is an even integer and n leaves if n is an odd integer. The situation for $r^2 = \cos n\theta$ is reversed. The graph has $2n$ leaves if n is an odd integer (provided negative values of r are allowed), and it has n leaves if n is even.

32. $r = \cos\theta\cos(m\theta)$
For odd m this flower has $2m$ petals, 2 large ones and 4 each of $(m-1)/2$ smaller sizes.
For even m the flower has $m+1$ petals, one large and each of $m/2$ smaller sizes.

34. $r = \sin(2\theta)\sin(m\theta)$
For odd m there are $m+1$ petals, 2 each of $(m + $ different sizes.
For even m there are always $2m$ petals. They ar different sizes if $m = 4n - 2$ or $m = 4n$.

36. $r = C + \cos\theta\cos(2\theta)$
The curve always has 3 bulges, one larger th other two. For $C = 0$ these are 3 distinct $0 < C < 1$ there is a fourth supplementar the large one. For $C = 1$ the curve has a gin. For $C > 1$ the curve does not appr and the petals become less distinct as C

38.

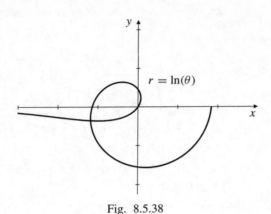

Fig. 8.5.38

We will have $[\ln\theta_1, \theta_1] = [\ln\theta_2, \theta_2]$ if

$$\theta_2 = \theta_1 + \pi \quad \text{and} \quad \ln\theta_1 = -\ln\theta_2,$$

that is, if $\ln\theta_1 + \ln(\theta_1 + \pi) = 0$. This equation has solution $\theta_1 \approx 0.29129956$. The corresponding intersection point has Cartesian coordinates $(\ln\theta_1 \cos\theta_1, \ln\theta_1 \sin\theta_1) \approx (-1.181442, -0.354230)$.

Section 8.6   Slopes, Areas, and Arc Lengths for Polar Curves   (page 468)

2.   Area $= \dfrac{1}{2}\displaystyle\int_0^{2\pi} \theta^2\, d\theta = \dfrac{\theta^3}{6}\Big|_0^{2\pi} = \dfrac{4}{3}\pi^3$ sq. units.

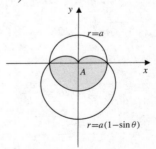

Fig. 8.6.2

4.   Area $= \dfrac{1}{2}\displaystyle\int_0^{\pi/3} \sin^2 3\theta\, d\theta = \dfrac{1}{4}\displaystyle\int_0^{\pi/3} (1 - \cos 6\theta)\, d\theta$

$$= \dfrac{1}{4}\left(\theta - \dfrac{1}{6}\sin 6\theta\right)\Big|_0^{\pi/3} = \dfrac{\pi}{12} \text{ sq. units.}$$

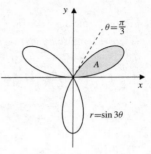

Fig. 8.6.4

6.   The circles $r = a$ and $r = 2a\cos\theta$ intersect at $\theta = \pm\pi/3$. By symmetry, the common area is $4 \times$ (area of sector $-$ area of right triangle) (see the figure), i.e.,

$$4 \times \left[\left(\dfrac{1}{6}\pi a^2\right) - \left(\dfrac{1}{2}\dfrac{a}{2}\dfrac{\sqrt{3}a}{2}\right)\right] = \dfrac{4\pi - 3\sqrt{3}}{6}a^2 \text{ sq. units.}$$

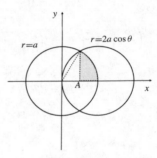

Fig. 8.6.6

8.   Area $= \dfrac{1}{2}\pi a^2 + 2 \times \dfrac{1}{2}\displaystyle\int_0^{\pi/2} a^2(1 - \sin\theta)^2\, d\theta$

$$= \dfrac{\pi a^2}{2} + a^2 \displaystyle\int_0^{\pi/2} \left(1 - 2\sin\theta + \dfrac{1 - \cos 2\theta}{2}\right) d\theta$$

$$= \dfrac{\pi a^2}{2} + a^2 \left(\dfrac{3}{2}\theta + 2\cos\theta - \dfrac{1}{4}\sin 2\theta\right)\Big|_0^{\pi/2}$$

$$= \left(\dfrac{5\pi}{4} - 2\right)a^2 \text{ sq. units.}$$

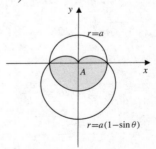

Fig. 8.6.8

10. Since $r^2 = 2\cos 2\theta$ meets $r = 1$ at $\theta = \pm\dfrac{\pi}{6}$ and $\pm\dfrac{5\pi}{6}$, the area inside the lemniscate and outside the circle is

$$4 \times \frac{1}{2}\int_0^{\pi/6}\Big[2\cos 2\theta - 1^2\Big]\,d\theta$$

$$= 2\sin 2\theta\,\Big|_0^{\pi/6} - \frac{\pi}{3} = \sqrt{3} - \frac{\pi}{3}\ \text{sq. units.}$$

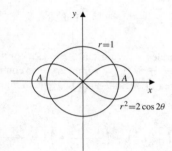

Fig. 8.6.10

12.
$$s = \int_0^{\pi}\sqrt{\left(\frac{dr}{d\theta}\right)^2 + r^2}\,d\theta = \int_0^{\pi}\sqrt{4\theta^2 + \theta^4}\,d\theta$$

$$= \int_0^{\pi}\theta\sqrt{4 + \theta^2}\,d\theta \quad \text{Let } u = 4 + \theta^2$$
$$du = 2\theta\,d\theta$$

$$= \frac{1}{2}\int_4^{4+\pi^2}\sqrt{u}\,du = \frac{1}{3}u^{3/2}\,\Big|_4^{4+\pi^2}$$

$$= \frac{1}{3}\Big[(4 + \pi^2)^{3/2} - 8\Big]\ \text{units.}$$

14.
$$s = \int_0^{2\pi}\sqrt{a^2 + a^2\theta^2}\,d\theta$$

$$= a\int_0^{2\pi}\sqrt{1 + \theta^2}\,d\theta \quad \text{Let } \theta = \tan u$$
$$d\theta = \sec^2 u\,du$$

$$= a\int_{\theta=0}^{\theta=2\pi}\sec^3 u\,du$$

$$= \frac{a}{2}\Big(\sec u\,\tan u + \ln|\sec u + \tan u|\Big)\Big|_{\theta=0}^{\theta=2\pi}$$

$$= \frac{a}{2}\Big[\theta\sqrt{1 + \theta^2} + \ln|\sqrt{1 + \theta^2} + \theta|\Big]\Big|_{\theta=0}^{\theta=2\pi}$$

$$= \frac{a}{2}\Big[2\pi\sqrt{1 + 4\pi^2} + \ln(2\pi + \sqrt{1 + 4\pi^2})\Big]\ \text{units.}$$

16. If $r^2 = \cos 2\theta$, then

$$2r\frac{dr}{d\theta} = -2\sin 2\theta \Rightarrow \frac{dr}{d\theta} = -\frac{\sin 2\theta}{\sqrt{\cos 2\theta}}$$

and

$$ds = \sqrt{\cos 2\theta + \frac{\sin^2 2\theta}{\cos 2\theta}}\,d\theta = \frac{d\theta}{\sqrt{\cos 2\theta}}.$$

a) Area of the surface generated by rotation about the x- axis is

$$S_x = 2\pi\int_0^{\pi/4} r\sin\theta\,ds$$

$$= 2\pi\int_0^{\pi/4}\sqrt{\cos 2\theta}\,\sin\theta\,\frac{d\theta}{\sqrt{\cos 2\theta}}$$

$$= -2\pi\cos\theta\,\Big|_0^{\pi/4} = (2 - \sqrt{2})\pi\ \text{sq. units.}$$

b) Area of the surface generated by rotation about the y- axis is

$$S_y = 2\pi\int_{-\pi/4}^{\pi/4} r\cos\theta\,ds$$

$$= 4\pi\int_0^{\pi/4}\sqrt{\cos 2\theta}\,\cos\theta\,\frac{d\theta}{\sqrt{\cos 2\theta}}$$

$$= 4\pi\sin\theta\,\Big|_0^{\pi/4} = 2\sqrt{2}\pi\ \text{sq. units.}$$

18. The two curves $r^2 = 2\sin 2\theta$ and $r = 2\cos\theta$ intersect where

$$2\sin 2\theta = 4\cos^2\theta$$
$$4\sin\theta\cos\theta = 4\cos^2\theta$$
$$(\sin\theta - \cos\theta)\cos\theta = 0$$
$$\Leftrightarrow \quad \sin\theta = \cos\theta \text{ or } \cos\theta = 0,$$

i.e., at $P_1 = \left[\sqrt{2}, \dfrac{\pi}{4}\right]$ and $P_2 = (0, 0)$.

For $r^2 = 2\sin 2\theta$ we have $2r\dfrac{dr}{d\theta} = 4\cos 2\theta$. At P_1 we have $r = \sqrt{2}$ and $dr/d\theta = 0$. Thus the angle ψ between the curve and the radial line $\theta = \pi/4$ is $\psi = \pi/2$.
For $r = 2\cos\theta$ we have $dr/d\theta = -2\sin\theta$, so the angle between this curve and the radial line $\theta = \pi/4$ satisfies $\tan\psi = \dfrac{r}{dr/d\theta}\Big|_{\theta=\pi/4} = -1$, and $\psi = 3\pi/4$. The two curves intersect at P_1 at angle $\dfrac{3\pi}{4} - \dfrac{\pi}{2} = \dfrac{\pi}{4}$.
The Figure shows that at the origin, P_2, the circle meets the lemniscate twice, at angles 0 and $\pi/2$.

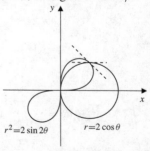

Fig. 8.6.18

20. We have $r = \cos\theta + \sin\theta$. For horizontal tangents:

$$0 = \frac{dy}{d\theta} = \frac{d}{d\theta}\left(\cos\theta\sin\theta + \sin^2\theta\right)$$

$$= \cos^2\theta - \sin^2\theta + 2\sin\theta\cos\theta$$

$$\Leftrightarrow \quad \cos 2\theta = -\sin 2\theta \quad \Leftrightarrow \quad \tan 2\theta = -1.$$

Thus $\theta = -\frac{\pi}{8}$ or $\frac{3\pi}{8}$. The tangents are horizontal at

$$\left[\cos\left(\frac{\pi}{8}\right) - \sin\left(\frac{\pi}{8}\right), -\frac{\pi}{8}\right] \text{ and}$$

$$\left[\cos\left(\frac{3\pi}{8}\right) + \sin\left(\frac{3\pi}{8}\right), \frac{3\pi}{8}\right].$$

For vertical tangent:

$$0 = \frac{dx}{d\theta} = \frac{d}{d\theta}\left(\cos^2\theta + \cos\theta\sin\theta\right)$$

$$= -2\cos\theta\sin\theta + \cos^2\theta - \sin^2\theta$$

$$\Leftrightarrow \quad \sin 2\theta = \cos 2\theta \quad \Leftrightarrow \quad \tan 2\theta = 1.$$

Thus $\theta = \pi/8$ of $5\pi/8$. There are vertical tangents at

$$\left[\cos\left(\frac{\pi}{8}\right) + \sin\left(\frac{\pi}{8}\right), \frac{\pi}{8}\right] \text{ and}$$

$$\left[\cos\left(\frac{5\pi}{8}\right) + \sin\left(\frac{5\pi}{8}\right), \frac{5\pi}{8}\right].$$

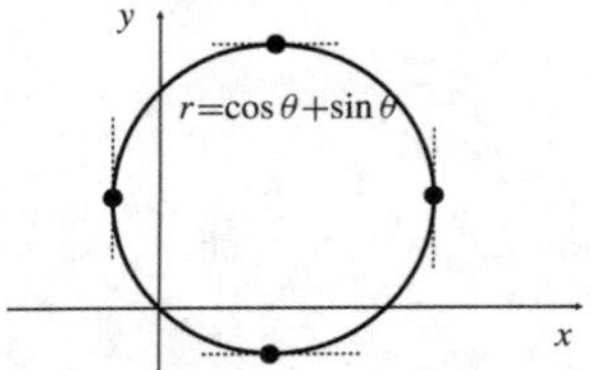

Fig. 8.6.20

22. We have $r^2 = \cos 2\theta$, and $2r\frac{dr}{d\theta} = -2\sin 2\theta$. For horizontal tangents:

$$0 = \frac{d}{d\theta}r\sin\theta = r\cos\theta + \sin\theta\left(-\frac{\sin 2\theta}{r}\right)$$

$$\Leftrightarrow \quad \cos 2\theta\cos\theta = \sin 2\theta\sin\theta$$

$$\Leftrightarrow \quad (\cos^2\theta - \sin^2\theta)\cos\theta = 2\sin^2\theta\cos\theta$$

$$\Leftrightarrow \quad \cos\theta = 0 \quad \text{or} \quad \cos^2\theta = 3\sin^2\theta.$$

There are no points on the curve where $\cos\theta = 0$. Therefore, horizontal tangents occur only where $\tan^2\theta = 1/3$. There are horizontal tangents at

$$\left[\frac{1}{\sqrt{2}}, \pm\frac{\pi}{6}\right] \text{ and } \left[\frac{1}{\sqrt{2}}, \pm\frac{5\pi}{6}\right].$$

For vertical tangents:

$$0 = \frac{d}{d\theta}r\cos\theta = -r\sin\theta + \cos\theta\left(-\frac{\sin 2\theta}{r}\right)$$

$$\Leftrightarrow \quad \cos 2\theta\sin\theta = -\sin 2\theta\cos\theta$$

$$\Leftrightarrow \quad (\cos^2\theta - \sin^2\theta)\sin\theta = -2\sin\theta\cos^2\theta$$

$$\Leftrightarrow \quad \sin\theta = 0 \quad \text{or} \quad 3\cos^2\theta = \sin^2\theta.$$

There are no points on the curve where $\tan^2\theta = 3$, so the only vertical tangents occur where $\sin\theta = 0$, that is, at the points with polar coordinates $[1, 0]$ and $[1, \pi]$.

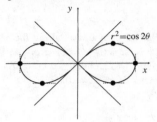

Fig. 8.6.22

24. We have $r = e^\theta$ and $\frac{dr}{d\theta} = e^\theta$. For horizontal tangents:

$$0 = \frac{d}{d\theta}r\sin\theta = e^\theta\cos\theta + e^\theta\sin\theta$$

$$\Leftrightarrow \quad \tan\theta = -1 \quad \Leftrightarrow \quad \theta = -\frac{\pi}{4} + k\pi,$$

where $k = 0, \pm 1, \pm 2, \ldots$. At the points $[e^{k\pi - \pi/4}, k\pi - \pi/4]$ the tangents are horizontal. For vertical tangents:

$$0 = \frac{d}{d\theta}r\cos\theta = e^\theta\cos\theta - e^\theta\sin\theta$$

$$\Leftrightarrow \quad \tan\theta = 1 \quad \leftrightarrow \quad \theta = \frac{\pi}{4} + k\pi.$$

At the points $[e^{k\pi + \pi/4}, k\pi + \pi/4]$ the tangents are vertical.

26. $x = r\cos\theta = f(\theta)\cos\theta$, $y = r\sin\theta = f(\theta)\sin\theta$.

$$\frac{dx}{d\theta} = f'(\theta)\cos\theta - f(\theta)\sin\theta, \quad \frac{dy}{d\theta} = f'(\theta)\sin\theta + f(\theta)\cos\theta$$

$$ds = \sqrt{\left(f'(\theta)\cos\theta - f(\theta)\sin\theta\right)^2 + \left(f'(\theta)\sin\theta + f(\theta)\cos\theta\right)^2}\, d\theta$$

$$= \left[\left(f'(\theta)\right)^2\cos^2\theta - 2f'(\theta)f(\theta)\cos\theta\sin\theta + \left(f(\theta)\right)^2\sin^2\theta\right.$$

$$\left. + \left(f'(\theta)\right)^2\sin^2\theta + 2f'(\theta)f(\theta)\sin\theta\cos\theta + \left(f(\theta)\right)^2\cos^2\theta\right]^{1/2} d\theta$$

$$= \sqrt{\left(f'(\theta)\right)^2 + \left(f(\theta)\right)^2}\, d\theta.$$

Review Exercises 8   (page 469)

2. $9x^2 - 4y^2 = 36 \quad \Leftrightarrow \quad \dfrac{x^2}{4} - \dfrac{y^2}{9} = 1$
Hyperbola, transverse axis along the x-axis.
Semi-transverse axis $a = 2$, semi-conjugate axis $b = 3$.
$c^2 = a^2 + b^2 = 13$. Foci: $(\pm\sqrt{13}, 0)$.
Asymptotes: $3x \pm 2y = 0$.

4. $2x^2 + 8y^2 = 4x - 48y$
$2(x^2 - 2x + 1) + 8(y^2 + 6y + 9) = 74$

$$\frac{(x-1)^2}{37} + \frac{(y+3)^2}{37/4} = 1.$$

Ellipse, centre $(1, -3)$, major axis along $y = -3$.
$a = \sqrt{37}$, $b = \sqrt{37}/2$, $c^2 = a^2 - b^2 = 111/4$.
Foci: $(1 \pm \sqrt{111}/2, -3)$.

6. $x = 2\sin(3t)$, $y = 2\cos(3t)$, $(0 \le t \le 2)$
Part of a circle of radius 2 centred at the origin from the
point $(0, 2)$ clockwise to $(2\sin 6, 2\cos 6)$.

8. $x = e^t$, $y = e^{-2t}$, $(-1 \le t \le 1)$.
Part of the curve $x^2 y = 1$ from $(1/e, e^2)$ to $(e, 1/e^2)$.

10. $x = \cos t + \sin t$, $y = \cos t - \sin t$, $(0 \le t \le 2\pi)$
The circle $x^2 + y^2 = 2$, traversed clockwise, starting and
ending at $(1, 1)$.

12. $\quad x = t^3 - 3t \qquad y = t^3 + 3t$
$\dfrac{dx}{dt} = 3(t^2 - 1) \qquad \dfrac{dy}{dt} = 3(t^2 + 1)$
Horizontal tangent: none.
Vertical tangent at $t = \pm 1$, i.e., at $(2, -4)$ and $(-2, 4)$.

Slope $\dfrac{dy}{dx} = \dfrac{t^2 + 1}{t^2 - 1} \quad \begin{cases} > 0 & \text{if } |t| > 1 \\ < 0 & \text{if } |t| < 1 \end{cases}$
Slope $\to 1$ as $t \to \pm\infty$.

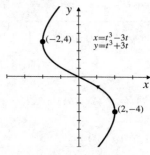

Fig. R-8.12

14. $\quad x = t^3 - 3t \qquad y = t^3 - 12t$
$\dfrac{dx}{dt} = 3(t^2 - 1) \qquad \dfrac{dy}{dt} = 3(t^2 - 4)$
Horizontal tangent at $t = \pm 2$, i.e., at $(2, -16)$ and
$(-2, 16)$.
Vertical tangent at $t = \pm 1$, i.e., at $(2, 11)$ and $(-2, -11)$.

Slope $\dfrac{dy}{dx} = \dfrac{t^2 - 4}{t^2 - 1} \quad \begin{cases} > 0 & \text{if } |t| > 2 \text{ or } |t| < 1 \\ < 0 & \text{if } 1 < |t| < 2 \end{cases}$
Slope $\to 1$ as $t \to \pm\infty$.

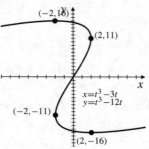

Fig. R-8.14

16. The volume of revolution about the y-axis is

$$V = \pi \int_{t=0}^{t=1} x^2 \, dy$$
$$= \pi \int_0^1 (t^6 - 2t^4 + t^2) 3t^2 \, dt$$
$$= 3\pi \int_0^1 (t^8 - 2t^6 + t^4) \, dt$$
$$= 3\pi \left(\frac{1}{9} - \frac{2}{7} + \frac{1}{5} \right) = \frac{8\pi}{105} \text{ cu. units.}$$

18. Area of revolution about the x-axis is

$$S = 2\pi \int 4e^{t/2}(e^t + 1) \, dt$$
$$= 8\pi \left(\frac{2}{3} e^{3t/2} + 2e^{t/2} \right) \Big|_0^2$$
$$= \frac{16\pi}{3} (e^3 + 3e - 4) \text{ sq. units.}$$

20. $r = |\theta|,\quad (-2\pi \le \theta \le 2\pi)$

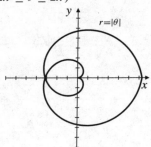

Fig. R-8.20

22. $r = 2 + \cos(2\theta)$

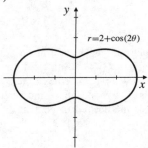

Fig. R-8.22

24. $r = 1 - \sin(3\theta)$

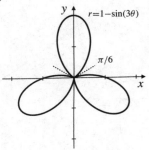

Fig. R-8.24

26. Area of a small loop:

$$A = 2 \times \frac{1}{2} \int_{\pi/3}^{\pi/2} (1 + 2\cos(2\theta))^2 \, d\theta$$

$$= \int_{\pi/3}^{\pi/2} [1 + 4\cos(2\theta) + 2(1 + \cos(4\theta))] \, d\theta$$

$$= \left(3\theta + 2\sin(2\theta) + \frac{1}{2}\sin(4\theta) \right) \Big|_{\pi/3}^{\pi/2}$$

$$= \frac{\pi}{2} - \frac{3\sqrt{3}}{4} \text{ sq. units.}$$

28. $r\cos\theta = x = 1/4$ and $r = 1 + \cos\theta$ intersect where

$$1 + \cos\theta = \frac{1}{4\cos\theta}$$

$$4\cos^2\theta + 4\cos\theta - 1 = 0$$

$$\cos\theta = \frac{-4 \pm \sqrt{16 + 16}}{8} = \frac{\pm\sqrt{2} - 1}{2}.$$

Only $(\sqrt{2} - 1)/2$ is between -1 and 1, so is a possible value of $\cos\theta$. Let $\theta_0 = \cos^{-1}\dfrac{\sqrt{2} - 1}{2}$. Then

$$\sin\theta_0 = \sqrt{1 - \left(\frac{\sqrt{2} - 1}{2} \right)^2} = \frac{\sqrt{1 + 2\sqrt{2}}}{2}.$$

By symmetry, the area inside $r = 1 + \cos\theta$ to the left of the line $x = 1/4$ is

$$A = 2 \times \frac{1}{2} \int_{\theta_0}^{\pi} \left(1 + 2\cos\theta + \frac{1 + \cos(2\theta)}{2} \right) d\theta + \cos\theta_0 \sin\theta_0$$

$$= \frac{3}{2}(\pi - \theta_0) + \left(2\sin\theta + \frac{1}{4}\sin(2\theta) \right) \Big|_{\theta_0}^{\pi}$$

$$+ \frac{(\sqrt{2} - 1)\sqrt{1 + 2\sqrt{2}}}{4}$$

$$= \frac{3}{2}\left(\pi - \cos^{-1}\frac{\sqrt{2} - 1}{2} \right) + \sqrt{1 + 2\sqrt{2}}\left(\frac{\sqrt{2} - 9}{8} \right) \text{ sq. units.}$$

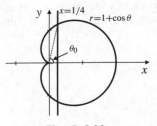

Fig. R-8.28

Challenging Problems 8   (page 469)

2. Let S_1 and S_2 be two spheres inscribed in the cylinder, one on each side of the plane that intersects the cylinder in the curve C that we are trying to show is an ellipse. Let the spheres be tangent to the cylinder around the circles C_1 and C_2, and suppose they are also tangent to the plane at the points F_1 and F_2, respectively, as shown in the figure.

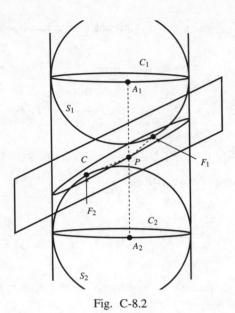

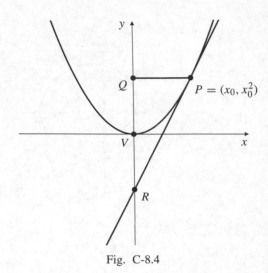

Fig. C-8.4

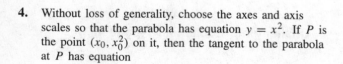

Fig. C-8.2

To construct the tangent at a given point P on a parabola with given vertex V and principal axis L, drop a perpendicular from P to L, meeting L at Q. Then find R on L on the side of V opposite Q and such that $QV = VR$. Then PR is the desired tangent.

Let P be any point on C. Let A_1A_2 be the line through P that lies on the cylinder, with A_1 on C_1 and A_2 on C_2. Then $PF_1 = PA_1$ because both lengths are of tangents drawn to the sphere S_1 from the same exterior point P. Similarly, $PF_2 = PA_2$. Hence

$$PF_1 + PF_2 = PA_1 + PA_2 = A_1A_2,$$

which is constant, the distance between the centres of the two spheres. Thus C must be an ellipse, with foci at F_1 and F_2.

6.

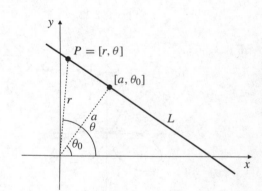

Fig. C-8.6

a) Let L be a line not passing through the origin, and let $[a, \theta_0]$ be the polar coordinates of the point on L that is closest to the origin. If $P = [r, \theta]$ is an point on the line, then, from the triangle in the figure,

$$\frac{a}{r} = \cos(\theta - \theta_0), \quad \text{or} \quad r = \frac{a}{\cos(\theta - \theta_0}$$

b) As shown in part (a), any line not passing t the origin has equation of the form

$$r = g(\theta) = \frac{a}{\cos(\theta - \theta_0)} = a\sec(\theta$$

4. Without loss of generality, choose the axes and axis scales so that the parabola has equation $y = x^2$. If P is the point (x_0, x_0^2) on it, then the tangent to the parabola at P has equation

$$y = x_0^2 + 2x_0(x - x_0),$$

which intersects the principal axis $x = 0$ at $(0, -x_0^2)$. Thus $R = (0, -x_0^2)$ and $Q = (0, x_0^2)$. Evidently the vertex $V = (0, 0)$ bisects RQ.

for some constants a and θ_0. We have

$$g'(\theta) = a\sec(\theta - \theta_0)\tan(\theta - \theta_0)$$

$$g''(\theta) = a\sec(\theta - \theta_0)\tan^2(\theta - \theta_0)$$
$$+ a\sec^3(\theta - \theta_0)\Big(g(\theta)\Big)^2 + 2\Big(g'(\theta)\Big)^2 - g(\theta)g''(\theta)$$
$$= a^2\sec^2(\theta - \theta_0) + 2a^2\sec^2(\theta - \theta_0)\tan^2(\theta - \theta_0)$$
$$- a^2\sec^2(\theta - \theta_0)\tan^2(\theta - \theta_0) - a^2\sec^4(\theta - \theta_0)$$
$$= a^2\Big[\sec^2(\theta - \theta_0)\Big(1 + \tan^2(\theta - \theta_0)\Big) - \sec^4(\theta - \theta_0)\Big]$$
$$= 0.$$

c) If $r = g(\theta)$ is the polar equation of the tangent to $r = f(\theta)$ at $\theta = \alpha$, then $g(\alpha) = f(\alpha)$ and $g'(\alpha) = f'(\alpha)$. Suppose that

$$\Big(f(\alpha)\Big)^2 + 2\Big(f'(\alpha)\Big)^2 - f(\alpha)f''(\alpha) > 0.$$

By part (b) we have

$$\Big(g(\alpha)\Big)^2 + 2\Big(g'(\alpha)\Big)^2 - g(\alpha)g''(\alpha) = 0.$$

Subtracting, and using $g(\alpha) = f(\alpha)$ and $g'(\alpha) = f'(\alpha)$, we get $f''(\alpha) < g''(\alpha)$. It follows that $f(\theta) < g(\theta)$ for values of θ near α; that is, the graph of $r = f(\theta)$ is curving to the origin side of its tangent at α. Similarly, if

$$\Big(f(\alpha)\Big)^2 + 2\Big(f'(\alpha)\Big)^2 - f(\alpha)f''(\alpha) < 0,$$

then the graph is curving to the opposite side of the tangent, away from the origin.

8. Take the origin at station O as shown in the figure. Both of the lines L_1 and L_2 pass at distance $100\cos\epsilon$ from the origin. Therefore, by Problem 6(a), their equations are

$$L_1: \quad r = \frac{100\cos\epsilon}{\cos\left[\theta - \left(\frac{\pi}{2} - \epsilon\right)\right]} = \frac{100\cos\epsilon}{\sin(\theta + \epsilon)}$$

$$L_2: \quad r = \frac{100\cos\epsilon}{\cos\left[\theta - \left(\frac{\pi}{2} + \epsilon\right)\right]} = \frac{100\cos\epsilon}{\sin(\theta - \epsilon)}.$$

The search area $A(\epsilon)$ is, therefore,

$$A(\epsilon) = \frac{1}{2}\int_{\frac{\pi}{4}-\epsilon}^{\frac{\pi}{4}+\epsilon}\left(\frac{100^2\cos^2\epsilon}{\sin^2(\theta - \epsilon)} - \frac{100^2\cos^2\epsilon}{\sin^2(\theta + \epsilon)}\right)d\theta$$

$$= 5,000\cos^2\epsilon\int_{\frac{\pi}{4}-\epsilon}^{\frac{\pi}{4}+\epsilon}\Big(\csc^2(\theta - \epsilon) - \csc^2(\theta + \epsilon)\Big)d\theta$$

$$= 5,000\cos^2\epsilon\left[\cot\left(\frac{\pi}{4} + 2\epsilon\right) - 2\cot\frac{\pi}{4} + \cot\left(\frac{\pi}{4} - 2\epsilon\right)\right]$$

$$= 5,000\cos^2\epsilon\left[\frac{\cos\left(\frac{\pi}{4} + 2\epsilon\right)}{\sin\left(\frac{\pi}{4} + 2\epsilon\right)} + \frac{\sin\left(\frac{\pi}{4} + 2\epsilon\right)}{\cos\left(\frac{\pi}{4} + 2\epsilon\right)} - 2\right]$$

$$= 10,000\cos^2\epsilon\left[\csc\left(\frac{\pi}{2} + 4\epsilon\right) - 1\right]$$

$$= 10,000\cos^2\epsilon(\sec(4\epsilon) - 1)\ \text{mi}^2.$$

For $\epsilon = 3° = \pi/60$, we have $A(\epsilon) \approx 222.8$ square miles. Also

$$A'(\epsilon) = -20,000\cos\epsilon\sin\epsilon(\sec(4\epsilon) - 1)$$
$$+ 40,000\cos^2\epsilon\sec(4\epsilon)\tan(4\epsilon)$$
$$A'(\pi/60) \approx 8645.$$

When $\epsilon = 3°$, the search area increases at about $8645(\pi/180) \approx 151$ square miles per degree increase in ϵ.

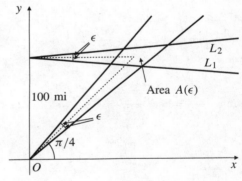

Fig. C-8.8

CHAPTER 9. SEQUENCES, SERIES, AND POWER SERIES

Section 9.1 Sequences and Convergence (page 478)

2. $\left\{\dfrac{2n}{n^2+1}\right\} = \left\{1, \dfrac{4}{5}, \dfrac{3}{5}, \dfrac{8}{17}, \dots\right\}$ is bounded, positive, decreasing, and converges to 0.

4. $\left\{\sin\dfrac{1}{n}\right\} = \left\{\sin 1, \sin\left(\dfrac{1}{2}\right), \sin\left(\dfrac{1}{3}\right), \dots\right\}$ is bounded, positive, decreasing, and converges to 0.

6. $\left\{\dfrac{e^n}{\pi^n}\right\} = \left\{\dfrac{e}{\pi}, \left(\dfrac{e}{\pi}\right)^2, \left(\dfrac{e}{\pi}\right)^3, \dots\right\}$ is bounded, positive, decreasing, and converges to 0, since $e < \pi$.

8. $\left\{\dfrac{(-1)^n n}{e^n}\right\} = \left\{\dfrac{-1}{e}, \dfrac{2}{e^2}, \dfrac{-3}{e^3}, \dots\right\}$ is bounded, alternating, and converges to 0.

10. $\dfrac{(n!)^2}{(2n)!} = \dfrac{1}{n+1}\dfrac{2}{n+2}\dfrac{3}{n+3}\cdots\dfrac{n}{2n} \le \left(\dfrac{1}{2}\right)^n$.

Also, $\dfrac{a_{n+1}}{a_n} = \dfrac{(n+1)^2}{(2n+2)(2n+1)} < \dfrac{1}{2}$. Thus the sequence $\left\{\dfrac{(n!)^2}{(2n)!}\right\}$ is positive, decreasing, bounded, and convergent to 0.

12. $\left\{\dfrac{\sin n}{n}\right\} = \left\{\sin 1, \dfrac{\sin 2}{2}, \dfrac{\sin 3}{3}, \dots\right\}$ is bounded and converges to 0.

14. $\lim\dfrac{5-2n}{3n-7} = \lim\dfrac{\dfrac{5}{n}-2}{3-\dfrac{7}{n}} = -\dfrac{2}{3}$.

16. $\lim\dfrac{n^2}{n^3+1} = \lim\dfrac{\dfrac{1}{n}}{1+\dfrac{1}{n^3}} = 0$.

18. $\lim\dfrac{n^2-2\sqrt{n}+1}{1-n-3n^2} = \lim\dfrac{1-\dfrac{2}{n\sqrt{n}}+\dfrac{1}{n^2}}{\dfrac{1}{n^2}-\dfrac{1}{n}-3} = -\dfrac{1}{3}$.

20. $\lim n\sin\dfrac{1}{n} = \lim_{x\to 0+}\dfrac{\sin x}{x} = \lim_{x\to 0+}\dfrac{\cos x}{1} = 1$.

22. $\lim\dfrac{n}{\ln(n+1)} = \lim_{x\to\infty}\dfrac{x}{\ln(x+1)}$

$= \lim_{x\to\infty}\dfrac{1}{\left(\dfrac{1}{x+1}\right)} = \lim_{x\to\infty} x+1 = \infty$.

24. $\lim\left(n - \sqrt{n^2-4n}\right) = \lim\dfrac{n^2-(n^2-4n)}{n+\sqrt{n^2-4n}}$

$= \lim\dfrac{4n}{n+\sqrt{n^2-4n}} = \lim\dfrac{4}{1+\sqrt{1-\dfrac{4}{n}}} = 2$.

26. If $a_n = \left(\dfrac{n-1}{n+1}\right)^n$, then

$\lim a_n = \lim\left(\dfrac{n-1}{n}\right)^n\left(\dfrac{n}{n+1}\right)^n$

$= \lim\left(1-\dfrac{1}{n}\right)^n \Big/ \lim\left(1+\dfrac{1}{n}\right)^n$

$= \dfrac{e^{-1}}{e} = e^{-2}$ (by Theorem 6 of Section 3.4).

28. We have $\lim\dfrac{n^2}{2^n} = 0$ since 2^n grows much faster than n^2 and $\lim\dfrac{4^n}{n!} = 0$ by Theorem 3(b). Hence,

$\lim\dfrac{n^2 2^n}{n!} = \lim\dfrac{n^2}{2^n}\cdot\dfrac{2^{2n}}{n!} = \left(\lim\dfrac{n^2}{2^n}\right)\left(\lim\dfrac{4^n}{n!}\right) = 0$.

30. Let $a_1 = 1$ and $a_{n+1} = \sqrt{1+2a_n}$ for $n = 1, 2, 3, \dots$. Then we have $a_2 = \sqrt{3} > a_1$. If $a_{k+1} > a_k$ for some k, then

$a_{k+2} = \sqrt{1+2a_{k+1}} > \sqrt{1+2a_k} = a_{k+1}$.

Thus, $\{a_n\}$ is increasing by induction. Observe that $a_1 < 3$ and $a_2 < 3$. If $a_k < 3$ then

$a_{k+1} = \sqrt{1+2a_k} < \sqrt{1+2(3)} = \sqrt{7} < \sqrt{9} = 3$.

Therefore, $a_n < 3$ for all n, by induction. Since $\{a_n\}$ is increasing and bounded above, it converges. Let $\lim a_n = a$. Then

$a = \sqrt{1+2a} \Rightarrow a^2 - 2a - 1 = 0 \Rightarrow a = 1 \pm \sqrt{2}$.

Since $a = 1 - \sqrt{2} < 0$, it is not appropriate. Hence, must have $\lim a_n = 1 + \sqrt{2}$.

32. Let $a_n = \left(1+\dfrac{1}{n}\right)^n$ so $\ln a_n = n\ln\left(1+\dfrac{1}{n}\right)$.

a) If $f(x) = x \ln\left(1 + \dfrac{1}{x}\right) = x \ln(x + 1) - x \ln x$, then

$$f'(x) = \ln(x + 1) + \frac{x}{x + 1} - \ln x - 1$$

$$= \ln\left(\frac{x + 1}{x}\right) - \frac{1}{x + 1}$$

$$= \int_x^{x+1} \frac{dt}{t} - \frac{1}{x + 1}$$

$$> \frac{1}{x + 1} \int_x^{x+1} dt - \frac{1}{x + 1}$$

$$= \frac{1}{x + 1} - \frac{1}{x + 1} = 0.$$

Since $f'(x) > 0$, $f(x)$ must be an increasing function. Thus, $\{a_n\} = \{e^{f(x_n)}\}$ is increasing.

b) Since $\ln x \le x - 1$,

$$\ln a_k = k \ln\left(1 + \frac{1}{k}\right) \le k\left(1 + \frac{1}{k} - 1\right) = 1$$

which implies that $a_k \le e$ for all k. Since $\{a_n\}$ is increasing, e is an upper bound for $\{a_n\}$.

34. If $\{|a_n|\}$ is bounded then it is bounded above, and there exists a constant K such that $|a_n| \le K$ for all n. Therefore, $-K \le a_n \le K$ for all n, and so $\{a_n\}$ is bounded above and below, and is therefore bounded.

36. a) "If $\lim a_n = \infty$ and $\lim b_n = L > 0$, then $\lim a_n b_n = \infty$" is TRUE. Let R be an arbitrary, large positive number. Since $\lim a_n = \infty$, and $L > 0$, it must be true that $a_n \ge \dfrac{2R}{L}$ for n sufficiently large. Since $\lim b_n = L$, it must also be that $b_n \ge \dfrac{L}{2}$ for n sufficiently large. Therefore $a_n b_n \ge \dfrac{2R}{L}\dfrac{L}{2} = R$ for n sufficiently large. Since R is arbitrary, $\lim a_n b_n = \infty$.

b) "If $\lim a_n = \infty$ and $\lim b_n = -\infty$, then $\lim(a_n + b_n) = 0$" is FALSE. Let $a_n = 1 + n$ and $b_n = -n$; then $\lim a_n = \infty$ and $\lim b_n = -\infty$ but $\lim(a_n + b_n) = 1$.

c) "If $\lim a_n = \infty$ and $\lim b_n = -\infty$, then $\lim a_n b_n = -\infty$" is TRUE. Let R be an arbitrary, large positive number. Since $\lim a_n = \infty$ and $\lim b_n = -\infty$, we must have $a_n \ge \sqrt{R}$ and $b_n \le -\sqrt{R}$, for all sufficiently large n. Thus $a_n b_n \le -R$, and $\lim a_n b_n = -\infty$.

d) "If neither $\{a_n\}$ nor $\{b_n\}$ converges, then $\{a_n b_n\}$ does not converge" is FALSE. Let $a_n = b_n = (-1)^n$; then $\lim a_n$ and $\lim b_n$ both diverge. But $a_n b_n = (-1)^{2n} = 1$ and $\{a_n b_n\}$ does converge (to 1).

e) "If $\{|a_n|\}$ converges, then $\{a_n\}$ converges" is FALSE. Let $a_n = (-1)^n$. Then $\lim_{n \to \infty} |a_n| = \lim_{n \to \infty} 1 = 1$, but $\lim_{n \to \infty} a_n$ does not exist.

Section 9.2   Infinite Series   (page 484)

2. $3 - \dfrac{3}{4} + \dfrac{3}{16} - \dfrac{3}{64} + \cdots = \displaystyle\sum_{n=1}^{\infty} 3\left(-\frac{1}{4}\right)^{n-1} = \dfrac{3}{1 + \frac{1}{4}} = \dfrac{12}{5}.$

4. $\displaystyle\sum_{n=0}^{\infty} \frac{5}{10^{3n}} = 5\left[1 + \frac{1}{1000} + \left(\frac{1}{1000}\right)^2 + \cdots\right]$

$$= \frac{5}{1 - \dfrac{1}{1000}} = \frac{5000}{999}.$$

6. $\displaystyle\sum_{n=0}^{\infty} \frac{1}{e^n} = 1 + \frac{1}{e} + \left(\frac{1}{e}\right)^2 + \cdots = \frac{1}{1 - \dfrac{1}{e}} = \frac{e}{e - 1}.$

8. $\sum_{j=1}^{\infty} \pi^{j/2} \cos(j\pi) = \sum_{j=2}^{\infty} (-1)^j \pi^{j/2}$ diverges because $\lim_{j \to \infty} (-1)^j \pi^{j/2}$ does not exist.

10. $\displaystyle\sum_{n=0}^{\infty} \frac{3 + 2^n}{3^{n+2}} = \frac{1}{3} \sum_{n=0}^{\infty} \left(\frac{1}{3}\right)^n + \frac{1}{9} \sum_{n=0}^{\infty} \left(\frac{2}{3}\right)^n$

$$= \frac{1}{3} \cdot \frac{1}{1 - \dfrac{1}{3}} + \frac{1}{9} \cdot \frac{1}{1 - \dfrac{2}{3}} = \frac{1}{2} + \frac{1}{3} = \frac{5}{6}.$$

12. Let

$$\sum_{n=1}^{\infty} \frac{1}{(2n - 1)(2n + 1)} = \frac{1}{1 \times 3} + \frac{1}{3 \times 5} + \frac{1}{5 \times 7} + \cdots.$$

Since $\dfrac{1}{(2n - 1)(2n + 1)} = \dfrac{1}{2}\left(\dfrac{1}{2n - 1} - \dfrac{1}{2n + 1}\right)$, the partial sum is

$$s_n = \frac{1}{2}\left(1 - \frac{1}{3}\right) + \frac{1}{2}\left(\frac{1}{3} - \frac{1}{5}\right) + \cdots$$

$$+ \frac{1}{2}\left(\frac{1}{2n - 3} - \frac{1}{2n - 1}\right) + \frac{1}{2}\left(\frac{1}{2n - 1} - \frac{1}{2n + 1}\right)$$

$$= \frac{1}{2}\left(1 - \frac{1}{2n + 1}\right).$$

Hence,

$$\sum_{n=1}^{\infty} \frac{1}{(2n - 1)(2n + 1)} = \lim s_n = \frac{1}{2}.$$

14. Since

$$\frac{1}{n(n + 1)(n + 2)} = \frac{1}{2}\left[\frac{1}{n} - \frac{2}{n + 1} + \frac{1}{n + 2}\right],$$

the partial sum is

$$s_n = \frac{1}{2}\left(1 - \frac{2}{2} + \frac{1}{3}\right) + \frac{1}{2}\left(\frac{1}{2} - \frac{2}{3} + \frac{1}{4}\right) + \cdots$$
$$+ \frac{1}{2}\left(\frac{1}{n-1} - \frac{2}{n} + \frac{1}{n+1}\right) + \frac{1}{2}\left(\frac{1}{n} - \frac{2}{n+1} + \frac{1}{n+2}\right)$$
$$= \frac{1}{2}\left(\frac{1}{2} - \frac{1}{n+1} + \frac{1}{n+2}\right).$$

Hence,

$$\sum_{n=1}^{\infty} \frac{1}{n(n+1)(n+2)} = \lim s_n = \frac{1}{4}.$$

16. $\displaystyle\sum_{n=1}^{\infty} \frac{n}{n+2}$ diverges to infinity since $\lim \dfrac{n}{n+2} = 1 > 0$.

18. $\displaystyle\sum_{n=1}^{\infty} \frac{2}{n+1} = 2\left(\frac{1}{2} + \frac{1}{3} + \frac{1}{4} + \cdots\right)$ diverges to infinity since it is just twice the harmonic series with the first term omitted.

20. Since $1 + 2 + 3 + \cdots + n = \dfrac{n(n+1)}{2}$, the given series is $\sum_{n=1}^{\infty} \dfrac{2}{n(n+1)}$ which converges to 2 by the result of Example 3 of this section.

22. The balance at the end of 8 years is

$$s_n = 1000\left[(1.1)^8 + (1.1)^7 + \cdots + (1.1)^2 + (1.1)\right]$$
$$= 1000(1.1)\left(\frac{(1.1)^8 - 1}{1.1 - 1}\right) \approx \$12,579.48.$$

24. If $\{a_n\}$ is ultimately positive, then the sequence $\{s_n\}$ of partial sums of the series must be ultimately increasing. By Theorem 2, if $\{s_n\}$ is ultimately increasing, then either it is bounded above, and therefore convergent, or else it is not bounded above and diverges to infinity. Since $\sum a_n = \lim s_n$, $\sum a_n$ must either converge when $\{s_n\}$ converges and $\lim s_n = s$ exists, or diverge to infinity when $\{s_n\}$ diverges to infinity.

26. "If $a_n = 0$ for every n, then $\sum a_n$ converge" is TRUE because $s_n = \sum_{k=0}^{n} 0 = 0$, for every n, and so $\sum a_n = \lim s_n = 0$.

28. "If $\sum a_n$ and $\sum b_n$ both diverge, then so does $\sum(a_n + b_n)$" is FALSE. Let $a_n = \dfrac{1}{n}$ and $b_n = -\dfrac{1}{n}$, then $\sum a_n = \infty$ and $\sum b_n = -\infty$ but $\sum(a_n + b_n) = \sum(0) = 0$.

30. "If $\sum a_n$ diverges and $\{b_n\}$ is bounded, then $\sum a_n b_n$ diverges" is FALSE. Let $a_n = \dfrac{1}{n}$ and $b_n = \dfrac{1}{n+1}$. Then $\sum a_n = \infty$ and $0 \le b_n \le 1/2$. But $\sum a_n b_n = \sum \dfrac{1}{n(n+1)}$ which converges by Example 3.

Section 9.3 Convergence Tests for Positive Series (page 494)

2. $\displaystyle\sum_{n=1}^{\infty} \frac{n}{n^4 - 2}$ converges by comparison with $\displaystyle\sum_{n=1}^{\infty} \frac{1}{n^3}$ since

$$\lim \frac{\left(\dfrac{n}{n^4 - 2}\right)}{\left(\dfrac{1}{n^3}\right)} = 1, \quad \text{and} \quad 0 < 1 < \infty.$$

4. $\displaystyle\sum_{n=1}^{\infty} \frac{\sqrt{n}}{n^2 + n + 1}$ converges by comparison with $\displaystyle\sum_{n=1}^{\infty} \frac{1}{n^{3/2}}$ since

$$\lim \frac{\left(\dfrac{\sqrt{n}}{n^2 + n + 1}\right)}{\left(\dfrac{1}{n^{3/2}}\right)} = 1, \quad \text{and} \quad 0 < 1 < \infty.$$

6. $\displaystyle\sum_{n=8}^{\infty} \frac{1}{\pi^n + 5}$ converges by comparison with the geometric series $\displaystyle\sum_{n=8}^{\infty} \left(\frac{1}{\pi}\right)^n$ since $0 < \dfrac{1}{\pi^n + 5} < \dfrac{1}{\pi^n}$.

8. $\displaystyle\sum_{n=1}^{\infty} \frac{1}{\ln(3n)}$ diverges to infinity by comparison with the harmonic series $\displaystyle\sum_{n=1}^{\infty} \frac{1}{3n}$ since $\dfrac{1}{\ln(3n)} > \dfrac{1}{3n}$ for $n \ge 1$.

10. $\displaystyle\sum_{n=0}^{\infty} \frac{1+n}{2+n}$ diverges to infinity since $\lim \dfrac{1+n}{2+n} = 1 > 0$.

12. $\displaystyle\sum_{n=1}^{\infty} \frac{n^2}{1 + n\sqrt{n}}$ diverges to infinity since

$$\lim \frac{n^2}{1 + n\sqrt{n}} = \infty.$$

14. $\displaystyle\sum_{n=2}^{\infty} \frac{1}{n \ln n (\ln \ln n)^2}$ converges by the integral test:

$$\int_a^{\infty} \frac{dt}{t \ln t (\ln \ln t)^2} = \int_{\ln \ln a}^{\infty} \frac{du}{u^2} < \infty \quad \text{if} \quad \ln \ln a > 0.$$

16. The series

$$\sum_{n=1}^{\infty} \frac{1 + (-1)^n}{\sqrt{n}} = 0 + \frac{2}{\sqrt{2}} + 0 + \frac{2}{\sqrt{4}} + 0 + \frac{2}{\sqrt{6}} + \cdots$$

$$= 2 \sum_{k=1}^{\infty} \frac{1}{\sqrt{2k}} = \sqrt{2} \sum_{k=1}^{\infty} \frac{1}{\sqrt{k}}$$

diverges to infinity.

18. $\sum_{n=1}^{\infty} \frac{n^4}{n!}$ converges by the ratio test since

$$\lim \frac{\dfrac{(n+1)^4}{(n+1)!}}{\dfrac{n^4}{n!}} = \lim \left(\frac{n+1}{n}\right)^4 \frac{1}{n+1} = 0.$$

20. $\sum_{n=1}^{\infty} \frac{(2n)! 6^n}{(3n)!}$ converges by the ratio test since

$$\lim \frac{(2n+2)! 6^{n+1}}{(3n+3)!} \bigg/ \frac{(2n)! 6^n}{(3n)!}$$

$$= \lim \frac{(2n+2)(2n+1)6}{(3n+3)(3n+2)(3n+1)} = 0.$$

22. $\sum_{n=0}^{\infty} \frac{n^{100} 2^n}{\sqrt{n!}}$ converges by the ratio test since

$$\lim \frac{(n+1)^{100} 2^{n+1}}{\sqrt{(n+1)!}} \bigg/ \frac{n^{100} 2^n}{\sqrt{n!}}$$

$$= \lim 2 \left(\frac{n+1}{n}\right)^{100} \frac{1}{\sqrt{n+1}} = 0.$$

24. $\sum_{n=1}^{\infty} \frac{1 + n!}{(1+n)!}$ diverges by comparison with the harmonic

series $\sum_{n=1}^{\infty} \frac{1}{n+1}$ since $\frac{1+n!}{(1+n)!} > \frac{n!}{(1+n)!} = \frac{1}{n+1}$.

26. $\sum_{n=1}^{\infty} \frac{n^n}{\pi^n n!}$ converges by the ratio test since

$$\lim \frac{(n+1)^{n+1}}{\pi^{(n+1)}(n+1)!} \bigg/ \frac{n^n}{\pi^n n!} = \frac{1}{\pi} \lim \left(1 + \frac{1}{n}\right)^n = \frac{e}{\pi} < 1.$$

Since $f(x) = \frac{1}{x^3}$ is positive, continuous and decreasing on $[1, \infty)$, for any $n = 1, 2, 3, \ldots$, we have

$$s_n + A_{n+1} \le s \le s_n + A_n$$

where $s_n = \sum_{k=1}^{n} \frac{1}{k^3}$ and $A_n = \int_n^{\infty} \frac{dx}{x^3} = \frac{1}{2n^2}$. If $s_n^* = s_n + \frac{1}{2}(A_{n+1} + A_n)$, then

$$|s_n - s_n^*| \le \frac{A_n - A_{n+1}}{2} = \frac{1}{4}\left[\frac{1}{n^2} - \frac{1}{(n+1)^2}\right]$$

$$= \frac{1}{4} \frac{2n+1}{n^2(n+1)^2} < 0.001$$

if $n = 8$. Thus, the error in the approximation $s \approx s_8^*$ is less than 0.001.

30. Again, we have $s_n + A_{n+1} \le s \le s_n + A_n$ where $s_n = \sum_{k=1}^{n} \frac{1}{k^2 + 4}$ and

$$A_n = \int_n^{\infty} \frac{dx}{x^2 + 4} = \frac{1}{2} \tan^{-1}\left(\frac{x}{2}\right) \bigg|_n^{\infty} = \frac{\pi}{4} - \frac{1}{2} \tan^{-1}\left(\frac{n}{2}\right).$$

If $s_n^* = s_n + \frac{1}{2}(A_{n+1} + A_n)$, then

$$|s_n - s_n^*| \le \frac{A_n - A_{n+1}}{2}$$

$$= \frac{1}{2}\left[\frac{\pi}{4} - \frac{1}{2}\tan^{-1}\left(\frac{n}{2}\right) - \frac{\pi}{4} + \frac{1}{2}\tan^{-1}\left(\frac{n+1}{2}\right)\right]$$

$$= \frac{1}{4}\left[\tan^{-1}\left(\frac{n+1}{2}\right) - \tan^{-1}\left(\frac{n}{2}\right)\right] = \frac{1}{4}(a - b),$$

where $a = \tan^{-1}\left(\frac{n+1}{2}\right)$ and $b = \tan^{-1}\left(\frac{n}{2}\right)$. Now

$$\tan(a - b) = \frac{\tan a - \tan b}{1 + \tan a \tan b}$$

$$= \frac{\left(\frac{n+1}{2}\right) - \left(\frac{n}{2}\right)}{1 + \left(\frac{n+1}{2}\right)\left(\frac{n}{2}\right)}$$

$$= \frac{2}{n^2 + n + 4}$$

$$\Leftrightarrow \quad a - b = \tan^{-1}\left(\frac{2}{n^2 + n + 4}\right).$$

We want error less than 0.001:

$$\frac{1}{4}(a - b) = \frac{1}{4} \tan^{-1}\left(\frac{2}{n^2 + n + 4}\right) < 0.001$$

$$\Leftrightarrow \quad \frac{2}{n^2 + n + 4} < \tan 0.004$$

$$\Leftrightarrow \quad n^2 + n > 2 \cot(0.004) - 4 \approx 496.$$

$n = 22$ will do. The approximation $s \approx s_{22}^*$ has error less than 0.001.

32. We have $s = \sum_{k=1}^{\infty} \frac{1}{(2k-1)!}$ and

$$s_n = \sum_{k=1}^{n} \frac{1}{(2k-1)!} = \frac{1}{1!} + \frac{1}{3!} + \frac{1}{5!} + \cdots + \frac{1}{(2n-1)!}.$$

Then

$$0 < s - s_n = \frac{1}{(2n+1)!} + \frac{1}{(2n+3)!} + \frac{1}{(2n+5)!} + \cdots$$

$$= \frac{1}{(2n+1)!}\left[1 + \frac{1}{(2n+2)(2n+3)} + \right.$$

$$\left. \frac{1}{(2n+2)(2n+3)(2n+4)(2n+5)} + \cdots\right]$$

$$< \frac{1}{(2n+1)!}\left[1 + \frac{1}{(2n+2)(2n+3)} + \right.$$

$$\left. \frac{1}{[(2n+2)(2n+3)]^2} + \cdots\right]$$

$$= \frac{1}{(2n+1)!}\left[\frac{1}{1 - \dfrac{1}{(2n+2)(2n+3)}}\right]$$

$$= \frac{1}{(2n+1)!} \frac{4n^2 + 10n + 6}{4n^2 + 10n + 5} < 0.001$$

if $n = 3$. Thus, $s \approx s_3 = 1 + \frac{1}{3!} + \frac{1}{5!} = 1.175$ with error less than 0.001.

34. We have $s = \sum_{k=1}^{\infty} \frac{1}{k^k}$ and

$$s_n = \sum_{k=1}^{n} \frac{1}{k^k} = \frac{1}{1} + \frac{1}{2^2} + \frac{1}{3^3} + \cdots + \frac{1}{n^n}.$$

Then

$$0 < s - s_n = \frac{1}{(n+1)^{n+1}} + \frac{1}{(n+2)^{n+2}} + \frac{1}{(n+3)^{n+3}} + \cdots$$

$$< \frac{1}{(n+1)^{n+1}}\left[1 + \frac{1}{n+1} + \frac{1}{(n+1)^2} + \cdots\right]$$

$$= \frac{1}{(n+1)^{n+1}}\left[\frac{1}{1 - \dfrac{1}{n+1}}\right]$$

$$= \frac{1}{n(n+1)^n} < 0.001$$

if $n = 4$. Thus, $s \approx s_4 = 1 + \frac{1}{2^2} + \frac{1}{3^3} + \frac{1}{4^4} = 1.291$ with error less than 0.001.

36. Let $u = \ln \ln t$, $du = \frac{dt}{t \ln t}$ and $\ln \ln a > 0$; then

$$\int_a^{\infty} \frac{dt}{t \ln t (\ln \ln t)^p} = \int_{\ln \ln a}^{\infty} \frac{du}{u^p}$$

will converge if and only if $p > 1$. Thus, $\sum_{n=3}^{\infty} \frac{1}{n \ln n (\ln \ln n)^p}$ will converge if and only if $p > 1$. Similarly,

$$\sum_{n=N}^{\infty} \frac{1}{n(\ln n)(\ln \ln n) \cdots (\ln_j n)(\ln_{j+1} n)^p}$$

converges if and only if $p > 1$, where N is large enough that $\ln_j N > 1$.

38. Let $a_n = 2^{n+1}/n^n$. Then

$$\lim_{n \to \infty} \sqrt[n]{a_n} = \lim_{n \to \infty} \frac{2 \times 2^{1/n}}{n} = 0.$$

Since this limit is less than 1, $\sum_{n=1}^{\infty} a_n$ converges by the root test.

40. Let $a_n = \frac{2^{n+1}}{n^n}$. Then

$$\frac{a_{n+1}}{a_n} = \frac{2^{n+2}}{(n+1)^{n+1}} \cdot \frac{n^n}{2^{n+1}}$$

$$= \frac{2}{(n+1)\left(\dfrac{n}{n+1}\right)^n} = \frac{2}{n+1} \cdot \frac{1}{\left(1 + \dfrac{1}{n}\right)^n}$$

$$\to 0 \times \frac{1}{e} = 0 \text{ as } n \to \infty.$$

Thus $\sum_{n=1}^{\infty} a_n$ converges by the ratio test.
(Remark: the question contained a typo. It was intended to ask that #33 be repeated, using the ratio test. That is a little harder.)

42. We have

$$a_n = \frac{(2n)!}{2^{2n}(n!)^2} = \frac{1 \times 2 \times 3 \times 4 \times \cdots \times 2n}{(2 \times 4 \times 6 \times 8 \times \cdots \times 2n)^2}$$

$$= \frac{1 \times 3 \times 5 \times \cdots \times (2n-1)}{2 \times 4 \times 6 \times \cdots \times (2n-2) \times 2n}$$

$$= 1 \times \frac{3}{2} \times \frac{5}{4} \times \frac{7}{6} \times \cdots \times \frac{2n-1}{2n-2} \times \frac{1}{2n} > \frac{1}{2n}$$

Therefore $\sum_{n=1}^{\infty} \frac{(2n)!}{2^{2n}(n!)^2}$ diverges to infinity by com

with the harmonic series $\sum_{n=1}^{\infty} \frac{1}{2n}$.

44. If $s = \sum_{k=1}^{\infty} c_k = \sum_{k=1}^{\infty} \frac{1}{k^2(k+1)}$, then we have

$$s_n + A_{n+1} \leq s \leq s_n + A_n$$

where $s_n = \sum_{k=1}^{n} \frac{1}{k^2(k+1)}$ and

$$A_n = \int_n^{\infty} \frac{dx}{x^2(x+1)} = \int_n^{\infty} \left(\frac{-1}{x} + \frac{1}{x^2} + \frac{1}{x+1} \right) dx$$

$$= -\ln x - \frac{1}{x} + \ln(x+1) \Big|_n^{\infty}$$

$$= \ln\left(1 + \frac{1}{x}\right) - \frac{1}{x} \Big|_n^{\infty}$$

$$= \frac{1}{n} - \ln\left(1 + \frac{1}{n}\right).$$

If $s_n^* = s_n + \frac{1}{2}(A_{n+1} + A_n)$, then

$$|s_n - s_n^*| \leq \frac{A_n - A_{n+1}}{2}$$

$$= \frac{1}{2}\left[\frac{1}{n} - \ln\left(1 + \frac{1}{n}\right) - \frac{1}{n+1} + \ln\left(1 + \frac{1}{n+1}\right) \right]$$

$$= \frac{1}{2}\left[\frac{1}{n(n+1)} + \ln\left(\frac{n^2 + 2n}{n^2 + 2n + 1} \right) \right]$$

$$\leq \frac{1}{2}\left[\frac{1}{n(n+1)} + \left(\frac{n^2 + 2n}{n^2 + 2n + 1} - 1 \right) \right]$$

$$= \frac{1}{2n(n+1)^2} < 0.001$$

if $n = 8$. Thus,

$$\sum_{n=1}^{\infty} \frac{1}{n^2} = 1 + s_8^* = 1 + s_8 + \frac{1}{2}(A_9 + A_8)$$

$$= 1 + \left[\frac{1}{2} + \frac{1}{2^2(3)} + \frac{1}{3^2(4)} + \cdots + \frac{1}{8^2(9)} \right] +$$

$$\frac{1}{2}\left[\left(\frac{1}{9} - \ln \frac{10}{9} \right) + \left(\frac{1}{8} - \ln \frac{9}{8} \right) \right]$$

$$= 1.6450$$

with error less than 0.001.

Section 9.4   Absolute and Conditional Convergence   (page 501)

2. $\sum_{n=1}^{\infty} \frac{(-1)^n}{n^2 + \ln n}$ converges absolutely since
$\left| \frac{(-1)^n}{n^2 + \ln n} \right| \leq \frac{1}{n^2}$ and $\sum_{n=1}^{\infty} \frac{1}{n^2}$ converges.

4. $\sum_{n=1}^{\infty} \frac{(-1)^{2n}}{2^n} = \sum_{n=1}^{\infty} \frac{1}{2^n}$ is a positive, convergent geometric series so must converge absolutely.

6. $\sum_{n=1}^{\infty} \frac{(-2)^n}{n!}$ converges absolutely by the ratio test since

$$\lim \left| \frac{(-2)^{n+1}}{(n+1)!} \cdot \frac{n!}{(-2)^n} \right| = 2 \lim \frac{1}{n+1} = 0.$$

8. $\sum_{n=0}^{\infty} \frac{-n}{n^2 + 1}$ diverges to $-\infty$ since all terms are negative
and $\sum_{n=0}^{\infty} \frac{n}{n^2 + 1}$ diverges to infinity by comparison with
$\sum_{n=0}^{\infty} \frac{1}{n}$.

10. $\sum_{n=1}^{\infty} \frac{100\cos(n\pi)}{2n+3} = \sum_{n=1}^{\infty} \frac{100(-1)^n}{2n+3}$ converges by the alternating series test but only conditionally since

$$\left| \frac{100(-1)^n}{2n+3} \right| = \frac{100}{2n+3}$$

and $\sum_{n=1}^{\infty} \frac{100}{2n+3}$ diverges to infinity.

12. $\sum_{n=10}^{\infty} \frac{\sin(n + \frac{1}{2})\pi}{\ln \ln n} = \sum_{n=10}^{\infty} \frac{(-1)^n}{\ln \ln n}$ converges by the alternating series test but only conditionally since $\sum_{n=10}^{\infty} \frac{1}{\ln \ln n}$
diverges to infinity by comparison with $\sum_{n=10}^{\infty} \frac{1}{n}$.
($\ln \ln n < n$ for $n \geq 10$.)

14. Since the terms of the series $s = \sum_{n=0}^{\infty} \frac{(-1)^n}{(2n)!}$ are alternating in sign and decreasing in size, the size of the error in the approximation $s \approx s_n$ does not exceed that of the first omitted term:

$$|s - s_n| \leq \frac{1}{(2n+2)!} < 0.001$$

if $n = 3$. Hence $s \approx 1 - \frac{1}{2!} + \frac{1}{4!} - \frac{1}{6!}$; four terms will approximate s with error less than 0.001 in absolute value.

16. Since the terms of the series $s = \sum_{n=0}^{\infty}(-1)^n\frac{3^n}{n!}$ are alternating in sign and ultimately decreasing in size (they decrease after the third term), the size of the error in the approximation $s \approx s_n$ does not exceed that of the first omitted term (provided $n \geq 3$):

$$|s - s_n| \leq \frac{3^{n+1}}{(n+1)!} < 0.001 \text{ if } n = 12. \text{ Thus twelve terms}$$

will suffice to approximate s with error less than 0.001 in absolute value.

18. Let $a_n = \frac{(x-2)^n}{n^2 2^{2n}}$. Apply the ratio test

$$\rho = \lim \left| \frac{(x-2)^{n+1}}{(n+1)^2 2^{2n+2}} \times \frac{n^2 2^{2n}}{(x-2)^n} \right| = \frac{|x-2|}{4} < 1$$

if and only if $|x-2| < 4$, that is $-2 < x < 6$. If $x = -2$, then $\sum_{n=1}^{\infty} a_n = \sum_{n=1}^{\infty} \frac{(-1)^n}{n^2}$, which converges absolutely. If $x = 6$, then $\sum_{n=1}^{\infty} a_n = \sum_{n=1}^{\infty} \frac{1}{n^2}$, which also converges absolutely. Thus, the series converges absolutely if $-2 \leq x \leq 6$ and diverges elsewhere.

20. Let $a_n = \frac{1}{2n-1}\left(\frac{3x+2}{-5}\right)^n$. Apply the ratio test

$$\rho = \lim \left| \frac{1}{2n+1}\left(\frac{3x+2}{-5}\right)^{n+1} \times \frac{2n-1}{1}\left(\frac{3x+2}{-5}\right)^{-n} \right|$$

$$= \left| \frac{3x+2}{5} \right| < 1$$

if and only if $\left| x + \frac{2}{3} \right| < \frac{5}{3}$, that is $-\frac{7}{3} < x < 1$. If $x = -\frac{7}{3}$, then $\sum_{n=1}^{\infty} a_n = \sum_{n=1}^{\infty} \frac{1}{2n-1}$, which diverges. If $x = 1$, then $\sum_{n=1}^{\infty} a_n = \sum_{n=1}^{\infty} \frac{(-1)^n}{2n-1}$, which converges conditionally. Thus, the series converges absolutely if $-\frac{7}{3} < x < 1$, converges conditionally if $x = 1$ and diverges elsewhere.

22. Let $a_n = \frac{(4x+1)^n}{n^3}$. Apply the ratio test

$$\rho = \lim \left| \frac{(4x+1)^{n+1}}{(n+1)^3} \times \frac{n^3}{(4x+1)^n} \right| = |4x+1| < 1$$

if and only if $-\frac{1}{2} < x < 0$. If $x = -\frac{1}{2}$, then $\sum_{n=1}^{\infty} a_n = \sum_{n=1}^{\infty} \frac{(-1)^n}{n^3}$, which converges absolutely. If $x = 0$, then $\sum_{n=1}^{\infty} a_n = \sum_{n=1}^{\infty} \frac{1}{n^3}$, which also converges absolutely. Thus, the series converges absolutely if $-\frac{1}{2} \leq x \leq 0$ and diverges elsewhere.

24. Let $a_n = \frac{1}{n}\left(1 + \frac{1}{x}\right)^n$. Apply the ratio test

$$\rho = \lim \left| \frac{1}{n+1}\left(1 + \frac{1}{x}\right)^{n+1} \times \frac{n}{1}\left(1 + \frac{1}{x}\right)^{-n} \right| = \left| 1 + \frac{1}{x} \right| < 1$$

if and only if $|x + 1| < |x|$, that is, $-2 < \frac{1}{x} < 0 \Rightarrow x < -\frac{1}{2}$. If $x = -\frac{1}{2}$, then $\sum_{n=1}^{\infty} a_n = \sum_{n=1}^{\infty} \frac{(-1)^n}{n}$, which converges conditionally. Thus, the series converges absolutely if $x < -\frac{1}{2}$, converges conditionally if $x = -\frac{1}{2}$ and diverges elsewhere. It is undefined at $x = 0$.

26. If

$$a_n = \begin{cases} \dfrac{10}{n^2}, & \text{if } n \text{ is even;} \\[2mm] \dfrac{-1}{10n^3}, & \text{if } n \text{ is odd;} \end{cases}$$

then $|a_n| \leq \frac{10}{n^2}$ for every $n \geq 1$. Hence, $\sum_{n=1}^{\infty} a_n$ converges absolutely by comparison with $\sum_{n=1}^{\infty} \frac{10}{n^2}$.

28. a) We have

$$\ln(n!) = \ln 1 + \ln 2 + \ln 3 + \cdots + \ln n$$
$$= \text{sum of area of the shaded rectangles}$$
$$> \int_1^n \ln t \, dt = (t \ln t - t)\Big|_1^n$$
$$= n \ln n - n + 1.$$

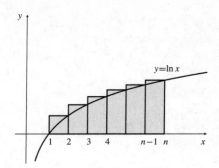

Fig. 9.4.28

b) Let $a_n = \dfrac{n!x^n}{n^n}$. Apply the ratio test

$$\rho = \lim \left| \frac{(n+1)!x^{n+1}}{(n+1)^{n+1}} \times \frac{n^n}{n!x^n} \right|$$

$$= \lim \frac{|x|}{\left(1 + \dfrac{1}{n}\right)^n} = \frac{|x|}{e} < 1$$

if and only if $-e < x < e$. If $x = \pm e$, then, by (a),

$$\ln \left| \frac{n!e^n}{n^n} \right| = \ln(n!) + \ln e^n - \ln n^n$$

$$> (n \ln n - n + 1) + n - n \ln n = 1.$$

$$\Rightarrow \left| \frac{n!e^n}{n^n} \right| > e.$$

Hence, $\displaystyle\sum_{n=1}^{\infty} a_n$ converges absolutely if $-e < x < e$
and diverges elsewhere.

30. Let $p_n = \dfrac{1}{2n - 1}$ and $q_n = -\dfrac{1}{2n}$. Then $\sum p_n$ diverges
to ∞ and $\sum q_n$ diverges to $-\infty$. Also, the alternating
harmonic series is the sum of all the p_ns and q_ns in a
specific order:

$$\sum_{n=1}^{\infty} \frac{(-1)^{n-1}}{n} = \sum_{n=1}^{\infty} (p_n + q_n).$$

a) Rearrange the terms as follows: first add terms of
$\sum p_n$ until the sum exceeds 2. Then add q_1. Then
add more terms of $\sum p_n$ until the sum exceeds 3.
Then add q_2. Continue in this way; at the nth stage,
add new terms from $\sum p_n$ until the sum exceeds
$n + 1$, and then add q_n. All partial sums after the
nth stage exceed n, so the rearranged series diverges
to infinity.

b) Rearrange the terms of the original alternating harmonic series as follows: first add terms of $\sum q_n$
until the sum is less than -2. Then add p_1. The
sum will now be greater than -2. (Why?) Then re-
sume adding new terms from $\sum q_n$ until the sum is
less than -2 again, and add p_2, which will raise the
sum above -2 again. Continue in this way. After
the nth stage, all succeeding partial sums will differ
from -2 by less than $1/n$, so the rearranged series
will converge to -2.

Section 9.5   Power Series   (page 511)

2. We have $\displaystyle\sum_{n=0}^{\infty} 3n(x + 1)^n$. The centre of convergence is
$x = -1$. The radius of convergence is

$$R = \lim \frac{3n}{3(n + 1)} = 1.$$

The series converges absolutely on $(-2, 0)$ and diverges
on $(-\infty, -2)$ and $(0, \infty)$. At $x = -2$, the series is
$\displaystyle\sum_{n=0}^{\infty} 3n(-1)^n$, which diverges. At $x = 0$, the series is
$\displaystyle\sum_{n=0}^{\infty} 3n$, which diverges to infinity. Hence, the interval of
convergence is $(-2, 0)$.

4. We have $\displaystyle\sum_{n=1}^{\infty} \frac{(-1)^n}{n^4 2^{2n}} x^n$. The centre of convergence is
$x = 0$. The radius of convergence is

$$R = \lim \left| \frac{(-1)^n}{n^4 2^{2n}} \cdot \frac{(n+1)^4 2^{2n+2}}{(-1)^{n+1}} \right|$$

$$= \lim \left| \left(\frac{n+1}{n}\right)^4 \cdot 4 \right| = 4.$$

At $x = 4$, the series is $\displaystyle\sum_{n=1}^{\infty} \frac{(-1)^n}{n^4}$, which converges.

At $x = -4$, the series is $\displaystyle\sum_{n=1}^{\infty} \frac{1}{n^4}$, which also converges.
Hence, the interval of convergence is $[-4, 4]$.

6. We have $\displaystyle\sum_{n=1}^{\infty} \frac{e^n}{n^3} (4 - x)^n$. The centre of convergence is
$x = 4$. The radius of convergence is

$$R = \lim \frac{e^n}{n^3} \cdot \frac{(n+1)^3}{e^{n+1}} = \frac{1}{e}.$$

At $x = 4 + \dfrac{1}{e}$, the series is $\displaystyle\sum_{n=1}^{\infty} \dfrac{(-1)^n}{n^3}$, which converges.

At $x = 4 - \dfrac{1}{e}$, the series is $\displaystyle\sum_{n=1}^{\infty} \dfrac{1}{n^3}$, which also converges.

Hence, the interval of convergence is $\left[4 - \dfrac{1}{e}, 4 + \dfrac{1}{e} \right]$.

8. We have $\displaystyle\sum_{n=1}^{\infty} \dfrac{(4x-1)^n}{n^n} = \sum_{n=1}^{\infty} \left(\dfrac{4}{n}\right)^n \left(x - \dfrac{1}{4}\right)^n$. The centre of convergence is $x = \frac{1}{4}$. The radius of convergence is

$$R = \lim \dfrac{4^n}{n^n} \cdot \dfrac{(n+1)^{n+1}}{4^{n+1}}$$
$$= \dfrac{1}{4} \lim \left(\dfrac{n+1}{n}\right)^n (n+1) = \infty.$$

Hence, the interval of convergence is $(-\infty, \infty)$.

10. We have

$$1 + x + x^2 + x^3 + \cdots = \dfrac{1}{1-x} = \sum_{n=0}^{\infty} x^n$$

and

$$1 - x + x^2 - x^3 + \cdots = \dfrac{1}{1+x} = \sum_{n=0}^{\infty} (-1)^n x^n$$

holds for $-1 < x < 1$. Since $a_n = 1$ and $b_n = (-1)^n$ for $n = 0, 1, 2, \ldots$, we have

$$C_n = \sum_{j=0}^{n} (-1)^{n-j} = \begin{cases} 0, & \text{if } n \text{ is odd;} \\ 1, & \text{if } n \text{ is even.} \end{cases}$$

Then the Cauchy product is

$$1 + x^2 + x^4 + \cdots = \sum_{n=0}^{\infty} x^{2n} = \dfrac{1}{1-x} \cdot \dfrac{1}{1+x} = \dfrac{1}{1-x^2}$$

for $-1 < x < 1$.

12. $\dfrac{1}{2-x} = \dfrac{1}{2} \dfrac{1}{\left(1 - \dfrac{x}{2}\right)} = \dfrac{1}{2} \sum_{n=0}^{\infty} \left(\dfrac{x}{2}\right)^n$

$$= \dfrac{1}{2} + \dfrac{x}{2^2} + \dfrac{x^2}{2^3} + \dfrac{x^3}{2^4} + \cdots \qquad (-2 < x < 2).$$

14. $\dfrac{1}{1+2x} = \displaystyle\sum_{n=0}^{\infty} (-2x)^n$

$$= 1 - 2x + 2^2 x^2 - 2^3 x^3 + \cdots \qquad \left(-\dfrac{1}{2} < x < \dfrac{1}{2}\right).$$

16. Let $y = x - 1$. Then $x = 1 + y$ and

$$\dfrac{1}{x} = \dfrac{1}{1+y} = \sum_{n=0}^{\infty} (-y)^n \qquad (-1 < y < 1)$$
$$= \sum_{n=0}^{\infty} \left[-(x-1)\right]^n$$
$$= 1 - (x-1) + (x-1)^2 - (x-1)^3 + (x-1)^4 - \cdots$$
$$(\text{for } 0 < x < 2).$$

18. $\dfrac{1-x}{1+x} = \dfrac{2}{1+x} - 1$

$$= 2(1 - x + x^2 - x^3 + \cdots) - 1$$
$$= 1 + 2 \sum_{n=1}^{\infty} (-x)^n \qquad (-1 < x < 1).$$

20. Let $y = x - 4$. Then $x = 4 + y$ and

$$\dfrac{1}{x} = \dfrac{1}{4+y} = \dfrac{1}{4} \dfrac{1}{\left(1 + \dfrac{y}{4}\right)} = \dfrac{1}{4} \sum_{n=0}^{\infty} \left(-\dfrac{y}{4}\right)^n$$
$$= \dfrac{1}{4} \sum_{n=0}^{\infty} \left[-\dfrac{(x-4)}{4}\right]^n$$
$$= \dfrac{1}{4} - \dfrac{(x-4)}{4^2} + \dfrac{(x-4)^2}{4^3} - \dfrac{(x-4)^3}{4^4} + \cdots$$

for $0 < x < 8$. Therefore,

$$\ln x = \int_1^x \dfrac{dt}{t} = \int_1^4 \dfrac{dt}{t} + \int_4^x \dfrac{dt}{t}$$
$$= \ln 4 + \int_4^x \left[\dfrac{1}{4} - \dfrac{(t-4)}{4^2} + \dfrac{(t-4)^2}{4^3} - \dfrac{(t-4)^3}{4^4} + \cdots\right] dt$$
$$= \ln 4 + \dfrac{x-4}{4} - \dfrac{(x-4)^2}{2 \cdot 4^2} + \dfrac{(x-4)^3}{3 \cdot 4^3} - \dfrac{(x-4)^4}{4 \cdot 4^4} + \cdots$$
$$(\text{for } 0 < x \leq 8).$$

22. We differentiate the series

$$\sum_{n=0}^{\infty} x^n = 1 + x + x^2 + x^3 + \cdots = \dfrac{1}{1-x}$$

and multiply by x to get

$$\sum_{n=0}^{\infty} nx^n = x + 2x^2 + 3x^3 + \cdots = \dfrac{x}{(1-x)^2}$$

for $-1 < x < 1$. Therefore,

$$\sum_{n=0}^{\infty} (n+3)x^n = \sum_{n=0}^{\infty} nx^n + 3 \sum_{n=0}^{\infty} x^n$$
$$= \dfrac{x}{(1-x)^2} + \dfrac{3}{1-x}$$
$$= \dfrac{3-2x}{(1-x)^2} \qquad (-1 < x < 1).$$

24. We start with

$$1 - x + x^2 - x^3 + x^4 - \cdots = \frac{1}{1+x}$$

and differentiate to get

$$-1 + 2x - 3x^3 + 4x^3 - \cdots = -\frac{1}{(1+x)^2}.$$

Now we multiply by $-x^3$:

$$x^3 - 2x^4 + 3x^5 - 4x^6 + \cdots = \frac{x^3}{(1+x)^2}.$$

Differentiating again we get

$$3x^2 - 2 \times 4x^3 + 3 \times 5x^4 - 4 \times 6x^5 + \cdots = \frac{x^3 + 3x^2}{(1+x)^3}.$$

Finally, we remove the factor x^2:

$$3 - 2 \times 4x + 3 \times 5x^2 - 4 \times 6x^3 + \cdots = \frac{x+3}{(1+x)^3}.$$

All steps are valid for $-1 < x < 1$.

26. Since $x - \dfrac{x^2}{2} + \dfrac{x^3}{3} - \dfrac{x^4}{4} + \cdots = \ln(1+x)$ for $-1 < x \le 1$,
therefore

$$x^2 - \frac{x^4}{2} + \frac{x^6}{3} - \frac{x^8}{4} + \cdots = \ln(1 + x^2)$$

for $-1 \le x \le 1$, and, dividing by x^2,

$$1 - \frac{x^2}{2} + \frac{x^4}{3} - \frac{x^6}{4} + \cdots = \begin{cases} \dfrac{\ln(1+x^2)}{x^2} & \text{if } -1 \le x \le 1, \, x \ne 0 \\ 1 & \text{if } x = 0. \end{cases}$$

28. From Example 5(a) with $x = 1/2$,

$$\sum_{n=0}^{\infty} \frac{n+1}{2^n} = \sum_{k=1}^{\infty} k\left(\frac{1}{2}\right)^{k-1} = \frac{1}{\left(1 - \frac{1}{2}\right)^2} = 4.$$

0. From Example 5(a),

$$\sum_{n=1}^{\infty} nx^{n-1} = \frac{1}{(1-x)^2}, \quad (-1 < x < 1).$$

Differentiate with respect to x and then replace n by $n + 1$:

$$\sum_{n=2}^{\infty} n(n-1)x^{n-2} = \frac{2}{(1-x)^3}, \quad (-1 < x < 1)$$

$$\sum_{n=1}^{\infty} (n+1)nx^{n-1} = \frac{2}{(1-x)^3}, \quad (-1 < x < 1).$$

Now let $x = -1/2$:

$$\sum_{n=1}^{\infty} (-1)^{n-1} \frac{n(n+1)}{2^{n-1}} = \frac{16}{27}.$$

Finally, multiply by $-1/2$:

$$\sum_{n=1}^{\infty} (-1)^n \frac{n(n+1)}{2^n} = -\frac{8}{27}.$$

32. In the series for $\ln(1+x)$ in Example 5(c), put $x = -1/2$ to get

$$\sum_{n=1}^{\infty} (-1)\frac{1}{n2^n} = \sum_{k=0}^{\infty} \frac{(-1)^k}{k+1}\left(-\frac{1}{2}\right)^{k+1} = \ln\left(1 - \frac{1}{2}\right) = -\ln 2.$$

Therefore

$$\sum_{n=1}^{\infty} \frac{1}{n2^n} = \ln 2$$

$$\sum_{n=3}^{\infty} \frac{1}{n2^n} = \ln 2 - \frac{1}{2} - \frac{1}{8} = \ln 2 - \frac{5}{8}.$$

Section 9.6 Taylor and Maclaurin Series (page 520)

2.
$$\cos(2x^3) = 1 - \frac{(2x^3)^2}{2!} + \frac{(2x^3)^4}{4!} - \frac{(2x^3)^6}{6!} + \cdots$$
$$= 1 - \frac{2^2 x^6}{2!} + \frac{2^4 x^{12}}{4!} - \frac{2^6 x^{18}}{6!} + \cdots$$
$$= \sum_{n=0}^{\infty} \frac{(-1)^n 4^n}{(2n)!} x^{6n} \quad \text{(for all } x\text{).}$$

4.
$$\cos(2x - \pi) = -\cos(2x)$$
$$= -1 + \frac{2^2 x^2}{2!} - \frac{2^4 x^4}{4!} + \frac{2^6 x^6}{6!} - \cdots$$
$$= -\sum_{n=0}^{\infty} \frac{(-1)^n}{(2n)!}(2x)^{2n}$$
$$= \sum_{n=0}^{\infty} \frac{(-1)^{n+1}}{(2n)!} 4^n (x)^{2n} \quad \text{(for all } x\text{).}$$

6. $\cos^2\left(\dfrac{x}{2}\right) = \dfrac{1}{2}(1 + \cos x)$

$\qquad = \dfrac{1}{2}\left(1 + 1 - \dfrac{x^2}{2!} + \dfrac{x^4}{4!} - \dfrac{x^6}{6!} + \cdots\right)$

$\qquad = 1 + \dfrac{1}{2}\sum_{n=1}^{\infty}\dfrac{(-1)^n}{(2n)!}x^{2n}\qquad\text{(for all }x\text{)}.$

8. $\tan^{-1}(5x^2) = (5x^2) - \dfrac{(5x^2)^3}{3} + \dfrac{(5x^2)^5}{5} - \dfrac{(5x^2)^7}{7} + \cdots$

$\qquad = \sum_{n=0}^{\infty}\dfrac{(-1)^n}{(2n+1)}(5x^2)^{2n+1}$

$\qquad = \sum_{n=0}^{\infty}\dfrac{(-1)^n 5^{2n+1}}{(2n+1)}x^{4n+2}$

$\qquad\left(\text{for } -\dfrac{1}{\sqrt{5}} \le x \le \dfrac{1}{\sqrt{5}}\right).$

10. $\ln(2 + x^2) = \ln 2\left(1 + \dfrac{x^2}{2}\right)$

$\qquad = \ln 2 + \ln\left(1 + \dfrac{x^2}{2}\right)$

$\qquad = \ln 2 + \left[\dfrac{x^2}{2} - \dfrac{1}{2}\left(\dfrac{x^2}{2}\right)^2 + \dfrac{1}{3}\left(\dfrac{x^2}{2}\right)^3 - \cdots\right]$

$\qquad = \ln 2 + \sum_{n=1}^{\infty}\dfrac{(-1)^{n-1}}{n}\cdot\dfrac{x^{2n}}{2^n}$

$\qquad\text{(for }-\sqrt{2} \le x \le \sqrt{2}).$

12. $\dfrac{e^{2x^2} - 1}{x^2} = \dfrac{1}{x^2}\left(e^{2x^2} - 1\right)$

$\qquad = \dfrac{1}{x^2}\left(1 + 2x^2 + \dfrac{(2x^2)^2}{2!} + \dfrac{(2x^2)^3}{3!} + \cdots - 1\right)$

$\qquad = 2 + \dfrac{2^2 x^2}{2!} + \dfrac{2^3 x^4}{3!} + \dfrac{2^4 x^6}{4!} + \cdots$

$\qquad = \sum_{n=0}^{\infty}\dfrac{2^{n+1}}{(n+1)!}x^{2n}\qquad\text{(for all }x \neq 0).$

14. $\sinh x - \sin x = \sum_{n=0}^{\infty}\left[1 - (-1)^n\right]\dfrac{x^{2n+1}}{(2n+1)!}$

$\qquad = 2\left(\dfrac{x^2}{2!} + \dfrac{x^6}{6!} + \dfrac{x^{10}}{10!} + \cdots\right)$

$\qquad = 2\sum_{n=0}^{\infty}\dfrac{x^{4n+3}}{(4n+3)!}\qquad\text{(for all }x\text{)}.$

16. Let $y = x - \dfrac{\pi}{2}$; then $x = y + \dfrac{\pi}{2}$. Hence,

$\qquad \sin x = \sin\left(y + \dfrac{\pi}{2}\right) = \cos y$

$\qquad = 1 - \dfrac{y^2}{2!} + \dfrac{y^4}{4!} - \cdots\qquad\text{(for all }y\text{)}$

$\qquad = 1 - \dfrac{1}{2!}\left(x - \dfrac{\pi}{2}\right)^2 + \dfrac{1}{4!}\left(x - \dfrac{\pi}{2}\right)^4 - \cdots$

$\qquad = \sum_{n=0}^{\infty}\dfrac{(-1)^n}{(2n)!}\left(x - \dfrac{\pi}{2}\right)^{2n}\qquad\text{(for all }x\text{)}.$

18. Let $y = x - 3$; then $x = y + 3$. Hence,

$\qquad \ln x = \ln(y + 3) = \ln 3 + \ln\left(1 + \dfrac{y}{3}\right)$

$\qquad = \ln 3 + \dfrac{y}{3} - \dfrac{1}{2}\left(\dfrac{y}{3}\right)^2 + \dfrac{1}{3}\left(\dfrac{y}{3}\right)^3 - \dfrac{1}{4}\left(\dfrac{y}{3}\right)^4 + \cdots$

$\qquad = \ln 3 + \dfrac{(x-3)}{3} - \dfrac{(x-3)^2}{2\cdot 3^2} + \dfrac{(x-3)^3}{3\cdot 3^3} - \dfrac{(x-3)^4}{4\cdot 3^4} + \cdots$

$\qquad = \ln 3 + \sum_{n=1}^{\infty}\dfrac{(-1)^{n-1}}{n\cdot 3^n}(x-3)^n\qquad(0 < x \le 6).$

20. Let $t = x + 1$. Then $x = t - 1$, and

$\qquad e^{2x+3} = e^{2t+1} = e\,e^{2t}$

$\qquad = e\sum_{n=0}^{\infty}\dfrac{2^n t^n}{n!}\qquad\text{(for all }t\text{)}$

$\qquad = \sum_{n=0}^{\infty}\dfrac{e 2^n (x+1)^n}{n!}\qquad\text{(for all }x\text{)}.$

22. Let $y = x - \dfrac{\pi}{8}$; then $x = y + \dfrac{\pi}{8}$. Thus,

$\qquad \cos^2 x = \cos^2\left(y + \dfrac{\pi}{8}\right)$

$\qquad = \dfrac{1}{2}\left[1 + \cos\left(2y + \dfrac{\pi}{4}\right)\right]$

$\qquad = \dfrac{1}{2}\left[1 + \dfrac{1}{\sqrt{2}}\cos(2y) - \dfrac{1}{\sqrt{2}}\sin(2y)\right]$

$\qquad = \dfrac{1}{2} + \dfrac{1}{2\sqrt{2}}\left[1 - \dfrac{(2y)^2}{2!} + \dfrac{(2y)^4}{4!} - \cdots\right]$

$\qquad\quad - \dfrac{1}{2\sqrt{2}}\left[2y - \dfrac{(2y)^3}{3!} + \dfrac{(2y)^5}{5!} - \cdots\right]$

$\qquad = \dfrac{1}{2} + \dfrac{1}{2\sqrt{2}}\left[1 - 2y - \dfrac{(2y)^2}{2!} + \dfrac{(2y)^3}{3!}\right.$

$\qquad\quad\left. + \dfrac{(2y)^4}{4!} - \dfrac{(2y)^5}{5!} - \cdots\right]$

$$= \frac{1}{2} + \frac{1}{2\sqrt{2}}\left[1 - 2\left(x - \frac{\pi}{8}\right) - \frac{2^2}{2!}\left(x - \frac{\pi}{8}\right)^2\right.$$

$$\left. + \frac{2^3}{3!}\left(x - \frac{\pi}{8}\right)^3 + \frac{2^4}{4!}\left(x - \frac{\pi}{8}\right)^4 - \frac{2^5}{5!}\left(x - \frac{\pi}{8}\right)^5 - \cdots\right]$$

$$= \frac{1}{2} + \frac{1}{2\sqrt{2}} + \frac{1}{2\sqrt{2}}\sum_{n=1}^{\infty}(-1)^n\left[\frac{2^{2n-1}}{(2n-1)!}\left(x - \frac{\pi}{8}\right)^{2n-1}\right.$$

$$\left. + \frac{2^{2n}}{(2n)!}\left(x - \frac{\pi}{8}\right)^{2n}\right] \quad \text{(for all } x\text{)}.$$

24. Let $y = x - 1$; then $x = y + 1$. Thus,

$$\frac{x}{1+x} = \frac{1+y}{2+y} = 1 - \frac{1}{2\left(1 + \frac{y}{2}\right)}$$

$$= 1 - \frac{1}{2}\left[1 - \frac{y}{2} + \left(\frac{y}{2}\right)^2 - \left(\frac{y}{2}\right)^3 + \cdots\right]$$

$$= \frac{1}{2}\left[1 + \frac{y}{2} - \frac{y^2}{2^2} + \frac{y^3}{2^3} - \frac{y^4}{2^4} + \cdots\right] \quad (-1 < y < 1)$$

$$= \frac{1}{2} + \frac{1}{2^2}(x-1) - \frac{1}{2^3}(x-1)^2 + \frac{1}{2^4}(x-1)^3 - \cdots$$

$$= \frac{1}{2} + \sum_{n=1}^{\infty}\frac{(-1)^{n-1}}{2^{n+1}}(x-1)^n \quad \text{(for } 0 < x < 2\text{)}.$$

26. Let $u = x + 2$. Then $x = u - 2$, and

$$xe^x = (u-2)e^{u-2}$$

$$= (u-2)e^{-2}\sum_{n=0}^{\infty}\frac{u^n}{n!} \quad \text{(for all } u\text{)}$$

$$= \sum_{n=0}^{\infty}\frac{e^{-2}u^{n+1}}{n!} - \sum_{n=0}^{\infty}\frac{2e^{-2}u^n}{n!}.$$

In the first sum replace n by $n-1$.

$$xe^x = \sum_{n=1}^{\infty}\frac{e^{-2}u^n}{(n-1)!} - \sum_{n=0}^{\infty}\frac{2e^{-2}u^n}{n!}$$

$$= -\frac{2}{e^2} + \sum_{n=1}^{\infty}\frac{1}{e^2}\left(\frac{1}{(n-1)!} - \frac{2}{n!}\right)u^n$$

$$= -\frac{2}{e^2} + \sum_{n=1}^{\infty}\frac{1}{e^2}\left(\frac{1}{(n-1)!} - \frac{2}{n!}\right)(x+2)^n \quad \text{(for all } x\text{)}.$$

28. If we divide the first four terms of the series

$$\cos x = 1 - \frac{x^2}{2} + \frac{x^4}{24} - \frac{x^6}{720} + \cdots$$

into 1 we obtain

$$\sec x = 1 + \frac{x^2}{2} + \frac{5x^4}{24} + \frac{61x^6}{720} + \cdots.$$

Now we can differentiate and obtain

$$\sec x\tan x = x + \frac{5x^3}{6} + \frac{61x^5}{120} + \cdots.$$

(Note: the same result can be obtained by multiplying the first three nonzero terms of the series for $\sec x$ (from Exercise 25) and $\tan x$ (from Example 6(b)).)

30. We have

$$e^{\tan^{-1}x} - 1 = \exp\left[x - \frac{x^3}{3} + \frac{x^5}{5} - \frac{x^7}{7} + \cdots\right] - 1$$

$$= 1 + \left(x - \frac{x^3}{3} + \frac{x^5}{5} - \cdots\right) + \frac{1}{2!}\left(x - \frac{x^3}{3} + \cdots\right)^2$$

$$+ \frac{1}{3!}(x - \cdots)^3 + \cdots - 1$$

$$= x - \frac{x^3}{3} + \frac{x^2}{2} + \frac{x^3}{6} + \text{higher degree terms}$$

$$= x + \frac{x^2}{2} - \frac{x^3}{6} + \cdots.$$

32. $\csc x$ does not have a Maclaurin series because $\lim_{x\to 0}\csc x$ does not exist.

Let $y = x - \frac{\pi}{2}$. Then $x = y + \frac{\pi}{2}$ and $\sin x = \cos y$.

Therefore, using the result of Exercise 25,

$$\csc x = \sec y = 1 + \frac{y^2}{2} + \frac{5y^4}{24} + \cdots$$

$$= 1 + \frac{1}{2}\left(x - \frac{\pi}{2}\right)^2 + \frac{5}{24}\left(x - \frac{\pi}{2}\right)^4 + \cdots.$$

34. $x^3 - \frac{x^9}{3!\times 4} + \frac{x^{15}}{5!\times 16} - \frac{x^{21}}{7!\times 64} + \frac{x^{27}}{9!\times 256} - \cdots$

$$= 2\left[\frac{x^3}{2} - \frac{1}{3!}\left(\frac{x^3}{2}\right)^3 + \frac{1}{5!}\left(\frac{x^3}{2}\right)^5 - \cdots\right]$$

$$= 2\sin\left(\frac{x^3}{2}\right) \quad \text{(for all } x\text{)}.$$

36. $1 + \frac{1}{2\times 2!} + \frac{1}{4\times 3!} + \frac{1}{8\times 4!} + \cdots$

$$= 2\left[\frac{1}{2} + \frac{1}{2!}\left(\frac{1}{2}\right)^2 + \frac{1}{3!}\left(\frac{1}{2}\right)^3 + \cdots\right]$$

$$= 2\left(e^{1/2} - 1\right).$$

38. If $a \neq 0$ and $|x - a| < |a|$, then

$$\frac{1}{x} = \frac{1}{a + (x-a)} = \frac{1}{a}\frac{1}{1 + \frac{x-a}{a}}$$

$$= \frac{1}{a}\left[1 - \frac{x-a}{a} + \frac{(x-a)^2}{a^2} - \frac{(x-a)^3}{a^3} + \cdots\right].$$

The radius of convergence of this series is $|a|$, and the series converges to $1/x$ throughout its interval of convergence. Hence, $1/x$ is analytic at a.

40. If

$$f(x) = \begin{cases} e^{-1/x^2}, & \text{if } x \neq 0; \\ 0, & \text{if } x = 0; \end{cases}$$

then the Maclaurin series for $f(x)$ is the identically zero series $0 + 0x + 0x^2 + \cdots$ since $f^{(k)}(0) = 0$ for every k. The series converges for every x, but converges to $f(x)$ only at $x = 0$, since $f(x) \neq 0$ if $x \neq 0$. Hence, f cannot be analytic at 0.

42. We want to prove that $f(x) = P_n(x) + E_n(x)$, where P_n is the nth-order Taylor polynomial for f about c and

$$E_n(x) = \frac{1}{n!} \int_c^x (x - t)^n f^{(n+1)}(t)\, dt.$$

(a) The Fundamental Theorem of Calculus written in the form

$$f(x) = f(c) + \int_c^x f'(t)\, dt = P_0(x) + E_0(x)$$

is the case $n = 0$ of the above formula. We now apply integration by parts to the integral, setting

$$U = f'(t), \qquad dV = dt,$$
$$dU = f''(t)\, dt, \qquad V = -(x - t).$$

(We have broken our usual rule about not including a constant of integration with V. In this case we have included the constant $-x$ in V in order to have V vanish when $t = x$.) We have

$$f(x) = f(c) - f'(t)(x - t)\Big|_{t=c}^{t=x} + \int_c^x (x - t) f''(t)\, dt$$
$$= f(c) + f'(c)(x - c) + \int_c^x (x - t) f''(t)\, dt$$
$$= P_1(x) + E_1(x).$$

We have now proved the case $n = 1$ of the formula.

(b) We complete the proof for general n by mathematical induction. Suppose the formula holds for some $n = k$:

$$f(x) = P_k(x) + E_k(x)$$
$$= P_k(x) + \frac{1}{k!} \int_c^x (x - t)^k f^{(k+1)}(t)\, dt.$$

Again we integrate by parts. Let

$$U = f^{(k+1)}(t), \qquad dV = (x - t)^k\, dt,$$
$$dU = f^{(k+2)}(t)\, dt, \qquad V = \frac{-1}{k+1}(x - t)^{k+1}.$$

We have

$$f(x) = P_k(x) + \frac{1}{k!}\left(-\frac{f^{(k+1)}(t)(x - t)^{k+1}}{k+1}\bigg|_{t=c}^{t=x} \right.$$
$$\left. + \int_c^x \frac{(x - t)^{k+1} f^{(k+2)}(t)}{k+1}\, dt \right)$$
$$= P_k(x) + \frac{f^{(k+1)}(c)}{(k+1)!}(x - c)^{k+1}$$
$$+ \frac{1}{(k+1)!} \int_c^x (x - t)^{k+1} f^{(k+2)}(t)\, dt$$
$$= P_{k+1}(x) + E_{k+1}(x).$$

Thus the formula is valid for $n = k + 1$ if it is valid for $n = k$. Having been shown to be valid for $n = 0$ (and $n = 1$), it must therefore be valid for every positive integer n for which $E_n(x)$ exists.

44. We follow the steps outlined in the problem:

(a) Note that $\ln(j - 1) < \int_{j-1}^j \ln x\, dx < \ln j$, $j = 1, 2, \ldots$. For $j = 0$ the integral is improper but convergent. We have

$$n \ln n - n = \int_0^n \ln x\, dx < \ln(n!) < \int_1^{n+1} \ln x\, dx$$
$$= (n + 1)\ln(n + 1) - n - 1 < (n + 1)\ln(n + 1) -$$

(b) If $c_n = \ln(n!) - \left(n + \frac{1}{2}\right)\ln n + n$, then

$$c_n - c_{n+1} = \ln\frac{n!}{(n+1)!} - \left(n + \frac{1}{2}\right)\ln n$$
$$+ \left(n + \frac{3}{2}\right)\ln(n + 1) - 1$$
$$= \ln\frac{1}{n+1} - \left(n + \frac{1}{2}\right)\ln n$$
$$+ \left(n + \frac{1}{2}\right)\ln(n + 1) + \ln(n + 1) - 1$$
$$= \left(n + \frac{1}{2}\right)\ln\frac{n + 1}{n} - 1$$
$$= \left(n + \frac{1}{2}\right)\ln\frac{1 + \frac{1}{2n+1}}{1 - \frac{1}{2n+1}} - 1.$$

(c) $\ln \dfrac{1+t}{1-t} = 2\left(t + \dfrac{t^3}{3} + \dfrac{t^5}{5} + \cdots\right)$ for $-1 < t < 1$. Thus

$$0 < c_n - c_{n+1} = (2n+1)\left(\frac{1}{2n+1} + \frac{1}{3(2n+1)^3}\right.$$

$$\left. + \frac{1}{5(2n+1)^5} + \cdots\right) - 1$$

$$< \frac{1}{3}\left(\frac{1}{(2n+1)^2} + \frac{1}{(2n+1)^4} + \cdots\right)$$

(geometric)

$$= \frac{1}{3(2n+1)^2}\frac{1}{1 - \dfrac{1}{(2n+1)^2}}$$

$$= \frac{1}{12(n^2+n)}$$

$$= \frac{1}{12}\left(\frac{1}{n} - \frac{1}{n+1}\right).$$

These inequalities imply that $\{c_n\}$ is decreasing and $\{c_n - \frac{1}{12n}\}$ is increasing. Thus $\{c_n\}$ is bounded below by $c_1 - \frac{1}{12} = \frac{11}{12}$ and so $\lim_{n\to\infty} c_n = c$ exists. Since $e^{c_n} = n! n^{-(n+1/2)} e^n$, we have

$$\lim_{n\to\infty} \frac{n!}{n^{n+1/2}e^{-n}} = \lim_{n\to\infty} e^{c_n} = e^c$$

exists. It remains to show that $e^c = \sqrt{2\pi}$.

(d) The Wallis Product,

$$\lim_{n\to\infty} \frac{2}{1}\frac{2}{3}\frac{4}{3}\frac{4}{5}\frac{6}{5} \cdots \frac{2n}{2n-1}\frac{2n}{2n+1} = \frac{\pi}{2}$$

can be rewritten in the form

$$\lim_{n\to\infty} \frac{2^n n!}{1 \cdot 3 \cdot 5 \cdots (2n-1)\sqrt{2n+1}} = \sqrt{\frac{\pi}{2}},$$

or, equivalently,

$$\lim_{n\to\infty} \frac{2^{2n}(n!)^2}{(2n)!\sqrt{2n+1}} = \sqrt{\frac{\pi}{2}}.$$

Substituting $n! = n^{n+1/2}e^{-n}e^{c_n}$ and a similar expression for $(2n)!$, we obtain

$$\lim_{n\to\infty} \frac{2^{2n}n^{2n+1}e^{-2n}e^{2c_n}}{2^{2n+1/2}n^{2n+1/2}e^{-2n}e^{c_{2n}}\sqrt{2n}} = \frac{e^{2c}}{2e^c} = \frac{e^c}{2}.$$

Thus $e^c/2 = \sqrt{\pi}2$, and $e^c = \sqrt{2\pi}$, which completes the proof of Stirling's Formula.

Section 9.7   Applications of Taylor and Maclaurin Series   (page 524)

2. If $f(x) = \ln x$, then $f'(x) = 1/x$, $f''(x) = -1/x^2$, $f'''(x) = 2/x^3$, $f^{(4)}(x) = -6/x^4$, and $f^{(5)}(x) = 24/x^5$. If $P_4(x)$ is the Taylor polynomial for f about $x = 2$, then for some s between 1.95 and 2 we have (using Taylor's Theorem)

$$|f(1.95) - P_4(1.95)| = \frac{24}{s^5} \cdot \frac{(0.05)^5}{5!}$$

$$\leq \frac{24(0.05)^5}{(1.95)^5 120} < 2.22 \times 10^{-9}.$$

4. We have

$$\frac{1}{e} = e^{-1} = 1 - \frac{1}{1!} + \frac{1}{2!} - \frac{1}{3!} + \frac{1}{4!} - \cdots$$

which satisfies the conditions for the alternating series test, and the error incurred in using a partial sum to approximate e^{-1} is less than the first omitted term in absolute value. Now $\dfrac{1}{(n+1)!} < 5 \times 10^{-5}$ if $n = 7$, so

$$\frac{1}{e} \approx \frac{1}{2} - \frac{1}{6} + \frac{1}{24} - \frac{1}{120} + \frac{1}{720} - \frac{1}{5040} \approx 0.36786$$

with error less than 5×10^{-5} in absolute value.

6. We have

$$\sin(0.1) = 0.1 - \frac{(0.1)^3}{3!} + \frac{(0.1)^5}{5!} - \frac{(0.1)^7}{7!} + \cdots.$$

Since $\dfrac{(0.1)^5}{5!} = 8.33 \times 10^{-8} < 5 \times 10^{-5}$, therefore

$$\sin(0.1) = 0.1 - \frac{(0.1)^3}{3!} \approx 0.09983$$

with error less than 5×10^{-5} in absolute value.

8. We have

$$\ln\left(\frac{6}{5}\right) = \ln\left(1 + \frac{1}{5}\right)$$

$$= \frac{1}{5} - \frac{1}{2}\left(\frac{1}{5}\right)^2 + \frac{1}{3}\left(\frac{1}{5}\right)^3 - \frac{1}{4}\left(\frac{1}{5}\right)^4 + \cdots.$$

Since $\dfrac{1}{n}\left(\dfrac{1}{5}\right)^n < 5 \times 10^{-5}$ if $n = 6$, therefore

$$\ln\left(\frac{6}{5}\right) \approx \frac{1}{5} - \frac{1}{2}\left(\frac{1}{5}\right)^2 + \frac{1}{3}\left(\frac{1}{5}\right)^3 - \frac{1}{4}\left(\frac{1}{5}\right)^4 + \frac{1}{5}\left(\frac{1}{5}\right)^5$$

$$\approx 0.18233$$

with error less than 5×10^{-5} in absolute value.

10. We have

$$\sin 80° = \cos 10° = \cos\left(\frac{\pi}{18}\right)$$
$$= 1 - \frac{1}{2!}\left(\frac{\pi}{18}\right)^2 + \frac{1}{4!}\left(\frac{\pi}{18}\right)^4 - \cdots.$$

Since $\frac{1}{4!}\left(\frac{\pi}{18}\right)^4 < 5 \times 10^{-5}$, therefore

$$\sin 80° \approx 1 - \frac{1}{2!}\left(\frac{\pi}{18}\right)^2 \approx 0.98477$$

with error less than 5×10^{-5} in absolute value.

12. We have

$$\tan^{-1}(0.2) = 0.2 - \frac{(0.2)^3}{3} + \frac{(0.2)^5}{5} - \frac{(0.2)^7}{7} + \cdots.$$

Since $\frac{(0.2)^7}{7} < 5 \times 10^{-5}$, therefore

$$\tan^{-1}(0.2) \approx 0.2 - \frac{(0.2)^3}{3} + \frac{(0.2)^5}{5} \approx 0.19740$$

with error less than 5×10^{-5} in absolute value.

14. We have

$$\ln\left(\frac{3}{2}\right) = \ln\left(1 + \frac{1}{2}\right)$$
$$= \frac{1}{2} - \frac{1}{2}\left(\frac{1}{2}\right)^2 + \frac{1}{3}\left(\frac{1}{2}\right)^3 - \frac{1}{4}\left(\frac{1}{2}\right)^4 + \cdots.$$

Since $\frac{1}{n}\left(\frac{1}{2}\right)^n < \frac{1}{20000}$ if $n = 11$, therefore

$$\ln\left(\frac{3}{2}\right) \approx \frac{1}{2} - \frac{1}{2}\left(\frac{1}{2}\right)^2 + \frac{1}{3}\left(\frac{1}{2}\right)^3 - \cdots - \frac{1}{10}\left(\frac{1}{2}\right)^{10}$$
$$\approx 0.40543$$

with error less than 5×10^{-5} in absolute value.

16. $J(x) = \int_0^x \frac{e^t - 1}{t}\, dt$

$$= \int_0^x \left(1 + \frac{t}{2!} + \frac{t^2}{3!} + \frac{t^3}{4!} + \cdots\right) dt$$
$$= x + \frac{x^2}{2! \cdot 2} + \frac{x^3}{3! \cdot 3} + \frac{x^4}{4! \cdot 4} + \cdots$$
$$= \sum_{n=1}^{\infty} \frac{x^n}{n! \cdot n}.$$

18. $L(x) = \int_0^x \cos(t^2)\, dt$

$$= \int_0^x \left(1 - \frac{t^4}{2!} + \frac{t^8}{4!} - \frac{t^{12}}{6!} + \cdots\right) dt$$
$$= x - \frac{x^5}{2! \cdot 5} + \frac{x^9}{4! \cdot 9} - \frac{x^{13}}{6! \cdot 13} + \cdots$$
$$= \sum_{n=0}^{\infty} (-1)^n \frac{x^{4n+1}}{(2n)! \cdot (4n + 1)}.$$

20. We have

$$L(0.5) = 0.5 - \frac{(0.5)^5}{2! \cdot 5} + \frac{(0.5)^9}{4! \cdot 9} - \frac{(0.5)^{13}}{6! \cdot 13} + \cdots.$$

Since $\frac{(0.5)^{4n+1}}{(2n)! \cdot (4n + 1)} < 5 \times 10^{-4}$ if $n = 2$, therefore

$$L(0.5) \approx 0.5 - \frac{(0.5)^5}{2! \cdot 5} \approx 0.497$$

rounded to 3 decimal places.

22. $\displaystyle\lim_{x \to 0} \frac{\sin(x^2)}{\sinh x} = \lim_{x \to 0} \frac{x^2 - \dfrac{x^6}{3!} + \dfrac{x^{10}}{5!} - \cdots}{x + \dfrac{x^3}{3!} + \dfrac{x^5}{5!} + \cdots}$

$$= \lim_{x \to 0} \frac{x - \dfrac{x^5}{3!} + \dfrac{x^9}{5!} - \cdots}{1 + \dfrac{x^2}{3!} + \dfrac{x^4}{5!} + \cdots} = 0.$$

24. We have

$$\lim_{x \to 0} \frac{(e^x - 1 - x)^2}{x^2 - \ln(1 + x^2)} = \lim_{x \to 0} \frac{\left(\dfrac{x^2}{2!} + \dfrac{x^3}{3!} + \dfrac{x^4}{4!} + \cdots\right)^2}{\dfrac{x^4}{2} - \dfrac{x^6}{3} + \dfrac{x^8}{4} - \cdots}$$

$$= \lim_{x \to 0} \frac{\dfrac{x^4}{4}\left(1 + \dfrac{x}{3} + \dfrac{x^2}{12} + \cdots\right)^2}{\dfrac{x^4}{2} - \dfrac{x^6}{3} + \dfrac{x^8}{4} - \cdots} = \frac{\left(\dfrac{1}{4}\right)}{\left(\dfrac{1}{2}\right)} = \frac{1}{2}.$$

26. We have

$$\lim_{x \to 0} \frac{\sin(\sin x) - x}{x[\cos(\sin x) - 1]}$$

$$= \lim_{x \to 0} \frac{\left(\sin x - \dfrac{1}{3!}\sin^3 x + \dfrac{1}{5!}\sin^5 x - \cdots\right) - x}{x\left[1 - \dfrac{1}{2!}\sin^2 x + \dfrac{1}{4!}\sin^4 x - \cdots - 1\right]}$$

$$= \lim_{x \to 0} \frac{\left(x - \dfrac{x^3}{3!} + \cdots\right) - \dfrac{1}{3!}\left(x - \dfrac{x^3}{3!} + \cdots\right)^3 + \dfrac{1}{5!}\left(x - \cdots\right)^5 - \cdots}{x\left[-\dfrac{1}{2!}\left(x - \dfrac{x^3}{3!} + \cdots\right)^2 + \dfrac{1}{4!}\left(x - \cdots\right)^4 - \cdots\right]}$$

$$= \lim_{x \to 0} \frac{-\dfrac{2}{3!}x^3 + \text{higher degree terms}}{-\dfrac{1}{2!}x^3 + \text{higher degree terms}} = \frac{\dfrac{2}{3!}}{\dfrac{1}{2!}} = \frac{2}{3}.$$

Section 9.8 The Binomial Theorem and Binomial Series (page 528)

2. $x\sqrt{1-x} = x(1-x)^{1/2}$

$$= x - \frac{x^2}{2} + \frac{1}{2}\left(-\frac{1}{2}\right)\frac{(-1)^2 x^3}{2!}$$

$$+ \frac{1}{2}\left(-\frac{1}{2}\right)\left(-\frac{3}{2}\right)\frac{(-1)^3 x^4}{3!} + \cdots$$

$$= x - \frac{x^2}{2} - \sum_{n=2}^{\infty} \frac{1\cdot 3\cdot 5\cdots(2n-3)}{2^n n!} x^{n+1}$$

$$= x - \frac{x^2}{2} - \sum_{n=2}^{\infty} (-1)^{n-1}\frac{(2n-2)!}{2^{2n-1}(n-1)!n!} x^{n+1} \quad (-1 < x < 1).$$

4. $\dfrac{1}{\sqrt{4+x^2}} = \dfrac{1}{2\sqrt{1+\left(\frac{x}{2}\right)^2}} = \dfrac{1}{2}\left[1+\left(\frac{x}{2}\right)^2\right]^{-1/2}$

$$= \frac{1}{2}\left[1+\left(-\frac{1}{2}\right)\left(\frac{x}{2}\right)^2 + \frac{1}{2!}\left(-\frac{1}{2}\right)\left(-\frac{3}{2}\right)\left(\frac{x}{2}\right)^4 + \right.$$

$$\left. \frac{1}{3!}\left(-\frac{1}{2}\right)\left(-\frac{3}{2}\right)\left(-\frac{5}{2}\right)\left(\frac{x}{2}\right)^6 + \cdots\right]$$

$$= \frac{1}{2} - \frac{1}{2^4}x^2 + \frac{3}{2^7 2!}x^4 - \frac{3\times 5}{2^{10}3!}x^6 + \cdots$$

$$= \frac{1}{2} + \sum_{n=1}^{\infty} (-1)^n \frac{1\times 2\times 3\times\cdots\times(2n-1)}{2^{3n+1}n!} x^{2n}$$

$$(-2 \le x \le 2).$$

6. $(1+x)^{-3} = 1 - 3x + \dfrac{(-3)(-4)}{2!}x^2 + \dfrac{(-3)(-4)(-5)}{3!}x^3 + \cdots$

$$= 1 - 3x + \frac{(3)(4)}{2}x^2 - \frac{(4)(5)}{2}x^3 + \cdots$$

$$= \sum_{n=0}^{\infty} (-1)^n \frac{(n+2)(n+1)}{2} x^n \quad (-1 < x < 1).$$

8. The formula $(a+b)^n = \sum_{k=0}^{n}\binom{n}{k}a^{n-k}b^k$

holds for $n = 1$; it says $a + b = a + b$ in this case. Suppose the formula holds for $n = m$, where m is some positive integer. Then

$$(a+b)^{m+1} = (a+b)\sum_{k=0}^{m}\binom{m}{k}a^{m-k}b^k$$

$$= \sum_{k=0}^{m}\binom{m}{k}a^{m+1-k}b^k + \sum_{k=0}^{m}\binom{m}{k}a^{m-k}b^{k+1}$$

(replace k by $k-1$ in the latter sum)

$$= \sum_{k=0}^{m}\binom{m}{k}a^{m+1-k}b^k + \sum_{k=1}^{m+1}\binom{m}{k-1}a^{m+1-k}b^k$$

$$= a^{m+1} + \sum_{k=1}^{m}\left[\binom{m}{k}+\binom{m}{k-1}\right]a^{m+1-k}b^k + b^{m+1}$$

(by #13(i))

$$= a^{m+1} + \sum_{k=1}^{m}\binom{m+1}{k}a^{m+1-k}b^k + b^{m+1} \quad \text{(by #13(ii))}$$

$$= \sum_{k=0}^{m+1}\binom{m+1}{k}a^{m+1-k}b^k \quad \text{(by #13(i) again)}.$$

Thus the formula holds for $n = m + 1$. By induction it holds for all positive integers n.

Section 9.9 Fourier Series (page 534)

2. $g(t) = \cos(3 + \pi t)$ has fundamental period 2 since $\cos t$ has fundamental period 2π:

$$g(t+2) = \cos\left(3 + \pi(t+2)\right) = \cos(3 + \pi t + 2\pi)$$

$$= \cos(3 + \pi t) = g(t).$$

4. Since $\sin 2t$ has periods π, 2π, 3π, $\ldots$, and $\cos 3t$ has periods $\frac{2\pi}{3}$, $\frac{4\pi}{3}$, $\frac{6\pi}{3} = 2\pi$, $\frac{8\pi}{3}$, $\ldots$, the sum $k(t) = \sin(2t) + \cos(3t)$ has periods 2π, 4π, $\ldots$. Its fundamental period is 2π.

6. $f(t) = \begin{cases} 0 & \text{if } 0 \le t < 1 \\ 1 & \text{if } 1 \le t < 2 \end{cases}$, f has period 2.

The Fourier coefficients of f are as follows:

$$\frac{a_0}{2} = \frac{1}{2}\int_0^2 f(t)\,dt = \frac{1}{2}\int_1^2 dt = \frac{1}{2}$$

$$a_n = \int_0^2 f(t)\cos(n\pi t)\,dt = \int_1^2 \cos(n\pi t)\,dt$$

$$= \frac{1}{n\pi}\sin(n\pi t)\Big|_1^2 = 0, \quad (n \ge 1)$$

$$b_n = \int_1^2 \sin(n\pi t)\,dt = -\frac{1}{n\pi}\cos(n\pi t)\Big|_1^2$$

$$= -\frac{1-(-1)^n}{n\pi} = \begin{cases} -\dfrac{2}{n\pi} & \text{if } n \text{ is odd} \\ 0 & \text{if } n \text{ is even} \end{cases}$$

The Fourier series of f is

$$\frac{1}{2} - \sum_{n=1}^{\infty} \frac{2}{(2n-1)\pi}\sin\left((2n-1)\pi t\right).$$

8. $f(t) = \begin{cases} t & \text{if } 0 \leq t < 1 \\ 1 & \text{if } 1 \leq t < 2, \quad f \text{ has period 3.} \\ 3 - t & \text{if } 2 \leq t < 3 \end{cases}$

f is even, so its Fourier sine coefficients are all zero. Its cosine coefficients are

$$\frac{a_0}{2} = \frac{1}{2} \cdot \frac{2}{3} \int_0^3 f(t)\, dt = \frac{2}{3}(2) = \frac{2}{3}$$

$$a_n = \frac{2}{3} \int_0^3 f(t) \cos \frac{2n\pi t}{3}\, dt$$

$$= \frac{2}{3} \left[\int_0^1 t \cos \frac{2n\pi t}{3}\, dt + \int_1^2 \cos \frac{2n\pi t}{3}\, dt \right.$$

$$\left. + \int_2^3 (3 - t) \cos \frac{2n\pi t}{3}\, dt \right]$$

$$= \frac{3}{2n^2\pi^2} \left[\cos \frac{2n\pi}{3} - 1 - \cos(2n\pi) + \cos \frac{4n\pi}{3} \right].$$

The latter expression was obtained using Maple to evaluate the integrals. If $n = 3k$, where k is an integer, then $a_n = 0$. For other integers n we have $a_n = -9/(2\pi^2 n^2)$. Thus the Fourier series of f is

$$\frac{2}{3} - \frac{9}{2\pi^2} \sum_{n=1}^{\infty} \frac{1}{n^2} \cos \frac{2n\pi t}{3} + \frac{1}{2\pi^2} \sum_{n=1}^{\infty} \frac{1}{n^2} \cos(2n\pi t).$$

10. The Fourier sine series of $g(t) = \pi - t$ on $[0, \pi]$ has coefficients

$$b_n = \frac{2}{\pi} \int_0^\pi (\pi - t) \sin nt\, dt = \frac{2}{n}.$$

The required Fourier sine series is

$$\sum_{n=1}^{\infty} \frac{2}{n} \sin nt.$$

12. The Fourier cosine series of $f(t) = t$ on $[0, 1]$ has coefficients

$$\frac{a_0}{2} = \int_0^1 t\, dt = \frac{1}{2}$$

$$a_n = 2 \int_0^1 t \cos(n\pi t)\, dt$$

$$= \frac{2(-1)^n - 2}{n^2\pi^2} = \begin{cases} 0 & \text{if } n \text{ is even} \\ \frac{-4}{n^2\pi^2} & \text{if } n \text{ is odd.} \end{cases}$$

The required Fourier cosine series is

$$\frac{1}{2} - \frac{4}{\pi^2} \sum_{n=1}^{\infty} \frac{\cos\big((2n - 1)\pi t\big)}{(2n - 1)^2}.$$

14. If f is even and has period T, then

$$b_n = \frac{2}{T} \int_{-T/2}^{T/2} f(t) \sin \frac{2n\pi t}{T}\, dt$$

$$= \frac{2}{T} \left[\int_{-T/2}^{0} f(t) \sin \frac{2n\pi t}{T}\, dt + \int_0^{T/2} f(t) \sin \frac{2n\pi t}{T}\, dt \right].$$

In the first integral in the line above replace t with $-t$. Since $f(-t) = f(t)$ and sine is odd, we get

$$b_n = \frac{2}{T} \left[\int_{T/2}^{0} f(t) \left(-\sin \frac{2n\pi t}{T} \right)(-dt) \right.$$

$$\left. + \int_0^{T/2} f(t) \sin \frac{2n\pi t}{T}\, dt \right]$$

$$= \frac{2}{T} \left[-\int_0^{T/2} f(t) \sin \frac{2n\pi t}{T}\, dt + \int_0^{T/2} f(t) \sin \frac{2n\pi t}{T}\, dt \right]$$

$$= 0.$$

Similarly,

$$a_n = \frac{2}{T} \left[\int_{-T/2}^{0} f(t) \cos \frac{2n\pi t}{T}\, dt + \int_0^{T/2} f(t) \cos \frac{2n\pi t}{T}\, dt \right]$$

$$= \frac{2}{T} \left[\int_{T/2}^{0} f(t) \cos \frac{2n\pi t}{T}\,(-dt) + \int_0^{T/2} f(t) \cos \frac{2n\pi t}{T}\, dt \right]$$

$$= \frac{4}{T} \int_0^{T/2} f(t) \cos \frac{2n\pi t}{T}\, dt.$$

The corresponding result for an odd function f states that $a_n = 0$ and

$$b_n = \frac{4}{T} \int_0^{T/2} f(t) \sin \frac{2n\pi t}{T}\, dt,$$

and is proved similarly.

Review Exercises 9 (page 534)

2. $\displaystyle \lim_{n \to \infty} \frac{n^{100} + 2^n \pi}{2^n} = \lim_{n \to \infty} \left(\pi + \frac{n^{100}}{2^n} \right) = \pi.$
The sequence converges.

4. $\displaystyle \lim_{n \to \infty} \frac{(-1)^n n^2}{\pi n(n - \pi)} = \lim_{n \to \infty} \frac{(-1)^n}{1 - (\pi/n)}$ does not exist.
The sequence diverges (oscillates).

6. By l'Hôpital's Rule,

$$\lim_{x \to \infty} \frac{\ln(x + 1)}{\ln x} = \lim_{x \to \infty} \frac{1/(x + 1)}{1/x} = \lim_{x \to \infty} \frac{x}{x + 1} = 1.$$

Thus

$$\lim_{n \to \infty} \Big(\ln \ln(n+1) - \ln \ln n \Big) = \lim_{n \to \infty} \ln \frac{\ln(n + 1)}{\ln n} = \ln 1 = 0.$$

8. $\displaystyle\sum_{n=0}^{\infty} \frac{4^{n-1}}{(\pi-1)^{2n}} = \frac{1}{4}\sum_{n=0}^{\infty}\left(\frac{4}{(\pi-1)^2}\right)^n$

$$= \frac{1}{4}\cdot\frac{1}{1-\dfrac{4}{(\pi-1)^2}} = \frac{(\pi-1)^2}{4(\pi-1)^2-16},$$

since $(\pi-1)^2 > 4$.

10. $\displaystyle\sum_{n=1}^{\infty}\frac{1}{n^2-\frac{9}{4}} = \sum_{n=1}^{\infty}\frac{1}{3}\left(\frac{1}{n-\frac{3}{2}}-\frac{1}{n+\frac{3}{2}}\right)$ (telescoping)

$$= \frac{1}{3}\left[\frac{1}{-1/2}-\frac{1}{5/2}+\frac{1}{1/2}-\frac{1}{7/2}\right.$$
$$\left.+\frac{1}{3/2}-\frac{1}{9/2}+\frac{1}{5/2}-\frac{1}{11/2}+\cdots\right]$$
$$= \frac{1}{3}\left[-2+2+\frac{2}{3}\right] = \frac{2}{9}.$$

12. $\displaystyle\sum_{n=1}^{\infty}\frac{n+2^n}{1+3^n}$ converges by comparison with the convergent

geometric series $\displaystyle\sum_{n=1}^{\infty}\left(\frac{2}{3}\right)^n$ because

$$\lim_{n\to\infty}\frac{\dfrac{n+2^n}{1+3^n}}{(2/3)^n} = \lim_{n\to\infty}\frac{(n/2^n)+1}{(1/3^n)+1} = 1.$$

14. $\displaystyle\sum_{n=1}^{\infty}\frac{n^2}{(1+2^n)(1+n\sqrt{n})}$ converges by comparison with

the convergent series $\displaystyle\sum_{n=1}^{\infty}\frac{\sqrt{n}}{2^n}$ (which converges by the

ratio test) because

$$\lim_{n\to\infty}\frac{\dfrac{n^2}{(1+2^n)(1+n\sqrt{n})}}{\dfrac{\sqrt{n}}{2^n}} = \lim_{n\to\infty}\frac{1}{\left(\dfrac{1}{2^n}+1\right)\left(\dfrac{1}{n^{3/2}+1}\right)} = 1.$$

16. $\displaystyle\sum_{n=1}^{\infty}\frac{n!}{(n+2)!+1}$ converges by comparison with the con-

vergent p-series $\displaystyle\sum_{n=1}^{\infty}\frac{1}{n^2}$, because

$$0 \le \frac{n!}{(n+2)!+1} < \frac{n!}{(n+2)!} = \frac{1}{(n+2)(n+1)} < \frac{1}{n^2}.$$

18. $\displaystyle\sum_{n=1}^{\infty}\frac{(-1)^n}{2^n-n}$ converges absolutely by comparison with the

convergent geometric series $\displaystyle\sum_{n=1}^{\infty}\frac{1}{2^n}$, because

$$\lim_{n\to\infty}\frac{\left|\dfrac{(-1)^n}{2^n-n}\right|}{\dfrac{1}{2^n}} = \lim_{n\to\infty}\frac{1}{1-\dfrac{n}{2^n}} = 1.$$

20. $\displaystyle\sum_{n=1}^{\infty}\frac{n^2\cos(n\pi)}{1+n^3}$ converges by the alternating series test
(note that $\cos(n\pi) = (-1)^n$), but the convergence is only
conditional because

$$\left|\frac{n^2\cos(n\pi)}{1+n^3}\right| = \frac{n^2}{1+n^3} \ge \frac{1}{2n}$$

for $n \ge 1$, and $\displaystyle\sum_{n=1}^{\infty}\frac{1}{2n}$ is a divergent harmonic series.

22. $\displaystyle\lim_{n\to\infty}\left|\frac{\dfrac{(5-2x)^{n+1}}{n+1}}{\dfrac{(5-2x)^n}{n}}\right| = \lim_{n\to\infty}|5-2x|\frac{n}{n+1} = |5-2x|.$

$\displaystyle\sum_{n=1}^{\infty}\frac{(5-2x)^n}{n}$ converges absolutely if $|5-2x| < 1$, that
is, if $2 < x < 3$, and diverges if $x < 2$ or $x > 3$.

If $x = 2$ the series is $\displaystyle\sum\frac{1}{n}$, which diverges.

If $x = 3$ the series is $\displaystyle\sum\frac{(-1)^n}{n}$, which converges conditionally.

24. Let $\displaystyle s = \sum_{k=1}^{\infty}\frac{1}{4+k^2}$ and $\displaystyle s_n = \sum_{k=1}^{n}\frac{1}{4+k^2}$. Then

$$\int_{n+1}^{\infty}\frac{dt}{4+t^2} < s-s_n < \int_n^{\infty}\frac{dt}{4+t^2}$$
$$s_n+\frac{\pi}{4}-\frac{1}{2}\tan^{-1}\frac{n+1}{2} < s < s_n+\frac{\pi}{4}-\frac{1}{2}\tan^{-1}\frac{n}{2}.$$

Let

$$s_n^* = s_n+\frac{\pi}{4}-\frac{1}{4}\left[\tan^{-1}\frac{n+1}{2}+\tan^{-1}\frac{n}{2}\right].$$

Then $s \approx s_n^*$ with error satisfying

$$|s-s_n^*| < \frac{1}{4}\left[\tan^{-1}\frac{n+1}{2}-\tan^{-1}\frac{n}{2}\right].$$

This error is less than 0.001 if $n \ge 22$. Hence

$$s \approx \sum_{k=1}^{22}\frac{1}{4+k^2}+\frac{\pi}{4}-\frac{1}{4}\left[\tan^{-1}\frac{23}{2}+\tan^{-1}(11)\right] \approx 0.6605$$

with error less than 0.001.

26. Replace x with x^2 in Exercise 25 and multiply by x to get

$$\frac{x}{3-x^2} = \sum_{n=0}^{\infty} \frac{x^{2n+1}}{3^{n+1}} \quad (-\sqrt{3} < x < \sqrt{3}).$$

28. $\dfrac{1-e^{-2x}}{x} = \dfrac{1}{x}\left(1 - 1 - \displaystyle\sum_{n=1}^{\infty} \dfrac{(-2x)^n}{n!}\right)$

$$= \sum_{n=1}^{\infty} (-1)^{n-1} \frac{2^n x^{n-1}}{n!} \quad \text{(for all } x \neq 0\text{)}.$$

30. $\sin\left(x + \dfrac{\pi}{3}\right) = \sin x \cos \dfrac{\pi}{3} + \cos x \sin \dfrac{\pi}{3}$

$$= \frac{1}{2}\sum_{n=0}^{\infty} (-1)^n \frac{x^{2n+1}}{(2n+1)!} + \frac{\sqrt{3}}{2}\sum_{n=0}^{\infty} (-1)^n \frac{x^{2n}}{(2n)!}$$

$$= \sum_{n=0}^{\infty} \frac{(-1)^n}{2}\left(\frac{\sqrt{3}x^{2n}}{(2n)!} + \frac{x^{2n+1}}{(2n+1)!}\right) \quad \text{(for all } x\text{)}.$$

32. $(1+x)^{1/3} = 1 + \dfrac{1}{3}x + \dfrac{\left(\dfrac{1}{3}\right)\left(-\dfrac{2}{3}\right)}{2!}x^2$

$$+ \frac{\left(\dfrac{1}{3}\right)\left(-\dfrac{2}{3}\right)\left(-\dfrac{5}{3}\right)}{3!}x^3 + \cdots$$

$$= 1 + \frac{x}{3} + \sum_{n=2}^{\infty} (-1)^{n-1} \frac{2 \cdot 5 \cdot 8 \cdots (3n-4)}{3^n n!} x^n \quad (-1 < x < 1).$$

(Remark: the series also converges at $x = 1$.)

34. Let $u = x - (\pi/4)$, so $x = u + (\pi/4)$. Then

$$\sin x + \cos x = \sin\left(u + \frac{\pi}{4}\right) + \cos\left(u + \frac{\pi}{4}\right)$$

$$= \frac{1}{\sqrt{2}}\big((\sin u + \cos u) + (\cos u - \sin u)\big)$$

$$= \sqrt{2}\cos u = \sqrt{2}\sum_{n=0}^{\infty} (-1)^n \frac{u^{2n}}{(2n)!}$$

$$= \sqrt{2}\sum_{n=0}^{\infty} \frac{(-1)^n}{(2n)!}\left(x - \frac{\pi}{4}\right)^{2n} \quad \text{(for all } x\text{)}.$$

36. $\sin(1+x) = \sin(1)\cos x + \cos(1)\sin x$

$$= \sin(1)\left(1 - \frac{x^2}{2!} + \cdots\right) + \cos(1)\left(x - \frac{x^3}{3!} + \cdots\right)$$

$$P_3(x) = \sin(1) + \cos(1)x - \frac{\sin(1)}{2}x^2 - \frac{\cos(1)}{6}x^3.$$

38. $\sqrt{1 + \sin x} = 1 + \dfrac{1}{2}\sin x + \dfrac{\left(\dfrac{1}{2}\right)\left(-\dfrac{1}{2}\right)}{2!}(\sin x)^2$

$$+ \frac{\left(\dfrac{1}{2}\right)\left(-\dfrac{1}{2}\right)\left(-\dfrac{3}{2}\right)}{3!}(\sin x)^3$$

$$+ \frac{\left(\dfrac{1}{2}\right)\left(-\dfrac{1}{2}\right)\left(-\dfrac{3}{2}\right)\left(-\dfrac{5}{2}\right)}{4!}(\sin x)^4 + \cdots$$

$$= 1 + \frac{1}{2}\left(x - \frac{x^3}{6} + \cdots\right) - \frac{1}{8}\left(x - \frac{x^3}{6} + \cdots\right)^2$$

$$+ \frac{1}{16}(x - \cdots)^3 - \frac{5}{128}(x - \cdots)^4 + \cdots$$

$$= 1 + \frac{x}{2} - \frac{x^3}{12} - \frac{x^2}{8} + \frac{x^4}{24} + \frac{x^3}{16} - \frac{5x^4}{128} + \cdots$$

$$P_4(x) = 1 + \frac{x}{2} - \frac{x^2}{8} - \frac{x^3}{48} + \frac{x^4}{384}.$$

40. Since

$$1 + \sum_{n=1}^{\infty} \frac{x^{2n}}{n^2} = \sum_{k=0}^{\infty} \frac{f^{(k)}(0)}{k!} x^k$$

for x near 0, we have, for $n = 1, 2, 3, \ldots$

$$f^{(2n)}(0) = \frac{(2n)!}{n^2}, \quad f^{(2n-1)}(0) = 0.$$

42. $\displaystyle\sum_{n=0}^{\infty} nx^n = \frac{x}{(1-x)^2}$ as in Exercise 23

$$\sum_{n=0}^{\infty} n^2 x^{n-1} = \frac{d}{dx}\frac{x}{(1-x)^2} = \frac{1+x}{(1-x)^3}$$

$$\sum_{n=0}^{\infty} n^2 x^n = \frac{x(1+x)}{(1-x)^3}$$

$$\sum_{n=0}^{\infty} \frac{n^2}{\pi^n} = \frac{\dfrac{1}{\pi}\left(1 + \dfrac{1}{\pi}\right)}{\left(1 - \dfrac{1}{\pi}\right)^3} = \frac{\pi(\pi+1)}{(\pi-1)^3}.$$

44. $\displaystyle\sum_{n=1}^{\infty} \frac{(-1)^{n-1}x^{2n-1}}{(2n-1)!} = \sin x$

$$\sum_{n=1}^{\infty} \frac{(-1)^n \pi^{2n-1}}{(2n-1)!} = -\sin \pi = 0$$

$$\sum_{n=2}^{\infty} \frac{(-1)^n \pi^{2n-4}}{(2n-1)!} = \frac{1}{\pi^3}\left(0 - \frac{(-1)\pi}{1!}\right) = \frac{1}{\pi^2}.$$

46. $\lim\limits_{x\to 0} \dfrac{(x-\tan^{-1}x)(e^{2x}-1)}{2x^2-1+\cos(2x)}$

$= \lim\limits_{x\to 0} \dfrac{\left(x - x + \dfrac{x^3}{3} - \dfrac{x^5}{5} + \cdots\right)\left(2x + \dfrac{4x^2}{2!} + \cdots\right)}{2x^2 - 1 + 1 - \dfrac{4x^2}{2!} + \dfrac{16x^4}{4!} - \cdots}$

$= \lim\limits_{x\to 0} \dfrac{x^4\left(\dfrac{2}{3} + \cdots\right)}{x^4\left(\dfrac{2}{3} + \cdots\right)} = 1.$

48. If $f(x) = \ln(\sin x)$, then calculation of successive derivatives leads to

$$f^{(5)}(x) = 24\csc^4 x \cot x - 8\csc^2 x \cot x.$$

Observe that $1.5 < \pi/2 \approx 1.5708$, that $\csc x \geq 1$ and $\cot x \geq 0$, and that both functions are decreasing on that interval. Thus

$$|f^{(5)}(x)| \leq 24\csc^4(1.5)\cot(1.5) \leq 2$$

for $1.5 \leq x \leq \pi/2$. Therefore, the error in the approximation

$$\ln(\sin 1.5) \approx P_4(x),$$

where P_4 is the 4th degree Taylor polynomial for $f(x)$ about $x = \pi/2$, satisfies

$$|\text{error}| \leq \frac{2}{5!}\left|1.5 - \frac{\pi}{2}\right|^5 \leq 3 \times 10^{-8}.$$

50. $f(t) = \begin{cases} 1 & \text{if } -\pi < t \leq 0 \\ t & \text{if } 0 < t \leq \pi \end{cases}$ has period 2π. Its Fourier coefficients are

$\dfrac{a_0}{2} = \dfrac{1}{2\pi}\displaystyle\int_{-\pi}^{\pi} f(t)\,dt$

$= \dfrac{1}{2\pi}\left[\displaystyle\int_{-\pi}^{0} dt + \int_{0}^{\pi} t\,dt\right] = \dfrac{1}{2} + \dfrac{\pi}{4}$

$a_n = \dfrac{1}{\pi}\left[\displaystyle\int_{-\pi}^{0}\cos(nt)\,dt + \int_{0}^{\pi} t\cos(nt)\,dt\right]$

$= \dfrac{1}{\pi}\displaystyle\int_{0}^{\pi}(1+t)\cos(nt)\,dt$

$= \dfrac{(-1)^n - 1}{\pi n^2} = \begin{cases} -2/(\pi n^2) & \text{if } n \text{ is odd} \\ 0 & \text{if } n \text{ is even} \end{cases}$

$b_n = \dfrac{1}{\pi}\left[\displaystyle\int_{-\pi}^{0}\sin(nt)\,dt + \int_{0}^{\pi} t\sin(nt)\,dt\right]$

$= \dfrac{1}{\pi}\displaystyle\int_{0}^{\pi}(t-1)\sin(nt)\,dt$

$= -\dfrac{1 + (-1)^n(\pi - 1)}{\pi n} = \begin{cases} (\pi - 2/(\pi n)) & \text{if } n \text{ is odd} \\ -(1/n) & \text{if } n \text{ is even.} \end{cases}$

The required Fourier series is, therefore,

$\dfrac{2+\pi}{4}$

$-\displaystyle\sum_{n=1}^{\infty}\left[\frac{2\cos\big((2n-1)t\big)}{\pi(2n-1)^2} + \frac{(2-\pi)\sin\big((2n-1)t\big)}{\pi(2n-1)} + \frac{\sin(2nt)}{2n}\right].$

Challenging Problems 9   (page 535)

2. a) If $s_n = \sum_{k=1}^{n} v_k$ for $n \geq 1$, and $s_0 = 0$, then $v_k = s_k - s_{k-1}$ for $k \geq 1$, and

$$\sum_{k=1}^{n} u_k v_k = \sum_{k=1}^{n} u_k s_k - \sum_{k=1}^{n} u_k s_{k-1}.$$

In the second sum on the right replace k with $k+1$:

$$\sum_{k=1}^{n} u_k v_k = \sum_{k=1}^{n} u_k s_k - \sum_{k=0}^{n-1} u_{k+1} s_k$$

$$= \sum_{k=1}^{n} (u_k - u_{k+1})s_k - u_1 s_0 + u_{n+1}s_n$$

$$= u_{n+1}s_n + \sum_{k=1}^{n} (u_k - u_{k+1})s_k.$$

b) If $\{u_n\}$ is positive and decreasing, and $\lim_{n\to\infty} u_n = 0$, then

$$\sum_{k=1}^{n} (u_k - u_{k+1}) = u_1 - u_2 + u_2 - u_3 + \cdots + u_n - u_{n+1}$$

$$= u_1 - u_{n+1} \to u_1 \text{ as } n \to \infty.$$

Thus $\sum_{k=1}^{n} (u_k - u_{k+1})$ is a convergent, positive, telescoping series.

If the partial sums s_n of $\{v_n\}$ are bounded, say $|s_n| \leq K$ for all n, then

$$|(u_n - u_{n+1})s_n| \leq K(u_n - u_{n+1}),$$

so $\sum_{n=1}^{\infty}(u_n - u_{n+1})s_n$ is absolutely convergent (and therefore convergent) by the comparison test. Therefore, by part (a),

$$\sum_{k=1}^{\infty} u_k v_k = \lim_{n\to\infty}\left(u_{n+1}s_n + \sum_{k=1}^{n}(u_k - u_{k+1})s_k\right)$$

$$= \sum_{k=1}^{\infty}(u_k - u_{k+1})s_k$$

converges.

4. Let a_n be the nth integer that has no zeros in its decimal representation. The number of such integers that have m digits is 9^m. (There are nine possible choices for each of the m digits.) Also, each such m-digit number is greater than 10^{m-1} (the smallest m-digit number). Therefore the sum of all the terms $1/a_n$ for which a_n has m digits is less than $9^m/(10^{m-1})$. Therefore,

$$\sum_{n=1}^{\infty} \frac{1}{a_n} < 9 \sum_{m=1}^{\infty} \left(\frac{9}{10}\right)^{m-1} = 90.$$

6.   a) Since $e = \sum_{j=0}^{\infty} \frac{1}{j!}$, we have

$$0 < e - \sum_{j=0}^{n} \frac{1}{j!} = \sum_{j=n+1}^{\infty} \frac{1}{j!}$$

$$= \frac{1}{(n+1)!} \left(1 + \frac{1}{n+2} + \frac{1}{(n+2)(n+3)} + \cdots\right)$$

$$\le \frac{1}{(n+1)!} \left(1 + \frac{1}{n+2} + \frac{1}{(n+2)^2} + \cdots\right)$$

$$= \frac{1}{(n+1)!} \cdot \frac{1}{1 - \dfrac{1}{n+2}} = \frac{n+2}{(n+1)!(n+1)} < \frac{1}{n!n}.$$

The last inequality follows from $\dfrac{n+2}{(n+1)^2} < \dfrac{1}{n}$, that is, $n^2 + 2n < n^2 + 2n + 1$.

   b) Suppose e is rational, say $e = M/N$ where M and N are positive integers. Then $N!e$ is an integer and $N! \sum_{j=0}^{N}(1/j!)$ is an integer (since each $j!$ is a factor of $N!$). Therefore the number

$$Q = N! \left(e - \sum_{j=0}^{N} \frac{1}{j!}\right)$$

is a difference of two integers and so is an integer.

   c) By part (a), $0 < Q < \dfrac{1}{N} \le 1$. By part (b), Q is an integer. This is not possible; there are no integers between 0 and 1. Therefore e cannot be rational.

8. Let f be a polynomial and let

$$g(x) = \sum_{j=0}^{\infty} (-1)^j f^{(2j)}(x).$$

This "series" is really just a polynomial since sufficiently high derivatives of f are all identically zero.

   a) By replacing j with $j - 1$, observe that

$$g''(x) = \sum_{j=0}^{\infty} (-1)^j f^{(2j+2)}(x)$$

196
$$= \sum_{j=1}^{\infty} (-1)^{j-1} f^{(2j)}(x) = -\Big(g(x) - f(x)\Big).$$

Also

$$\frac{d}{dx}\Big(g'(x)\sin x - g(x)\cos x\Big)$$

$$= g''(x)\sin x + g'(x)\cos x - g'(x)\cos x + g(x)\sin x$$

$$= \Big(g''(x) + g(x)\Big)\sin x = f(x)\sin x.$$

Thus

$$\int_0^{\pi} f(x)\sin x \, dx = \Big(g'(x)\sin x - g(x)\cos x\Big)\Big|_0^{\pi} = g(\pi) + g(0).$$

   b) Suppose that $\pi = m/n$, where m and n are positive integers. Since $\lim_{k\to\infty} x^k/k! = 0$ for any x, there exists an integer k such that $(\pi m)^k/k! < 1/2$. Let

$$f(x) = \frac{x^k(m - nx)^k}{k!} = \frac{1}{k!} \sum_{j=0}^{k} \binom{k}{j} m^{k-j}(-n)^j x^{j+k}.$$

The sum is just the binomial expansion. For $0 < x < \pi = m/n$ we have

$$0 < f(x) < \frac{\pi^k m^k}{k!} < \frac{1}{2}.$$

Thus $0 < \int_0^{\pi} f(x)\sin x \, dx < \dfrac{1}{2} \int_0^{\pi} \sin x \, dx = 1$, and so $0 < g(\pi) + g(0) < 1$.

   c) $f^{(i)}(x) = \dfrac{1}{k!} \sum_{j=0}^{k} \binom{k}{j} m^{k-j}(-n)^j$

$$\times (j+k)(j+k-1)\cdots(j+k-i+1)x^{j+k-i}$$

$$= \frac{1}{k!} \sum_{j=0}^{k} \binom{k}{j} m^{k-j}(-n)^j \frac{(j+k)!}{(j+k-i)!} x^{j+k-i}.$$

   d) Evidently $f^{(i)}(0) = 0$ if $i < k$ or if $i > 2k$. If $k \le i \le 2k$, the only term in the sum for $f^{(i)}(0)$ that is not zero is the term for which $j = i - k$. This term is the constant

$$\frac{1}{k!} \binom{k}{i-k} m^{k-j}(-n)^j \frac{i!}{0!}.$$

This constant is an integer because the binomial coefficient $\binom{k}{i-k}$ is an integer and $i!/k!$ is an integer. (The other factors are also integers.) Hence $f^{(i)}(0)$ is an integer, and so $g(0)$ is an integer.

   e) Observe that $f(\pi - x) = f((m/n) - x) = f(x)$ for all x. Therefore $f^{(i)}(\pi)$ is an integer (for each i), and so $g(\pi)$ is an integer. Thus $g(\pi) + g(0)$ is an integer, which contradicts the conclusion of part (b). (There is no integer between 0 and 1.) Therefore, π cannot be rational.

CHAPTER 17. ORDINARY DIFFEREN-TIAL EQUATIONS

Section 17.1 Classifying Differential Equations (page 902)

2. $\dfrac{d^2y}{dx^2} + x = y$: 2nd order, linear, nonhomogeneous.

4. $y''' + xy' = x\sin x$: 3rd order, linear, nonhomogeneous.

6. $y'' + 4y' - 3y = 2y^2$: 2nd order, nonlinear.

8. $\cos x \dfrac{dx}{dt} + x\sin t = 0$: 1st order, nonlinear, homogeneous.

10. $x^2 y'' + e^x y' = \dfrac{1}{y}$: 2nd order, nonlinear.

12. If $y = e^x$, then $y'' - y = e^x - e^x = 0$; if $y = e^{-x}$, then $y'' - y = e^{-x} - e^{-x} = 0$. Thus e^x and e^{-x} are both solutions of $y'' - y = 0$. Since $y'' - y = 0$ is linear and homogeneous, any function of the form

$$y = Ae^x + Be^{-x}$$

is also a solution. Thus $\cosh x = \frac{1}{2}(e^x + e^{-x})$ is a solution, but neither $\cos x$ nor x^e is a solution.

14. Given that $y_1 = e^{kx}$ is a solution of $y'' - k^2 y = 0$, we suspect that $y_2 = e^{-kx}$ is also a solution. This is easily verified since

$$y_2'' - k^2 y_2 = k^2 e^{-kx} - k^2 e^{-kx} = 0.$$

Since the DE is linear and homogeneous,

$$y = Ay_1 + By_2 = Ae^{kx} + Be^{-kx}$$

is a solution for any constants A and B. It will satisfy

$$0 = y(1) = Ae^k + Be^{-k}$$
$$2 = y'(1) = Ake^k - Bke^{-k},$$

provided $A = e^{-k}/k$ and $B = -e^k/k$. The required solution is

$$y = \frac{1}{k} e^{k(x-1)} - \frac{1}{k} e^{-k(x-1)}.$$

16. $y = e^{rx}$ is a solution of the equation $y'' - y' - 2y = 0$ if $r^2 e^{rx} - re^{rx} - 2e^{rx} = 0$, that is, if $r^2 - r - 2 = 0$. This quadratic has two roots, $r = 2$, and $r = -1$. Since the DE is linear and homogeneous, the function $y = Ae^{2x} + Be^{-x}$ is a solution for any constants A and B. This solution satisfies

$$1 = y(0) = A + B, \quad 2 = y'(0) = 2A - B,$$

provided $A = 1$ and $B = 0$. Thus, the required solution is $y = e^{2x}$.

18. If $y = y_1(x) = -e$, then $y_1' = 0$ and $y_1'' = 0$. Thus $y_1'' - y_1 = 0 + e = e$. By Exercise 12 we know that $y_2 = Ae^x + Be^{-x}$ satisfies the homogeneous DE $y'' - y = 0$. Therefore, by Theorem 2,

$$y = y_1(x) + y_2(x) = -e + Ae^x + Be^{-x}$$

is a solution of $y'' - y = e$. This solution satisfies

$$0 = y(1) = Ae + \frac{B}{e} - e, \quad 1 = y'(1) = Ae - \frac{B}{e},$$

provided $A = (e+1)/(2e)$ and $B = e(e-1)/2$. Thus the required solution is $y = -e + \frac{1}{2}(e+1)e^{x-1} + \frac{1}{2}(e-1)e^{1-x}$.

Section 17.2 Solving First-Order Equations (page 913)

2.
$$\frac{dy}{dx} = \frac{3y - 1}{x}$$
$$\int \frac{dy}{3y - 1} = \int \frac{dx}{x}$$
$$\frac{1}{3}\ln|3y - 1| = \ln|x| + \frac{1}{3}\ln C$$
$$\frac{3y - 1}{x^3} = C$$
$$\Rightarrow \quad y = \frac{1}{3}(1 + Cx^3).$$

4.
$$\frac{dy}{dx} = x^2 y^2$$
$$\int \frac{dy}{y^2} = \int x^2 \, dx$$
$$-\frac{1}{y} = \frac{1}{3}x^3 + \frac{1}{3}C$$
$$\Rightarrow \quad y = -\frac{3}{x^3 + C}.$$

6.
$$\frac{dx}{dt} = e^x \sin t$$
$$\int e^{-x} \, dx = \int \sin t \, dt$$
$$-e^{-x} = -\cos t - C$$
$$\Rightarrow \quad x = -\ln(\cos t + C).$$

8.
$$\frac{dy}{dx} = 1 + y^2$$
$$\int \frac{dy}{1 + y^2} = \int dx$$
$$\tan^{-1} y = x + C$$
$$\Rightarrow \quad y = \tan(x + C).$$

10. We have

$$\frac{dy}{dx} = y^2(1-y)$$

$$\int \frac{dy}{y^2(1-y)} = \int dx = x + K.$$

Expand the left side in partial fractions:

$$\frac{1}{y^2(1-y)} = \frac{A}{y} + \frac{B}{y^2} + \frac{C}{1-y}$$

$$= \frac{A(y-y^2) + B(1-y) + Cy^2}{y^2(1-y)}$$

$$\Rightarrow \begin{cases} -A + C = 0; \\ A - B = 0; \quad \Rightarrow A = B = C = 1. \\ B = 1. \end{cases}$$

Hence,

$$\int \frac{dy}{y^2(1-y)} = \int \left(\frac{1}{y} + \frac{1}{y^2} + \frac{1}{1-y} \right) dy$$

$$= \ln|y| - \frac{1}{y} - \ln|1-y|.$$

Therefore,

$$\ln\left| \frac{y}{1-y} \right| - \frac{1}{y} = x + K.$$

12. $\dfrac{dy}{dx} = \dfrac{xy}{x^2 + 2y^2}$ $\qquad$ Let $y = vx$

$$v + x\frac{dv}{dx} = \frac{vx^2}{(1 + 2v^2)x^2}$$

$$x\frac{dv}{dx} = \frac{v}{1 + 2v^2} - v = -\frac{2v^3}{1 + 2v^2}$$

$$\int \frac{1 + 2v^2}{v^3}\, dv = -2 \int \frac{dx}{x}$$

$$-\frac{1}{2v^2} + 2\ln|v| = -2\ln|x| + C_1$$

$$-\frac{x^2}{2y^2} + 2\ln|y| = C_1$$

$$x^2 - 4y^2 \ln|y| = Cy^2.$$

14. $\dfrac{dy}{dx} = \dfrac{x^3 + 3xy^2}{3x^2 y + y^3}$ $\qquad$ Let $y = vx$

$$v + x\frac{dv}{dx} = \frac{x^3(1 + 3v^2)}{x^3(3v + v^3)}$$

$$x\frac{dv}{dx} = \frac{1 + 3v^2}{3v + v^3} - v = \frac{1 - v^4}{v(3 + v^2)}$$

$$\int \frac{(3 + v^2)v\, dv}{1 - v^4} = \int \frac{dx}{x} \qquad \begin{aligned} &\text{Let } u = v^2 \\ &du = 2v\, dv \end{aligned}$$

$$\frac{1}{2} \int \frac{3 + u}{1 - u^2}\, du = \ln|x| + C_1$$

$$\frac{3}{4} \ln\left| \frac{u+1}{u-1} \right| - \frac{1}{4} \ln|1 - u^2| = \ln|x| + C_1$$

$$3 \ln\left| \frac{y^2 + x^2}{y^2 - x^2} \right| - \ln\left| \frac{x^4 - y^4}{x^4} \right| = 4\ln|x| + C_2$$

$$\ln\left| \left(\frac{x^2 + y^2}{x^2 - y^2} \right)^3 \frac{1}{x^4 - y^4} \right| = C_2$$

$$\ln\left| \frac{(x^2 + y^2)^2}{(x^2 - y^2)^4} \right| = C_2$$

$$x^2 + y^2 = C(x^2 - y^2)^2.$$

16. $\dfrac{dy}{dx} = \dfrac{y}{x} - e^{-y/x}$ $\quad$ (let $y = vx$)

$$v + x\frac{dv}{dx} = v - e^{-v}$$

$$e^v\, dv = -\frac{dx}{x}$$

$$e^v = -\ln|x| + \ln|C|$$

$$e^{y/x} = \ln\left| \frac{C}{x} \right|$$

$$y = x \ln\ln\left| \frac{C}{x} \right|.$$

18. We have $\dfrac{dy}{dx} + \dfrac{2y}{x} = \dfrac{1}{x^2}$. Let

$$\mu = \int \frac{2}{x}\, dx = 2\ln x = \ln x^2, \text{ then } e^\mu = x^2, \text{ and}$$

$$\frac{d}{dx}(x^2 y) = x^2 \frac{dy}{dx} + 2xy$$

$$= x^2 \left(\frac{dy}{dx} + \frac{2y}{x} \right) = x^2 \left(\frac{1}{x^2} \right) = 1$$

$$\Rightarrow \quad x^2 y = \int dx = x + C$$

$$\Rightarrow \quad y = \frac{1}{x} + \frac{C}{x^2}.$$

20. We have $\dfrac{dy}{dx} + y = e^x$. Let $\mu = \int dx = x$, then $e^\mu = e^x$, and

$$\frac{d}{dx}(e^x y) = e^x \frac{dy}{dx} + e^x y = e^x\left(\frac{dy}{dx} + y\right) = e^{2x}$$

$$\Rightarrow \quad e^x y = \int e^{2x}\, dx = \frac{1}{2}e^{2x} + C.$$

Hence, $y = \dfrac{1}{2}e^x + Ce^{-x}$.

22. We have $\dfrac{dy}{dx} + 2e^x y = e^x$. Let $\mu = \int 2e^x\, dx = 2e^x$, then

$$\frac{d}{dx}\left(e^{2e^x} y\right) = e^{2e^x}\frac{dy}{dx} + 2e^x e^{2e^x} y$$

$$= e^{2e^x}\left(\frac{dy}{dx} + 2e^x y\right) = e^{2e^x} e^x.$$

Therefore,

$$e^{2e^x} y = \int e^{2e^x} e^x\, dx \quad \text{Let } u = 2e^x$$
$$du = 2e^x\, dx$$
$$= \frac{1}{2}\int e^u\, du = \frac{1}{2}e^{2e^x} + C.$$

Hence, $y = \dfrac{1}{2} + Ce^{-2e^x}$.

24. $\dfrac{dy}{dx} + 3x^2 y = x^2, \qquad y(0) = 1$

$$\mu = \int 3x^2\, dx = x^3$$

$$\frac{d}{dx}(e^{x^3} y) = e^{x^3}\frac{dy}{dx} + 3x^2 e^{x^3} y = x^2 e^{x^3}$$

$$e^{x^3} y = \int x^2 e^{x^3}\, dx = \frac{1}{3}e^{x^3} + C$$

$$y(0) = 1 \Rightarrow 1 = \frac{1}{3} + C \Rightarrow C = \frac{2}{3}$$

$$y = \frac{1}{3} + \frac{2}{3}e^{-x^3}.$$

26. $y' + (\cos x)y = 2xe^{-\sin x}, \qquad y(\pi) = 0$

$$\mu = \int \cos x\, dx = \sin x$$

$$\frac{d}{dx}(e^{\sin x} y) = e^{\sin x}(y' + (\cos x)y) = 2x$$

$$e^{\sin x} y = \int 2x\, dx = x^2 + C$$

$$y(\pi) = 0 \Rightarrow 0 = \pi^2 + C \Rightarrow C = -\pi^2$$

$$y = (x^2 - \pi^2)e^{-\sin x}.$$

28. $y(x) = 1 + \displaystyle\int_0^x \frac{(y(t))^2}{1+t^2}\, dt \quad \Longrightarrow \quad y(0) = 1$

$$\frac{dy}{dx} = \frac{y^2}{1+x^2}, \qquad \text{i.e. } dy/y^2 = dx/(1+x^2)$$

$$-\frac{1}{y} = \tan^{-1}x + C$$

$$-1 = 0 + C \quad \Longrightarrow \quad C = -1$$

$$y = 1/(1 - \tan^{-1}x).$$

30. $y(x) = 3 + \displaystyle\int_0^x e^{-y}\, dt \quad \Longrightarrow \quad y(0) = 3$

$$\frac{dy}{dx} = e^{-y}, \qquad \text{i.e. } e^y\, dy = dx$$

$$e^y = x + C \quad \Longrightarrow \quad y = \ln(x + C)$$

$$3 = y(0) = \ln C \quad \Longrightarrow \quad C = e^3$$

$$y = \ln(x + e^3).$$

32. $\dfrac{dy}{dx} = 1 + \dfrac{2y}{x} \qquad \text{Let } y = vx$

$$v + x\frac{dv}{dx} = 1 + 2v$$

$$x\frac{dv}{dx} = 1 + v$$

$$\int \frac{dv}{1+v} = \int \frac{dx}{x}$$

$$\ln|1+v| = \ln|x| + C_1$$

$$1 + \frac{y}{x} = Cx \quad \Rightarrow \quad x + y = Cx^2.$$

Since $(1, 3)$ lies on the curve, $4 = C$. Thus the curve has equation $x + y = 4x^2$.

34. The system $x_0 + 2y_0 - 4 = 0$, $2x_0 - y_0 - 3 = 0$ has solution $x_0 = 2$, $y_0 = 1$. Thus, if $\xi = x - 2$ and $\eta = y - 1$, where

$$\frac{dy}{dx} = \frac{x + 2y - 4}{2x - y - 3},$$

then

$$\frac{d\eta}{d\xi} = \frac{\xi + 2\eta}{2\xi - \eta} \qquad \text{Let } \eta = v\xi$$

$$v + \xi\frac{dv}{d\xi} = \frac{1 + 2v}{2 - v}$$

$$\xi\frac{dv}{d\xi} = \frac{1 + 2v}{2 - v} - v = \frac{1 + v^2}{2 - v}$$

$$\int \left(\frac{2 - v}{1 + v^2}\right) dv = \int \frac{d\xi}{\xi}$$

$$2\tan^{-1}v - \frac{1}{2}\ln(1 + v^2) = \ln|\xi| + C_1$$

$$4\tan^{-1}\frac{\eta}{\xi} - \ln(\xi^2 + \eta^2) = C.$$

Hence the solution of the original equation is

$$4\tan^{-1}\frac{y - 1}{x - 2} - \ln\left((x - 2)^2 + (y - 1)^2\right) = C.$$

36. $(e^x \sin y + 2x)\, dx + (e^x \cos y + 2y)\, dy = 0$

$d(e^x \sin y + x^2 + y^2) = 0$

$e^x \sin y + x^2 + y^2 = C.$

38. $\left(2x + 1 - \dfrac{y^2}{x^2}\right) dx + \dfrac{2y}{x}\, dy = 0$

$d\left(x^2 + x + \dfrac{y^2}{x}\right) = 0$

$x^2 + x + \dfrac{y^2}{x} = C.$

40. $(xe^x + x \ln y + y)\, dx + \left(\dfrac{x^2}{y} + x \ln x + x \sin y\right) dy = 0$

$M = xe^x + x \ln y + y, \qquad N = \dfrac{x^2}{y} + x \ln x + x \sin y$

$\dfrac{\partial M}{\partial y} = \dfrac{x}{y} + 1, \qquad \dfrac{\partial N}{\partial x} = \dfrac{2x}{y} + \ln x + 1 + \sin y$

$\dfrac{1}{N}\left(\dfrac{\partial M}{\partial y} - \dfrac{\partial N}{\partial x}\right) = \dfrac{1}{N}\left(-\dfrac{x}{y} - \ln x - \sin y\right) = -\dfrac{1}{x}$

$\dfrac{d\mu}{\mu} = -\dfrac{1}{x}\, dx \quad \Rightarrow \quad \mu = \dfrac{1}{x}$

$\left(e^x + \ln y + \dfrac{y}{x}\right) dx + \left(\dfrac{x}{y} + \ln x + \sin y\right) dy$

$d\left(e^x + x \ln y + y \ln x - \cos y\right) = 0$

$e^x + x \ln y + y \ln x - \cos y = C.$

42. Since $b > a > 0$ and $k > 0$,

$$\lim_{t \to \infty} x(t) = \lim_{t \to \infty} \dfrac{ab\left(e^{(b-a)kt} - 1\right)}{be^{(b-a)kt} - a}$$

$$= \lim_{t \to \infty} \dfrac{ab\left(1 - e^{(a-b)kt}\right)}{b - ae^{(a-b)kt}}$$

$$= \dfrac{ab(1 - 0)}{b - 0} = a.$$

44. Given that $m\dfrac{dv}{dt} = mg - kv$, then

$$\int \dfrac{dv}{g - \dfrac{k}{m}v} = \int dt$$

$$-\dfrac{m}{k} \ln\left|g - \dfrac{k}{m}v\right| = t + C.$$

Since $v(0) = 0$, therefore $C = -\dfrac{m}{k} \ln g$. Also, $g - \dfrac{k}{m}v$ remains positive for all $t > 0$, so

$$\dfrac{m}{k} \ln \dfrac{g}{g - \dfrac{k}{m}v} = t$$

$$\dfrac{g - \dfrac{k}{m}v}{g} = e^{-kt/m}$$

$$\Rightarrow \quad v = v(t) = \dfrac{mg}{k}\left(1 - e^{-kt/m}\right).$$

Note that $\lim_{t \to \infty} v(t) = \dfrac{mg}{k}$. This limiting velocity can be obtained directly from the differential equation by setting $\dfrac{dv}{dt} = 0$.

46. The balance in the account after t years is $y(t)$ and $y(0) = 1000$. The balance must satisfy

$$\dfrac{dy}{dt} = 0.1y - \dfrac{y^2}{1,000,000}$$

$$\dfrac{dy}{dt} = \dfrac{10^5 y - y^2}{10^6}$$

$$\int \dfrac{dy}{10^5 y - y^2} = \int \dfrac{dt}{10^6}$$

$$\dfrac{1}{10^5} \int \left(\dfrac{1}{y} + \dfrac{1}{10^5 - y}\right) dy = \dfrac{t}{10^6} - \dfrac{C}{10^5}$$

$$\ln|y| - \ln|10^5 - y| = \dfrac{t}{10} - C$$

$$\dfrac{10^5 - y}{y} = e^{C - (t/10)}$$

$$y = \dfrac{10^5}{e^{C - (t/10)} + 1}.$$

Since $y(0) = 1000$, we have

$$1000 = y(0) = \dfrac{10^5}{e^C + 1} \quad \Rightarrow \quad C = \ln 99,$$

and

$$y = \dfrac{10^5}{99e^{-t/10} + 1}.$$

The balance after 1 year is

$$y = \dfrac{10^5}{99e^{-1/10} + 1} \approx \$1,104.01.$$

As $t \to \infty$, the balance can grow to

$$\lim_{t \to \infty} y(t) = \lim_{t \to \infty} \dfrac{10^5}{e^{(4.60 - 0.1t)} + 1} = \dfrac{10^5}{0 + 1} = \$100,000.$$

For the account to grow to \$50,000, t must satisfy

$$50,000 = y(t) = \dfrac{100,000}{99e^{-t/10} + 1}$$

$$\Rightarrow \quad 99e^{-t/10} + 1 = 2$$

$$\Rightarrow \quad t = 10 \ln 99 \approx 46 \text{ years}.$$

48. Let $x(t)$ be the number of kg of salt in the solution in the tank after t minutes. Thus, $x(0) = 50$. Salt is coming into the tank at a rate of 10 g/L × 12 L/min = 0.12 kg/min. Since the contents flow out at a rate of 10 L/min, the volume of the solution is increasing at 2 L/min and thus, at any time t, the volume of the solution is $1000 + 2t$ L. Therefore the concentration of salt is $\dfrac{x(t)}{1000 + 2t}$ L. Hence, salt is being removed at a rate

$$\frac{x(t)}{1000 + 2t} \text{ kg/L} \times 10 \text{ L/min} = \frac{5x(t)}{500 + t} \text{ kg/min}.$$

Therefore,

$$\frac{dx}{dt} = 0.12 - \frac{5x}{500 + t}$$
$$\frac{dx}{dt} + \frac{5}{500 + t}x = 0.12.$$

Let $\mu = \displaystyle\int \frac{5}{500 + t} dt = 5\ln|500 + t| = \ln(500 + t)^5$ for $t > 0$. Then $e^\mu = (500 + t)^5$, and

$$\frac{d}{dt}\left[(500 + t)^5 x\right] = (500 + t)^5 \frac{dx}{dy} + 5(500 + t)^4 x$$
$$= (500 + t)^5 \left(\frac{dx}{dy} + \frac{5x}{500 + t}\right)$$
$$= 0.12(500 + t)^5.$$

Hence,

$$(500 + t)^5 x = 0.12 \int (500 + t)^5 dt = 0.02(500 + t)^6 + C$$
$$\Rightarrow x = 0.02(500 + t) + C(500 + t)^{-5}.$$

Since $x(0) = 50$, we have $C = 1.25 \times 10^{15}$ and

$$x = 0.02(500 + t) + (1.25 \times 10^{15})(500 + t)^{-5}.$$

After 40 min, there will be

$$x = 0.02(540) + (1.25 \times 10^{15})(540)^{-5} = 38.023 \text{ kg}$$

of salt in the tank.

50. $2y^2(x + y^2)\,dx + xy(x + 6y^2)\,dy = 0$
$(2xy^2 + 2y^4)\mu(y)\,dx + (x^2y + 6xy^3)\mu(y)\,dy = 0$
$\dfrac{\partial M}{\partial y} = (4xy + 8y^3)\mu(y) + (2xy^2 + 2y^4)\mu'(y)$
$\dfrac{\partial N}{\partial x} = (2xy + 6y^3)\mu(y).$
For exactness we require
$(2xy^2 + 2y^4)\mu'(y) = [(2xy + 6y^3) - (4xy + 8y^3)]\mu(y)$
$y(2xy + 2y^3)\mu'(y) = -(2xy + 2y^3)\mu(y)$
$y\mu'(y) = -\mu(y) \quad \Rightarrow \quad \mu(y) = \dfrac{1}{y}$
$(2xy + 2y^3)\,dx + (x^2 + 6xy^2)\,dy = 0$
$d(x^2y + 2xy^3) = 0 \quad \Rightarrow \quad x^2y + 2xy^3 = C.$

52. If $\mu(xy)$ is an integrating factor for $M\,dx + N\,dy = 0$, then

$$\frac{\partial}{\partial y}(\mu M) = \frac{\partial}{\partial x}(\mu N), \qquad \text{or}$$
$$x\mu'(xy)M + \mu(xy)\frac{\partial M}{\partial y} = y\mu'(xy)N + \mu(xy)\frac{\partial N}{\partial x}.$$

Thus M and N will have to be such that the right-hand side of the equation

$$\frac{\mu'(xy)}{\mu(xy)} = \frac{1}{xM - yN}\left(\frac{\partial N}{\partial x} - \frac{\partial M}{\partial y}\right)$$

depends only on the product xy.

Section 17.3 Existence, Uniqueness, and Numerical Methods (page 921)

A computer spreadsheet was used in Exercises 1–12. The intermediate results appearing in the spreadsheet are not shown in these solutions.

2. We start with $x_0 = 1$, $y_0 = 0$, and calculate

$$x_{n+1} = x_n + h, \qquad u_{n+1} = y_n + h(x_n + y_n)$$
$$y_{n+1} = y_n + \frac{h}{2}(x_n + y_n + x_{n+1} + u_{n+1}).$$

a) For $h = 0.2$ we get $x_5 = 2$, $y_5 = 2.405416$.
b) For $h = 0.1$ we get $x_{10} = 2$, $y_{10} = 2.428162$.
c) For $h = 0.05$ we get $x_{20} = 2$, $y_{20} = 2.434382$.

4. We start with $x_0 = 0$, $y_0 = 0$, and calculate

$$x_{n+1} = x_n + h, \qquad y_{n+1} = hx_n e^{-y_n}.$$

a) For $h = 0.2$ we get $x_{10} = 2$, $y_{10} = 1.074160$.
b) For $h = 0.1$ we get $x_{20} = 2$, $y_{20} = 1.086635$.

6. We start with $x_0 = 0$, $y_0 = 0$, and calculate

$$x_{n+1} = x_n + h$$
$$p_n = x_n e^{-y_n}$$
$$q_n = \left(x_n + \frac{h}{2}\right) e^{-(y_n + (h/2)p_n)}$$
$$r_n = \left(x_n + \frac{h}{2}\right) e^{-(y_n + (h/2)q_n)}$$
$$s_n = (x_n + h)e^{-(y_n + hr_n)}$$
$$y_{n+1} = y_n + \frac{h}{6}(p_n + 2q_n + 2r_n + s_n).$$

a) For $h = 0.2$ we get $x_{10} = 2$, $y_{10} = 1.098614$.

b) For $h = 0.1$ we get $x_{20} = 2$, $y_{20} = 1.098612$.

8. We start with $x_0 = 0$, $y_0 = 0$, and calculate

$$x_{n+1} = x_n + h, \qquad u_{n+1} = y_n + h \cos y_n$$
$$y_{n+1} = y_n + \frac{h}{2}(\cos y_n + \cos u_{n+1}).$$

a) For $h = 0.2$ we get $x_5 = 1$, $y_5 = 0.862812$.

b) For $h = 0.1$ we get $x_{10} = 1$, $y_{10} = 0.865065$.

c) For $h = 0.05$ we get $x_{20} = 1$, $y_{20} = 0.865598$.

10. We start with $x_0 = 0$, $y_0 = 0$, and calculate

$$x_{n+1} = x_n + h, \qquad y_{n+1} = y_n + h \cos(x_n^2).$$

a) For $h = 0.2$ we get $x_5 = 1$, $y_5 = 0.944884$.

b) For $h = 0.1$ we get $x_{10} = 1$, $y_{10} = 0.926107$.

c) For $h = 0.05$ we get $x_{20} = 1$, $y_{20} = 0.915666$.

12. We start with $x_0 = 0$, $y_0 = 0$, and calculate

$$x_{n+1} = x_n + h$$
$$p_n = \cos(x_n^2)$$
$$q_n = \cos((x_n + (h/2))^2)$$
$$r_n = \cos((x_n + (h/2))^2)$$
$$q_n = \cos((x_n + h)^2)$$
$$y_{n+1} = y_n + \frac{h}{6}(p_n + 2q_n + 2r_n + s_n).$$

a) For $h = 0.2$ we get $x_5 = 1$, $y_5 = 0.904524$.

b) For $h = 0.1$ we get $x_{10} = 1$, $y_{10} = 0.904524$.

c) For $h = 0.05$ we get $x_{20} = 1$, $y_{20} = 0.904524$.

14. $u(x) = 1 + 3 \int_2^x t^2 u(t)\, dt$

$$\frac{du}{dx} = 3x^2 u(x), \qquad u(2) = 1 + 0 = 1$$
$$\frac{du}{u} = 3x^2\, dx \;\Rightarrow\; \ln u = x^3 + C$$
$$0 = \ln 1 = \ln u(2) = 2^3 + C \;\Rightarrow\; C = -8$$
$$u = e^{x^3 - 8}.$$

16. If $\phi(0) = A \geq 0$ and $\phi'(x) \geq k\phi(x)$ on an interval $[0, X]$, where $k > 0$ and $X > 0$, then

$$\frac{d}{dx}\left(\frac{\phi(x)}{e^{kx}}\right) = \frac{e^{kx}\phi'(x) - ke^{kx}\phi(x)}{e^{2kx}} \geq 0.$$

Thus $\phi(x)/e^{kx}$ is increasing on $[0, X]$. Since its value at $x = 0$ is $\phi(0) = A \geq 0$, therefore $\phi(x)/e^{kx} \geq A$ on $[0, X]$, and $\phi(x) \geq Ae^{kx}$ there.

Section 17.4 Differential Equations of Second Order   (page 925)

2. If $y_1 = e^{-2x}$, then $y_1'' - y_1' - 6y_1 = e^{-2x}(4 + 2 - 6) = 0$, so y_1 is a solution of the DE $y'' - y' - 6y = 0$. Let $y = e^{-2x}v$. Then

$$y' = e^{-2x}(v' - 2v), \qquad y'' = e^{-2x}(v'' - 4v' + 4v)$$
$$y'' - y' - 6y = e^{-2x}(v'' - 4v' + 4v - v' + 2v - 6v)$$
$$= e^x(v'' - 5v').$$

y satisfies $y'' - y' - 6y = 0$ provided $w = v'$ satisfies $w' - 5w = 0$. This equation has solution $v' = w = (C_1/5)e^{5x}$, so $v = C_1 e^{5x} + C_2$. Thus the given DE has solution $y = e^{-2x}v = C_1 e^{3x} + C_2 e^{-2x}$.

4. If $y_1 = x^2$ on $(0, \infty)$, then

$$x^2 y_1'' - 3xy_1' + 4y_1 = 2x^2 - 6x^2 + 4x^2 = 0,$$

so y_1 is a solution of the DE $x^2 y'' - 3xy' + 4y = 0$. Let $y = x^2 v(x)$. Then

$$y' = x^2 v' + 2xv, \qquad y'' = x^2 v'' + 4xv' + 2v$$
$$x^2 y'' - 3xy' + 4y = x^4 v'' + 4x^3 v' + 2x^2 v$$
$$- 3x^3 v' - 6x^2 v + 4x^2 v$$
$$= x^3(xv'' + v').$$

y satisfies $x^2 y'' - 3xy' + 4y = 0$ provided $w = v'$ satisfies $xw' + w = 0$. This equation has solution $v' = w = C_1/x$ (obtained by separation of variables), so $v = C_1 \ln x + C_2$. Thus the given DE has solution $y = x^2 v = C_1 x^2 \ln x + C_2 x^2$.

6. If $y = x^{-1/2} \cos x$, then

$$y' = -\frac{1}{2}x^{-3/2}\cos x - x^{-1/2}\sin x$$
$$y'' = \frac{3}{4}x^{-5/2}\cos x + x^{-3/2}\sin x - x^{-1/2}\cos x.$$

Thus

$$x^2 y'' + xy' + \left(x^2 - \frac{1}{4}\right)y$$
$$= \frac{3}{4}x^{-1/2}\cos x + x^{1/2}\sin x - x^{3/2}\cos x$$
$$- \frac{1}{2}x^{-1/2}\cos x - x^{1/2}\sin x + x^{3/2}\cos x - \frac{1}{4}x^{-1/2}\cos x$$
$$= 0.$$

Therefore $y = x^{-1/2}\cos x$ is a solution of the Bessel equation

$$x^2 y'' + xy' + \left(x^2 - \frac{1}{4}\right)y = 0. \qquad (*)$$

Now let $y = x^{-1/2}(\cos x)v(x)$. Then

$$y' = -\frac{1}{2}x^{-3/2}(\cos x)v - x^{-1/2}(\sin x)v + x^{-1/2}(\cos x)v'$$

$$y'' = \frac{3}{4}x^{-5/2}(\cos x)v + x^{-3/2}(\sin x)v - x^{-3/2}(\cos x)v'$$
$$- x^{-1/2}(\cos x)v - 2x^{-1/2}(\sin x)v' + x^{-1/2}(\cos x)v''.$$

If we substitute these expressions into the equation $(*)$, many terms cancel out and we are left with the equation

$$(\cos x)v'' - 2(\sin x)v' = 0.$$

Substituting $u = v'$, we rewrite this equation in the form

$$(\cos x)\frac{du}{dx} = 2(\sin x)u$$

$$\int \frac{du}{u} = 2\int \tan x\, dx \;\Rightarrow\; \ln|u| = 2\ln|\sec x| + C_0.$$

Thus $v' = u = C_1 \sec^2 x$, from which we obtain

$$v = C_1 \tan x + C_2.$$

Thus the general solution of the Bessel equation $(*)$ is

$$y = x^{-1/2}(\cos x)v = C_1 x^{-1/2}\sin x + C_2 x^{-1/2}\cos x.$$

8. If y satisfies

$$y^{(n)} + a_{n-1}(x)y^{(n-1)} + \cdots + a_1(x)y' + a_0(x)y = f(x),$$

then let

$$y_1 = y,\quad y_2 = y',\quad y_3 = y'',\quad \ldots \quad y_n = y^{(n-1)}.$$

Therefore

$$y_1' = y_2,\quad y_2' = y_3,\quad \ldots \quad y_{n-2}' = y_{n-1},\quad \text{and}$$
$$y_n' = -a_0 y_1 - a_1 y_2 - a_2 y_3 - \cdots - a_{n-1}y_n + f,$$

and we have

$$\frac{d}{dx}\begin{pmatrix} y_1 \\ y_2 \\ \vdots \\ y_n \end{pmatrix} = \begin{pmatrix} 0 & 1 & 0 & \ldots & 0 \\ 0 & 0 & 1 & \ldots & 0 \\ \vdots & \vdots & \vdots & & \vdots \\ 0 & 0 & 0 & \ldots & 1 \\ -a_0 & -a_1 & -a_2 & \ldots & -a_n \end{pmatrix}\begin{pmatrix} y_1 \\ y_2 \\ \vdots \\ y_n \end{pmatrix}$$
$$+ \begin{pmatrix} 0 \\ 0 \\ \vdots \\ 0 \\ f \end{pmatrix}.$$

10.
$$\begin{vmatrix} 2-\lambda & 1 \\ 2 & 3-\lambda \end{vmatrix} = 6 - 5\lambda + \lambda^2 - 2$$
$$= \lambda^2 - 5\lambda + 4$$
$$= (\lambda - 1)(\lambda - 4) = 0$$

if $\lambda = 1$ or $\lambda = 4$.

Let $\mathcal{A} = \begin{pmatrix} 2 & 1 \\ 2 & 3 \end{pmatrix}$.

If $\lambda = 1$ and $\mathcal{A}\mathbf{v} = \mathbf{v}$, then

$$\mathcal{A} = \begin{pmatrix} 2 & 1 \\ 2 & 3 \end{pmatrix}\begin{pmatrix} v_1 \\ v_2 \end{pmatrix} = \begin{pmatrix} v_1 \\ v_2 \end{pmatrix} \;\Leftrightarrow\; v_1 + v_2 = 0.$$

Thus we may take $\mathbf{v} = \mathbf{v}_1 = \begin{pmatrix} 1 \\ -1 \end{pmatrix}$.

If $\lambda = 4$ and $\mathcal{A}\mathbf{v} = 4\mathbf{v}$, then

$$\mathcal{A} = \begin{pmatrix} 2 & 1 \\ 2 & 3 \end{pmatrix}\begin{pmatrix} v_1 \\ v_2 \end{pmatrix} = 4\begin{pmatrix} v_1 \\ v_2 \end{pmatrix} \;\Leftrightarrow\; 2v_1 - v_2 = 0.$$

Thus we may take $\mathbf{v} = \mathbf{v}_2 = \begin{pmatrix} 1 \\ 2 \end{pmatrix}$.

By the result of Exercise 9, $\mathbf{y} = e^x\mathbf{v}_1$ and $\mathbf{y} = e^{4x}\mathbf{v}_2$ are solutions of the homogeneous linear system $\mathbf{y}' = \mathcal{A}\mathbf{y}$. Therefore the general solution of the system is

$$\mathbf{y} = C_1 e^x \mathbf{v}_1 + C_2 e^{4x}\mathbf{v}_2,$$

that is

$$\begin{pmatrix} y_1 \\ y_2 \end{pmatrix} = C_1 e^x \begin{pmatrix} 1 \\ -1 \end{pmatrix} + C_2 e^{4x}\begin{pmatrix} 1 \\ 2 \end{pmatrix}, \quad \text{or}$$
$$y_1 = C_1 e^x + C_2 e^{4x}$$
$$y_2 = -C_1 e^x + 2C_2 e^{4x}.$$

Section 17.5   Linear Differential Equations with Constant Coefficients   (page 935)

2.
$$y'' - 2y' - 3y = 0$$
auxiliary eqn $r^2 - 2r - 3 = 0 \;\Rightarrow\; r = -1,\ r = 3$
$$y = Ae^{-t} + Be^{3t}$$

4. $4y'' - 4y' - 3y = 0$
$4r^2 - 4r - 3 = 0 \Rightarrow (2r+1)(2r-3) = 0$
Thus, $r_1 = -\frac{1}{2}$, $r_2 = \frac{3}{2}$, and $y = Ae^{-(1/2)t} + Be^{(3/2)t}$.

6. $y'' - 2y' + y = 0$
$r^2 - 2r + 1 = 0 \Rightarrow (r-1)^2 = 0$
Thus, $r = 1,\ 1$, and $y = Ae^t + Bte^t$.

8. $9y'' + 6y' + y = 0$

$9r^2 + 6r + 1 = 0 \Rightarrow (3r + 1)^2 = 0$

Thus, $r = -\frac{1}{3}, -\frac{1}{3}$, and $y = Ae^{-(1/3)t} + Bte^{-(1/3)t}$.

10. For $y'' - 4y' + 5y = 0$ the auxiliary equation is $r^2 - 4r + 5 = 0$, which has roots $r = 2 \pm i$. Thus, the general solution of the DE is $y = Ae^{2t}\cos t + Be^{2t}\sin t$.

12. Given that $y'' + y' + y = 0$, hence $r^2 + r + 1 = 0$. Since $a = 1$, $b = 1$ and $c = 1$, the discriminant is $D = b^2 - 4ac = -3 < 0$ and $-(b/2a) = -\frac{1}{2}$ and $\omega = \sqrt{3}/2$. Thus, the general solution is

$$y = Ae^{-(1/2)t}\cos\left(\frac{\sqrt{3}}{2}t\right) + Be^{-(1/2)t}\sin\left(\frac{\sqrt{3}}{2}t\right).$$

14. Given that $y'' + 10y' + 25y = 0$, hence $r^2 + 10r + 25 = 0 \Rightarrow (r + 5)^2 = 0 \Rightarrow r = -5$. Thus,

$$y = Ae^{-5t} + Bte^{-5t}$$
$$y' = -5e^{-5t}(A + Bt) + Be^{-5t}.$$

Since

$$0 = y(1) = Ae^{-5} + Be^{-5}$$
$$2 = y'(1) = -5e^{-5}(A + B) + Be^{-5},$$

we have $A = -2e^5$ and $B = 2e^5$.
Thus, $y = -2e^5 e^{-5t} + 2te^5 e^{-5t} = 2(t - 1)e^{-5(t-1)}$.

16. The auxiliary equation $r^2 - (2 + \epsilon)r + (1 + \epsilon)$ factors to $(r - 1 - \epsilon)(r - 1) = 0$ and so has roots $r = 1 + \epsilon$ and $r = 1$. Thus the DE $y'' - (2 + \epsilon)y' + (1 + \epsilon)y = 0$ has general solution $y = Ae^{(1+\epsilon)t} + Be^t$. The function $y_\epsilon(t) = \dfrac{e^{(1+\epsilon)t} - e^t}{\epsilon}$ is of this form with $A = -B = 1/\epsilon$. We have, substituting $\epsilon = h/t$,

$$\lim_{\epsilon \to 0} y_\epsilon(t) = \lim_{\epsilon \to 0} \frac{e^{(1+\epsilon)t} - e^t}{\epsilon}$$
$$= t \lim_{h \to 0} \frac{e^{t+h} - e^t}{h}$$
$$= t\left(\frac{d}{dt}e^t\right) = te^t$$

which is, along with e^t, a solution of the CASE II DE $y'' - 2y' + y = 0$.

18. The auxiliary equation $ar^2 + br + c = 0$ has roots

$$r_1 = \frac{-b - \sqrt{D}}{2a}, \quad r_2 = \frac{-b + \sqrt{D}}{2a},$$

where $D = b^2 - 4ac$. Note that $a(r_2 - r_1) = \sqrt{D} = -(2ar_1 + b)$. If $y = e^{r_1 t}u$, then $y' = e^{r_1 t}(u' + r_1 u)$, and $y'' = e^{r_1 t}(u'' + 2r_1 u' + r_1^2 u)$. Substituting these expressions into the DE $ay'' + by' + cy = 0$, and simplifying, we obtain

$$e^{r_1 t}(au'' + 2ar_1 u' + bu') = 0,$$

or, more simply, $u'' - (r_2 - r_1)u' = 0$. Putting $v = u'$ reduces this equation to first order:

$$v' = (r_2 - r_1)v,$$

which has general solution $v = Ce^{(r_2 - r_1)t}$. Hence

$$u = \int Ce^{(r_2-r_1)t}\,dt = Be^{(r_2-r_1)t} + A,$$

and $y = e^{r_1 t}u = Ae^{r_1 t} + Be^{r_2 t}$.

20. $y^{(4)} - 2y'' + y = 0$
Auxiliary: $r^4 - 2r^2 + 1 = 0$
$$(r^2 - 1)^2 = 0 \Rightarrow r = -1, -1, 1, 1$$
General solution: $y = C_1 e^{-t} + C_2 te^{-t} + C_3 e^t + C_4 te^t$.

22. $y^{(4)} + 4y^{(3)} + 6y'' + 4y' + y = 0$
Auxiliary: $r^4 + 4r^3 + 6r^2 + 4r + 1 = 0$
$$(r + 1)^4 = 0 \Rightarrow r = -1, -1, -1, -1$$
General solution: $y = e^{-t}(C_1 + C_2 t + C_3 t^2 + C_4 t^3)$.

24. Aux. eqn: $(r^2 - r - 2)^2(r^2 - 4)^2 = 0$
$$(r + 1)^2(r - 2)^2(r - 2)^2(r + 2)^2 = 0$$
$$r = 2, 2, 2, 2, -1, -1, -2, -2.$$
The general solution is

$$y = e^{2t}(C_1 + C_2 t + C_3 t^2 + C_4 t^3) + e^{-t}(C_5 + C_6 t)$$
$$+ e^{-2t}(C_7 + C_8 t).$$

26. $x^2 y'' - xy' - 3y = 0$
$r(r - 1) - r - 3 = 0 \Rightarrow r^2 - 2r - 3 = 0$
$\Rightarrow (r - 3)(r + 1) = 0 \Rightarrow r_1 = -1$ and $r_2 = 3$
Thus, $y = Ax^{-1} + Bx^3$.

28. Consider $x^2 y'' - xy' + 5y = 0$. Since $a = 1$, $b = -1$, and $c = 5$, therefore $(b-a)^2 < 4ac$. Then $k = (a-b)/2a = 1$ and $\omega^2 = 4$. Thus, the general solution is $y = Ax\cos(2\ln x) + Bx\sin(2\ln x)$.

30. Given that $x^2 y'' + xy' + y = 0$. Since $a = 1$, $b = 1$, $c = 1$ therefore $(b - a)^2 < 4ac$. Then $k = (a - b)/2a = 0$ and $\omega^2 = 1$. Thus, the general solution is $y = A\cos(\ln x) + B\sin(\ln x)$.

32. Because $y'' + 4y = 0$, therefore $y = A\cos 2t + B\sin 2t$. Now

$$y(0) = 2 \Rightarrow A = 2,$$
$$y'(0) = -5 \Rightarrow B = -\tfrac{5}{2}.$$

Thus, $y = 2\cos 2t - \frac{5}{2}\sin 2t$.
circular frequency $= \omega = 2$, frequency $=$
$\frac{\omega}{2\pi} = \frac{1}{\pi} \approx 0.318$
period $= \frac{2\pi}{\omega} = \pi \approx 3.14$
amplitude $= \sqrt{(2)^2 + (-\frac{5}{2})^2} \simeq 3.20$

34. For $y'' + y = 0$, we have $y = A\sin t + B\cos t$. Since,

$$y(2) = 3 = A\sin 2 + B\cos 2$$
$$y'(2) = -4 = A\cos 2 - B\sin 2,$$

therefore

$$A = 3\sin 2 - 4\cos 2$$
$$B = 4\sin 2 + 3\cos 2.$$

Thus,

$$y = (3\sin 2 - 4\cos 2)\sin t + (4\sin 2 + 3\cos 2)\cos t$$
$$= 3\cos(t-2) - 4\sin(t-2).$$

36. $y = \mathcal{A}\cos\big(\omega(t-c)\big) + \mathcal{B}\sin\big(\omega(t-c)\big)$
(easy to calculate $y'' + \omega^2 y = 0$)
$y = \mathcal{A}\big(\cos(\omega t)\cos(\omega c) + \sin(\omega t)\sin(\omega c)\big)$
$\quad + \mathcal{B}\big(\sin(\omega t)\cos(\omega c) - \cos(\omega t)\sin(\omega c)\big)$
$= \big(\mathcal{A}\cos(\omega c) - \mathcal{B}\sin(\omega c)\big)\cos\omega t$
$\quad + \big(\mathcal{A}\sin(\omega c) + \mathcal{B}\cos(\omega c)\big)\sin\omega t$
$= A\cos\omega t + B\sin\omega t$
where $A = \mathcal{A}\cos(\omega c) - \mathcal{B}\sin(\omega c)$ and
$B = \mathcal{A}\sin(\omega c) + \mathcal{B}\cos(\omega c)$

37. If $y = A\cos\omega t + B\sin\omega t$ then

$$y'' + \omega^2 y = -A\omega^2\cos\omega t - B\omega^2\sin\omega t$$
$$+ \omega^2(A\cos\omega t + B\sin\omega t) = 0$$

for all t. So y is a solution of (†).

38. If $f(t)$ is any solution of (†) then $f''(t) = -\omega^2 f(t)$ for all t. Thus,

$$\frac{d}{dt}\Big[\omega^2\big(f(t)\big)^2 + \big(f'(t)\big)^2\Big]$$
$$= 2\omega^2 f(t)f'(t) + 2f'(t)f''(t)$$
$$= 2\omega^2 f(t)f'(t) - 2\omega^2 f(t)f'(t) = 0$$

for all t. Thus, $\omega^2\big(f(t)\big)^2 + \big(f'(t)\big)^2$ is constant. (This can be interpreted as a conservation of energy statement.)

39. If $g(t)$ satisfies (†) and also $g(0) = g'(0) = 0$, then by Exercise 20,

$$\omega^2\big(g(t)\big)^2 + \big(g'(t)\big)^2$$
$$= \omega^2\big(g(0)\big)^2 + \big(g'(0)\big)^2 = 0.$$

Since a sum of squares cannot vanish unless each term vanishes, $g(t) = 0$ for all t.

40. If $f(t)$ is any solution of (†), let
$g(t) = f(t) - A\cos\omega t - B\sin\omega t$ where $A = f(0)$ and $B\omega = f'(0)$. Then g is also solution of (†). Also $g(0) = f(0) - A = 0$ and $g'(0) = f'(0) - B\omega = 0$. Thus, $g(t) = 0$ for all t by Exercise 24, and therefore $f(x) = A\cos\omega t + B\sin\omega t$. Thus, it is proved that every solution of (†) is of this form.

42. From Example 9, the spring constant is
$k = 9 \times 10^4$ gm/sec^2. For a frequency of 10 Hz (i.e., a circular frequency $\omega = 20\pi$ rad/sec.), a mass m satisfying $\sqrt{k/m} = 20\pi$ should be used. So,

$$m = \frac{k}{400\pi^2} = \frac{9 \times 10^4}{400\pi^2} = 22.8 \text{ gm.}$$

The motion is determined by

$$\begin{cases} y'' + 400\pi^2 y = 0 \\ y(0) = -1 \\ y'(0) = 2 \end{cases}$$

therefore, $y = A\cos 20\pi t + B\sin 20\pi t$ and

$$y(0) = -1 \Rightarrow A = -1$$
$$y'(0) = 2 \Rightarrow B = \frac{2}{20\pi} = \frac{1}{10\pi}.$$

Thus, $y = -\cos 20\pi t + \frac{1}{10\pi}\sin 20\pi t$, with y in cm and t in second, gives the displacement at time t. The amplitude is $\sqrt{(-1)^2 + (\frac{1}{10\pi})^2} \approx 1.0005$ cm.

44. Using the addition identities for cosine and sine,

$$y = e^{kt}[A\cos\omega(t-t_0)B\sin\omega(t-t_0)]$$
$$= e^{kt}[A\cos\omega t\cos\omega t_0 + A\sin\omega t\sin\omega t_0$$
$$+ B\sin\omega t\cos\omega t_0 - B\cos\omega t\sin\omega t_0]$$
$$= e^{kt}[A_1\cos\omega t + B_1\sin\omega t],$$

where $A_1 = A\cos\omega t_0 - B\sin\omega t_0$ and $B_1 = A\sin\omega t_0 + B\cos\omega t_0$. Under the conditions of this problem we know that $e^{kt}\cos\omega t$ and $e^{kt}\sin\omega t$ are independent solutions of $ay'' + by' + cy = 0$, so our function y must also be a solution, and, since it involves two arbitrary constants, it is a general solution.

46. $\begin{cases} y'' + 2y' + 5y = 0 \\ y(3) = 2 \\ y'(3) = 0 \end{cases}$

The DE has auxiliary equation $r^2 + 2r + 5 = 0$ with roots $r = -1 \pm 2i$. By the second previous problem, a general solution can be expressed in the form $y = e^{-t}[A\cos 2(t-3) + B\sin 2(t-3)]$ for which

$$y' = -e^{-t}[A\cos 2(t-3) + B\sin 2(t-3)]$$
$$+ e^{-t}[-2A\sin 2(t-3) + 2B\cos 2(t-3)].$$

The initial conditions give

$$2 = y(3) = e^{-3}A$$
$$0 = y'(3) = -e^{-3}(A + 2B)$$

Thus $A = 2e^3$ and $B = -A/2 = -e^3$. The IVP has solution

$$y = e^{3-t}[2\cos 2(t-3) - \sin 2(t-3)].$$

48. Let $u(x) = c - k^2 y(x)$. Then $u(0) = c - k^2 a$. Also $u'(x) = -k^2 y'(x)$, so $u'(0) = -k^2 b$. We have

$$u''(x) = -k^2 y''(x) = -k^2\left(c - k^2 y(x)\right) = -k^2 u(x)$$

This IVP for the equation of simple harmonic motion has solution

$$u(x) = (c - k^2 a)\cos(kx) - kb\sin(kx)$$

so that

$$y(x) = \frac{1}{k^2}\left(c - u(x)\right)$$
$$= \frac{c}{k^2}\left(c - (c - k^2 a)\cos(kx) + kb\sin(kx)\right)$$
$$= \frac{c}{k^2}(1 - \cos(kx)) + a\cos(kx) + \frac{b}{k}\sin(kx).$$

Section 17.6 Nonhomogeneous Linear Equations (page 942)

2. $y'' + y' - 2y = x$.
The complementary function is $y_h = C_1 e^{-2x} + C_2 e^x$, as shown in Exercise 1. For a particular solution try $y = Ax + B$. Then $y' = A$ and $y'' = 0$, so y satisfies the given equation if

$$x = A - 2(Ax + B) = A - 2B - 2Ax.$$

We require $A - 2B = 0$ and $-2A = 1$, so $A = -1/2$ and $B = -1/4$. The general solution of the given equation is

$$y = -\frac{2x+1}{4} + C_1 e^{-2x} + C_2 e^x.$$

4. $y'' + y' - 2y = e^x$.
The complementary function is $y_h = C_1 e^{-2x} + C_2 e^x$, as shown in Exercise 1. For a particular solution try $y = Axe^x$. Then

$$y' = Ae^x(1 + x), \qquad y'' = Ae^x(2 + x),$$

so y satisfies the given equation if

$$e^x = Ae^x(2 + x + 1 + x - 2x) = 3Ae^x.$$

We require $A = 1/3$. The general solution of the given equation is

$$y = \frac{1}{3}xe^x + C_1 e^{-2x} + C_3 e^x.$$

6. $y'' + 4y = x^2$. The complementary function is $y = C_1 \cos(2x) + C_2 \sin(2x)$. For the given equation, try $y = Ax^2 + Bx + C$. Then

$$x^2 = y'' + 4y = 2A + 4Ax^2 + 4Bx + 4C$$

Thus $2A + 4C = 0$, $4A = 1$, $4B = 0$, and we have $A = \frac{1}{4}$, $B = 0$, and $C = -\frac{1}{8}$. The given equation has general solution

$$y = \frac{1}{4}x^2 - \frac{1}{8} + C_1 \cos(2x) + C_2 \sin(2x).$$

8. $y'' + 4y' + 4y = e^{-2x}$.
The homogeneous equation has auxiliary equation $r^2 + 4r + 4 = 0$ with roots $r = -2, -2$. Thus the complementary function is

$$y_h = C_1 e^{-2x} + C_2 xe^{-2x}.$$

For a particular solution, try $y = Ax^2 e^{-2x}$. Then $y' = e^{-2x}(2Ax - 2Ax^2)$ and $y'' = e^{-2x}(2A - 8Ax + 4Ax^2)$. We have

$$e^{-2x} = y'' + 4y' + 4y$$
$$= e^{-2x}(2A - 8Ax + 4Ax^2 + 8Ax - 8Ax^2 + 4Ax^2)$$
$$= 2Ae^{-2x}.$$

Thus we require $A = 1/2$. The given equation has general solution

$$y = e^{-2x}\left(\frac{x^2}{2} + C_1 + C_2 x\right).$$

APPENDICES

Appendix I.   Complex Numbers
(page A-10)

2. $z = 4 - i$, $\text{Re}(z) = 4$, $\text{Im}(z) = -1$

4. $z = -6$, $\text{Re}(z) = -6$, $\text{Im}(z) = 0$

6. $z = -2$, $|z| = 2$, $\text{Arg}(z) = \pi$
$z = 2(\cos \pi + i \sin \pi)$

8. $z = -5i$, $|z| = 5$, $\text{Arg}(z) = -\pi/2$
$z = 5(\cos(-\pi/2) + i \sin(-\pi/2))$

10. $z = -2 + i$, $|z| = \sqrt{5}$, $\theta = \text{Arg}(z) = \pi - \tan^{-1}(1/2)$
$z = \sqrt{5}(\cos \theta + i \sin \theta)$

12. $z = 3 - 4i$, $|z| = 5$, $\theta = \text{Arg}(z) = -\tan^{-1}(4/3)$
$z = 5(\cos \theta + i \sin \theta)$

14. $z = -\sqrt{3} - 3i$, $|z| = 2\sqrt{3}$, $\text{Arg}(z) = -2\pi/3$
$z = 2\sqrt{3}(\cos(-2\pi/3) + i \sin(-2\pi/3))$

16.   If $\text{Arg}(z) = \dfrac{3\pi}{4}$ and $\text{Arg}(w) = \dfrac{\pi}{2}$, then
$\arg(zw) = \dfrac{3\pi}{4} + \dfrac{\pi}{2} = \dfrac{5\pi}{4}$, so
$\text{Arg}(zw) = \dfrac{5\pi}{4} - 2\pi = \dfrac{-3\pi}{4}$.

18. $|z| = 2$, $\arg(z) = \pi \Rightarrow z = 2(\cos \pi + i \sin \pi) = -2$

20. $|z| = 1$, $\arg(z) = \dfrac{3\pi}{4} \Rightarrow z = \left(\cos \dfrac{3\pi}{4} + i \sin \dfrac{3\pi}{4} \right)$
$\Rightarrow z = -\dfrac{1}{\sqrt{2}} + \dfrac{1}{\sqrt{2}}i$

22. $|z| = 0 \Rightarrow z = 0$ for any value of $\arg(z)$

24. $\overline{5 + 3i} = 5 - 3i$

26. $\overline{4i} = -4i$

28. $|z| = 2$ represents all points on the circle of radius 2 centred at the origin.

30. $|z - 2i| \le 3$ represents all points in the closed disk of radius 3 centred at the point $2i$.

32. $\arg(z) = \pi/3$ represents all points on the ray from the origin in the first quadrant, making angle $60°$ with the positive direction of the real axis.

34. $(2 + 5i) + (3 - i) = 5 + 4i$

36. $(4 + i)(4 - i) = 16 - i^2 = 17$

38. $(a + bi)(\overline{2a - bi}) = (a + bi)(2a + bi) = 2a^2 - b^2 + 3abi$

40. $\dfrac{2 - i}{2 + i} = \dfrac{(2 - i)^2}{4 - i^2} = \dfrac{3 - 4i}{5}$

42. $\dfrac{1 + i}{i(2 + 3i)} = \dfrac{1 + i}{-3 + 2i} = \dfrac{(1 + i)(-3 - 2i)}{9 + 4} = \dfrac{-1 - 5i}{13}$

44.   If $z = x + yi$ and $w = u + vi$, where x, y, u, and v are real, then

$$\overline{z + w} = \overline{x + u + (y + v)i}$$
$$= x + u - (y + v)i = x - yi + u - vi = \overline{z} + \overline{w}.$$

46. $z = 3 + i\sqrt{3} = 2\sqrt{3} \left(\cos \dfrac{\pi}{6} + i \sin \dfrac{\pi}{6} \right)$
$w = -1 + i\sqrt{3} = 2 \left(\cos \dfrac{2\pi}{3} + i \sin \dfrac{2\pi}{3} \right)$
$zw = 4\sqrt{3} \left(\cos \dfrac{5\pi}{6} + i \sin \dfrac{5\pi}{6} \right)$
$\dfrac{z}{w} = \sqrt{3} \left(\cos \dfrac{-\pi}{2} + i \sin \dfrac{-\pi}{2} \right) = -i\sqrt{3}$

48. $\cos(3\theta) + i \sin(3\theta) = (\cos \theta + i \sin \theta)^3$
$= \cos^3 \theta + 3i \cos^2 \theta \sin \theta - 3 \cos \theta \sin^2 \theta - i \sin^3 \theta$
Thus

$$\cos(3\theta) = \cos^3 \theta - 3 \cos \theta \sin^2 \theta = 4 \cos^3 \theta - 3 \cos \theta$$
$$\sin(3\theta) = 3 \cos^2 \theta \sin \theta - \sin^3 \theta = 3 \sin \theta - 4 \sin^3 \theta.$$

50.   If $z = w = -1$, then $zw = 1$, so $\sqrt{zw} = 1$. But if we use $\sqrt{z} = \sqrt{-1} = i$ and the same value for $\sqrt{w}$, then $\sqrt{z}\sqrt{w} = i^2 = -1 \ne \sqrt{zw}$.

52.   The three cube roots of $-8i = 8 \left(\cos \dfrac{3\pi}{2} + i \sin \dfrac{3\pi}{2} \right)$
are of the form $2(\cos \theta + i \sin \theta)$ where $\theta = \pi/2$,
$\theta = 7\pi/6$, and
$\theta = 11\pi/6$. Thus they are

$$2i, \quad -\sqrt{3} - i, \quad \sqrt{3} - i.$$

54.   The four fourth roots of $4 = 4(\cos 0 + i \sin 0)$ are of the form $\sqrt{2}(\cos \theta + i \sin \theta)$ where $\theta = 0$, $\theta = \pi/2$, π, and $\theta = 3\pi/2$. Thus they are $\sqrt{2}$, $i\sqrt{2}$, $-\sqrt{2}$, and $-i\sqrt{2}$.

56.   The equation $z^5 + a^5 = 0$ $(a > 0)$ has solutions that are the five fifth roots of $-a^5 = a(\cos \pi + i \sin \pi)$; they are of the form $a(\cos \theta + i \sin \theta)$, where $\theta = \pi/5$, $3\pi/5$, π, $7\pi/5$, and $9\pi/5$.

Appendix II. Complex Functions (page A-19)

In Solutions 1–12, $z = x + yi$ and $w = u + vi$, where x, y, u, and v are real.

2. The function $w = \bar{z}$ transforms the line $x + y = 1$ to the line $u - v = 1$.

4. The function $w = z^3$ transforms the closed quarter-circular disk $0 \le |z| \le 2$, $0 \le \arg(z) \le \pi/2$ to the closed three-quarter disk $0 \le |w| \le 8$, $0 \le \arg(w) \le 3\pi/2$.

6. The function $w = -iz$ rotates the z-plane $-90°$, so transforms the wedge $\pi/4 \le \arg(z) \le \pi/3$ to the wedge $-\pi/4 \le \arg(z) \le -\pi/6$.

8. The function $w = z^2 = x^2 - y^2 + 2xyi$ transforms the line $x = 1$ to $u = 1 - y^2$, $v = 2y$, which is the parabola $v^2 = 4 - 4u$ with vertex at $w = 1$, opening to the left.

10. The function $w = 1/z = (x - yi)/(x^2 + y^2)$ transforms the line $x = 1$ to the curve given parametrically by

$$u = \frac{1}{1 + y^2}, \qquad v = \frac{-y}{1 + y^2}.$$

This curve is, in fact, a circle,

$$u^2 + v^2 = \frac{1 + y^2}{(1 + y^2)^2} = u,$$

with centre $w = 1/2$ and radius $1/2$.

12. The function $w = e^{iz} = e^{-y}(\cos x + i \sin x)$ transforms the vertical half-strip $0 < x < \pi/2$, $0 < y < \infty$ to the first-quadrant part of the unit open disk $|w| = e^{-y} < 1$, $0 < \arg(w) = x < \pi/2$, that is $u > 0$, $v > 0$, $u^2 + v^2 < 1$.

14. $f(z) = z^3 = (x + yi)^3 = x^3 - 3xy^2 + (3x^2y - y^3)i$

$u = x^3 - 3xy^2, \qquad v = 3x^2y - y^3$

$\dfrac{\partial u}{\partial x} = 3(x^2 - y^2) = \dfrac{\partial v}{\partial y}, \quad \dfrac{\partial u}{\partial y} = -6xy = -\dfrac{\partial v}{\partial x}$

$f'(z) = \dfrac{\partial u}{\partial x} + i\dfrac{\partial v}{\partial x} = 3(x^2 - y^2 + 2xyi) = 3z^2.$

16. $f(z) = e^{z^2} = e^{x^2 - y^2}(\cos(2xy) + i \sin(2xy))$

$u = e^{x^2 - y^2}\cos(2xy), \qquad v = e^{x^2 - y^2}\sin(2xy)$

$\dfrac{\partial u}{\partial x} = e^{x^2 - y^2}(2x\cos(2xy) - 2y\sin(2xy)) = \dfrac{\partial v}{\partial y}$

$\dfrac{\partial u}{\partial y} = -e^{x^2 - y^2}(2y\cos(2xy) + 2x\sin(2xy)) = -\dfrac{\partial v}{\partial x}$

$f'(z) = \dfrac{\partial u}{\partial x} + i\dfrac{\partial v}{\partial x}$

$= e^{x^2 - y^2}[2x\cos(2xy) - 2y\sin(2xy)$
$\qquad + i(2y\cos(2xy) + 2x\sin(2xy))]$

$= (2x + 2yi)e^{x^2 - y^2}(\cos(2xy) + i\sin(2xy)) = 2ze^{z^2}.$

18. $e^{z+2\pi i} = e^x(\cos(y + 2\pi) + i\sin(y + 2\pi))$
$\qquad = e^x(\cos y + i\sin y) = e^z.$

Thus e^z is periodic with period $2\pi i$. So is $e^{-z} = 1/e^z$. Since $e^{i(z+2\pi)} = e^{zi+2\pi i} = e^{zi}$, therefore e^{zi} and also e^{-zi} are periodic with period 2π. Hence

$$\cos z = \frac{e^{zi} + e^{-zi}}{2} \text{ and } \sin z = \frac{e^{zi} - e^{-zi}}{2i}$$

are periodic with period 2π, and

$$\cosh z = \frac{e^z + e^{-z}}{2} \text{ and } \sinh z = \frac{e^z - e^{-z}}{2}$$

are periodic with period $2\pi i$.

20. $\cosh(iz) = \dfrac{e^{iz} + e^{-iz}}{2} = \cosh z$

$-i\sinh(iz) = \dfrac{1}{i}\dfrac{e^{iz} - e^{-iz}}{2} = \sin z$

$\cos(iz) = \dfrac{e^{-z} + e^z}{2} = \cosh z$

$\sin(iz) = \dfrac{e^{-z} - e^z}{2i} = i\dfrac{-e^{-z} + e^z}{2} = i\sinh z$

22. $\sin z = 0 \Leftrightarrow e^{zi} = e^{-zi} \Leftrightarrow e^{2zi} = 1$
$\qquad \Leftrightarrow e^{-2y}[\cos(2x) + i\sin(2x)] = 1$
$\qquad \Leftrightarrow \sin(2x) = 0, \quad e^{-2y}\cos(2x) = 1$
$\qquad \Leftrightarrow y = 0, \ \cos(2x) = 1$
$\qquad =\Leftrightarrow y = 0, \ x = 0, \pm\pi, \ \pm 2\pi, \ldots$

Thus the only complex zeros of $\sin z$ are its real zeros at $z = n\pi$ for integers n.

24. $e^z = e^{x+yi} = e^x\cos y + ie^x\sin y$

$e^{-z} = e^{-x-yi} = e^{-x}\cos y - e^{-x}\sin y$

$\cosh z = \dfrac{e^z + e^{-z}}{2} = \dfrac{e^x + e^{-x}}{2}\cos y + i\dfrac{e^x - e^{-x}}{2}\sin y$

$\qquad = \cosh x\cos y + i\sinh x\sin y$

$\qquad \mathrm{Re}(\cosh z) = \cosh x\cos y, \quad \mathrm{Im}(\cosh z) = \sinh x\sin y.$

26. $e^{iz} = e^{-y+xi} = e^{-y}\cos x + ie^{-y}\sin x$

$e^{-iz} = e^{y-xi} = e^y\cos x - ie^y\sin x$

$\cos z = \dfrac{e^{iz} + e^{-iz}}{2} = \dfrac{e^{-y} + e^y}{2}\cos x + i\dfrac{e^{-y} - e^y}{2}\sin x$

$\qquad = \cos x\cosh y - i\sin x\sinh y$

$\qquad \mathrm{Re}(\cos z) = \cos x\cosh y, \quad \mathrm{Im}(\cos z) = -\sin x\sinh y$

$\sin z = \dfrac{e^{iz} - e^{-iz}}{2i} = \dfrac{e^{-y} - e^y}{2i}\cos x + i\dfrac{e^{-y} + e^y}{2i}\sin x$

$\qquad = \sin x\cosh y + i\cos x\sinh y$

$\qquad \mathrm{Re}(\sin z) = \sin x\cosh y, \quad \mathrm{Im}(\sin z) = \cos x\sinh y.$

28. $z^2 - 2z + i = 0 \Rightarrow (z-1)^2 = 1 - i$

$$= \sqrt{2}\left(\cos\frac{7\pi}{4} + i\sin\frac{7\pi}{4}\right)$$

$$\Rightarrow z = 1 \pm 2^{1/4}\left(\cos\frac{7\pi}{8} + i\sin\frac{7\pi}{8}\right)$$

30. $z^2 - 2iz - 1 = 0 \Rightarrow (z-i)^2 = 0$

$$\Rightarrow z = i \quad \text{(double root)}$$

32. $z^4 - 2z^2 + 4 = 0 \Rightarrow (z^2-1)^2 = -3$

$z^2 = 1 - i\sqrt{3} \quad \text{or} \quad z^2 = 1 + i\sqrt{3}$

$$z^2 = 2\left(\cos\frac{5\pi}{3} + i\sin\frac{5\pi}{3}\right), \quad z^2 = 2\left(\cos\frac{\pi}{3} + i\sin\frac{\pi}{3}\right)$$

$$z = \pm\sqrt{2}\left(\cos\frac{5\pi}{6} + i\sin\frac{5\pi}{6}\right), \quad \text{or}$$

$$z = \pm\sqrt{2}\left(\cos\frac{\pi}{6} + i\sin\frac{\pi}{6}\right)$$

$$z = \pm\left(\sqrt{\frac{3}{2}} - \frac{i}{\sqrt{2}}\right), \quad z = \pm\left(\sqrt{\frac{3}{2}} + \frac{i}{\sqrt{2}}\right)$$

34. Since $P(z) = z^4 - 4z^3 + 12z^2 - 16z + 16$ has real coefficients, if $z_1 = 1 - \sqrt{3}i$ is a zero of $P(z)$, then so is $\overline{z_1}$. Now

$$(z-z_1)(z-\overline{z_1}) = (z-1)^2 + 3 = z^2 - 2z + 4.$$

By long division (details omitted) we discover that

$$\frac{z^4 - 4z^3 + 12z^2 - 16z + 16}{z^2 - 2z + 4} = z^2 - 2z + 4.$$

Thus z_1 and $\overline{z_1}$ are both *double zeros* of $P(z)$. These are the only zeros.

36. Since $P(z) = z^5 - 2z^4 - 8z^3 + 8z^2 + 31z - 30$ has real coefficients, if $z_1 = -2 + i$ is a zero of $P(z)$, then so is $z_2 = -2 - i$. Now

$$(z-z_1)(z-z_2) = z^2 + 4z + 5.$$

By long division (details omitted) we discover that

$$\frac{z^5 - 2z^4 - 8z^3 + 8z^2 + 31z - 30}{z^2 + 4z + 5}$$

$$= z^3 - 6z^2 + 11z - 6.$$

Observe that $z_3 = 1$ is a zero of $z^3 - 6z^2 + 11z - 6$. By long division again:

$$\frac{z^3 - 6z^2 + 11z - 6}{z-1} = z^2 - 5z + 6 = (z-2)(z-3).$$

Hence $P(z)$ has the five zeros $-2+i$, $-2-i$, 1, 2, and 3.

Appendix III. Continuous Functions (page A-25)

2. To be proved: If $f(x) \le K$ on $[a, b)$ and $(b, c]$, and if $\lim_{x\to b} f(x) = L$, then $L \le K$.

Proof: If $L > K$, then let $\epsilon = (L - K)/2$; thus $\epsilon > 0$. There exists $\delta > 0$ such that $\delta < b - a$ and $\delta < c - b$, and such that if $0 < |x - b| < \delta$, then $|f(x) - L| < \epsilon$. In this case

$$f(x) > L - \epsilon = L - \frac{L-K}{2} > K,$$

which contradicts the fact that $f(x) \le K$ on $[a, b)$ and $(b, c]$. Therefore $L \le K$.

4. a) Let $f(x) = C$, $g(x) = x$. Let $\epsilon > 0$ be given and let $\delta = \epsilon$. For any real number x, if $|x - a| < \delta$, then

$$|f(x) - f(a)| = |C - C| = 0 < \epsilon,$$
$$|g(x) - g(a)| = |x - a| < \delta = \epsilon.$$

Thus $\lim_{x\to a} f(x) = f(a)$ and $\lim_{x\to a} g(x) = g(a)$, and f and g are both continuous at every real number a.

6. If P and Q are polynomials, they are continuous everywhere by Exercise 5. If $Q(a) \ne 0$, then $\lim_{x\to a} \dfrac{P(x)}{Q(x)} = \dfrac{P(a)}{Q(a)}$ by Theorem 1(a). Hence P/Q is continuous everywhere except at the zeros of Q.

8. By Exercise 5, x^m is continuous everywhere. By Exercise 7, $x^{1/n}$ is continuous at each $a > 0$. Thus for $a > 0$ we have

$$\lim_{x\to a} x^{m/n} = \lim_{x\to a}\left(x^{1/n}\right)^m = \left(\lim_{x\to a} x^{1/n}\right)^m$$
$$= (a^{1/n})^m = a^{m/n},$$

and $x^{m/n}$ is continuous at each positive number.

10. Let $\epsilon > 0$ be given. Let $\delta = \epsilon$. If a is any real number then

$$\Big||x| - |a|\Big| \le |x - a| < \epsilon \quad \text{if} \quad |x - a| < \delta.$$

Thus $\lim_{x\to a} |x| = |a|$, and the absolute value function is continuous at every real number.

12. The proof that cos is continuous everywhere is almost identical to that for sin in Exercise 11.

14. Let a be any real number, and let $\epsilon > 0$ be given. Assume (making ϵ smaller if necessary) that $\epsilon < e^a$. Since

$$\ln\left(1 - \frac{\epsilon}{e^a}\right) + \ln\left(1 + \frac{\epsilon}{e^a}\right) = \ln\left(1 - \frac{\epsilon^2}{e^{2a}}\right) < 0,$$

we have $\ln\left(1 + \dfrac{\epsilon}{e^a}\right) < -\ln\left(1 - \dfrac{\epsilon}{e^a}\right)$.

Let $\delta = \ln\left(1 + \dfrac{\epsilon}{e^a}\right)$. If $|x - a| < \delta$, then

$$\ln\left(1 - \frac{\epsilon}{e^a}\right) < x - a < \ln\left(1 + \frac{\epsilon}{e^a}\right)$$

$$1 - \frac{\epsilon}{e^a} < e^{x-a} < 1 + \frac{\epsilon}{e^a}$$

$$\left|e^{x-a} - 1\right| < \frac{\epsilon}{e^a}$$

$$|e^x - e^a| = e^a|e^{x-a} - 1| < \epsilon.$$

Thus $\lim_{x \to a} e^x = e^a$ and e^x is continuous at every point a in its domain.

16. Let $g(t) = \dfrac{t}{1 + |t|}$. For $t \neq 0$ we have

$$g'(t) = \frac{1 + |t| - t\,\text{sgn}\,t}{(1 + |t|)^2} = \frac{1 + |t| - |t|}{(1 + |t|)^2} = \frac{1}{(1 + |t|)^2} > 0.$$

If $t = 0$, g is also differentiable, and has derivative 1:

$$g'(0) = \lim_{h \to 0} \frac{g(h) - g(0)}{h} = \lim_{h \to 0} \frac{1}{1 + |h|} = 1.$$

Thus g is continuous and increasing on $\mathbb{R}$. If f is continuous on $[a, b]$, then

$$h(x) = g\Big(f(x)\Big) = \frac{f(x)}{1 + |f(x)|}$$

is also continuous there, being the composition of continuous functions. Also, $h(x)$ is bounded on $[a, b]$, since

$$\left|g\Big(f(x)\Big)\right| \leq \frac{|f(x)|}{1 + |f(x)|} \leq 1.$$

By assumption in this problem, $h(x)$ must assume maximum and minimum values; there exist c and d in $[a, b]$ such that

$$g\Big(f(c)\Big) \leq g\Big(f(x)\Big) \leq g\Big(f(d)\Big)$$

for all x in $[a, b]$. Since g is increasing, so is its inverse g^{-1}. Therefore

$$f(c) \leq f(x) \leq f(d)$$

for all x in $[a, b]$, and f is bounded on that interval.

342

Appendix IV. The Riemann Integral (page A-30)

2. $f(x) = \begin{cases} 1 & \text{if } x = 1/n \quad (n = 1, 2, 3, \ldots) \\ 0 & \text{otherwise} \end{cases}$

If P is any partition of $[0, 1]$ then $L(f, P) = 0$. Let $0 < \epsilon \leq 2$. Let N be an integer such that $N + 1 > \dfrac{2}{\epsilon} \geq N$. A partition P of $[0, 1]$ can be constructed so that the first two points of P are 0 and $\dfrac{\epsilon}{2}$, and such that each of the N points $\dfrac{1}{n}$ $(n = 1, 2, 3, \ldots, n)$ lies in a subinterval of P having length at most $\dfrac{\epsilon}{2N}$. Since every number $\dfrac{1}{n}$ with n a positive integer lies either in $\left[0, \dfrac{\epsilon}{2}\right]$ or one of these other N subintervals of P, and since $\max f(x) = 1$ for these subintervals and $\max f(x) = 0$ for all other subintervals of P, therefore $U(f, P) \leq \dfrac{\epsilon}{2} + N\dfrac{\epsilon}{2N} = \epsilon$. By Theorem 3, f is integrable on $[0, 1]$. Evidently

$$\int_0^1 f(x)\,dx = \text{least upper bound } L(f, P) = 0.$$

4. Suppose, to the contrary, that $I_* > I^*$. Let $\epsilon = \dfrac{I_* - I^*}{3}$, so $\epsilon > 0$. By the definition of I_* and I^*, there exist partitions P_1 and P_2 of $[a, b]$, such that $L(f, P_1) \geq I_* - \epsilon$ and $U(f, P_2) \leq I^* + \epsilon$. By Theorem 2, $L(f, P_1) \leq U(f, P_2)$, so

$$3\epsilon = I_* - I^* \leq L(f, P_1) + \epsilon - U(f, P_2) + \epsilon \leq 2\epsilon.$$

Since $\epsilon > 0$, it follows that $3 \leq 2$. This contradiction shows that we must have $I_* \leq I^*$.

6. Let $\epsilon > 0$ be given. Let $\delta = \epsilon^2/2$. Let $0 \leq x \leq 1$ and $0 \leq y \leq 1$. If $x < \epsilon^2/4$ and $y < \epsilon^2/4$ then $|\sqrt{x} - \sqrt{y}| \leq \sqrt{x} + \sqrt{y} < \epsilon$. If $|x - y| < \delta$ and either $x \geq \epsilon^2/4$ or $y \geq \epsilon^2/4$ then

$$|\sqrt{x} - \sqrt{y}| = \frac{|x - y|}{\sqrt{x} + \sqrt{y}} < \frac{2}{\epsilon} \times \frac{\epsilon^2}{2} = \epsilon.$$

Thus $f(x) = \sqrt{x}$ is uniformly continuous on $[0, 1]$.

8. Suppose that $|f(x)| \leq K$ on $[a, b]$ (where $K > 0$), and that f is integrable on $[a, b]$. Let $\epsilon > 0$ be given, and let $\delta = \epsilon/K$. If x and y belong to $[a, b]$ and $|x - y| < \delta$, then

$$|F(x) - F(y)| = \left|\int_a^x f(t)\,dt - \int_a^y f(t)\,dt\right|$$
$$= \left|\int_y^x f(t)\,dt\right| \leq K|x - y| < K\frac{\epsilon}{K} = \epsilon.$$

(See Theorem 3(f) of Section 6.4.) Thus F is uniformly continuous on $[a, b]$.